D0978892

PRINCE ALBERT

Also by A. N. Wilson

FICTION

The Sweets of Pimlico
Unguarded Hours
Kindly Light
The Healing Art
Who Was Oswald Fish?
Wise Virgin
Scandal: Or Priscilla's Kindness
Gentlemen in England
Love Unknown
Stray
The Vicar of Sorrows
Dream Children
My Name Is Legion
A Jealous Ghost
Winnie and Wolf
The Potter's Hand
Resolution
Aftershocks

THE LAMPITT CHRONICLES

Incline Our Hearts
A Bottle in the Smoke
Daughters of Albion
Hearing Voices
A Watch in the Night

NON-FICTION

The Laird of Abbotsford
A Life of John Milton
Hilaire Belloc
How Can We Know?
Tolstoy
Penfriends from Porlock
Eminent Victorians
C. S. Lewis
Against Religion
Jesus
The Rise and Fall of the House of Windsor
Paul
God's Funeral
The Victorians
Iris Murdoch As I Knew Her
London: A Short History
After the Victorians
Betjeman: A Life
Our Times
Dante in Love
The Elizabethans
Hitler
Victoria
The Book of the People: How to Read the Bible
The Queen
Charles Darwin: Victorian Mythmaker

PRINCE ALBERT

THE MAN WHO SAVED THE MONARCHY

A.N. WILSON

HARPER
An Imprint of HarperCollins*Publishers*

 For information, address HarperCollins Publishers, 195 Broadway, New York, NY 10007.

HarperCollins books may be purchased for educational, business, or sales promotional use. For information, please email the Special Markets Department at SPsales@harpercollins.com.

Originally published in the United Kingdom in 2019 by Atlantic Books.

FIRST U.S. EDITION

Image on p. viii: Queen Victoria's etching of Prince Albert, after Sir George Hayter's The Marriage of Queen Victoria, 1840. Courtesy of Royal Collection Trust / © Her Majesty the Queen Elizabeth 2019.

Library of Congress Cataloging-in-Publication Data has been applied for.

ISBN 978-0-06-274955-0

19 20 21 22 23 LSC 10 9 8 7 6 5 4 3 2 1

For

Matthew

and

Rebecca

'Here am I,' said Pancks, pursuing his argument with the weekly tenant. 'What else do you think I am made for? Nothing. Rattle me out of bed early, set me going, give me as short a time as you like to bolt my meals in, and keep me at it. Keep me always at it, and I'll keep you always at it, you keep somebody else always at it. There you are with the Whole Duty of Man in a commercial country.'

Charles Dickens, *Little Dorrit*

ἦ φῂς τοῦτο κάκιστον ἐν ἀνθρώποισι τετύχθαι;
οὐ μὲν γάρ τι κακὸν βασιλευέμεν
[Do you deny it is an honourable thing to be a king?]

Homer, *Odyssey*, Book 1, lines 391–2

Translation by Emily Wilson

Victoria's etching of Albert, after Hayter's *The Marriage of Queen Victoria, 1840.*

CONTENTS

ACKNOWLEDGEMENTS

GRATEFUL ACKNOWLEDGEMENT IS owing to Her Majesty the Queen, for graciously giving her consent for this biography to be written, and for allowing me full access to all the relevant material in the Royal Archives at Windsor. She let it be known that she thought a biography of Prince Albert 'long overdue'.

When I was halfway through my task, good fortune allowed me to encounter HRH the Duke of Edinburgh. He asked what I was writing. When he heard, he expressed the view, with characteristic forcefulness, that the world did not need 'yet another' book about Prince Albert. I had the temerity to reply that there had not, as it happened, been many, or indeed any, full biographies, since that of Sir Theodore Martin in the nineteenth century. He wandered off to speak to someone else, but then he returned. 'There is one good aspect,' he said. 'Prince Albert had a very short life, so you can write a very short book.'

I hope that His Royal Highness will not consider the finished result unduly prolix. In the course of its composition, Oliver Urquhart Irvine, the Royal Librarian and Archivist, could not have been more helpful, and the entire team in the Round Tower made each visit a pleasure, not least the austere, but jolly, lunch breaks. Julie Crocker has saved me from many howlers. Oliver Walton, a brilliant Germanist and expert on all Coburgian matters, was at hand to advise and help, especially when a letter in the old script seemed illegible. Bill Stockting had much good advice, and the entire team, of regulars and interns, with their unshowy willingness to share knowledge, made visits to the Royal Archives the high point of any month. In the Prints and Drawings Collection, I am especially grateful to Kate Heard and to Carly Collier for discussing Prince Albert's aesthetic, for making material available and for giving useful advice about the illustrations for this book. Helen Trompeteler of the Historic Photographs, Royal Collection Trust, and Leah Johnston, Archive Cataloguer at The Royal Household, were also very helpful.

I owe a great debt to Philip Mould for allowing me to use the superb watercolour by Alfred Chalon as the jacket of this book. This picture has subsequently been acquired by Historic Royal Palaces.

Richard Edgcumbe, Senior Curator in the Metalwork Collection at the Victoria and Albert Museum, made time to show me the sapphire and diamond coronet designed by Prince Albert and lately acquired by the museum.

Much of the primary material is in what Queen Victoria called 'the dear German language', and so I am especially grateful to my late mother, Jean, a good Germanist, who urged me in her latter years to start learning it, and to Ute Ormerod who has patiently taught me each week for many years.

I have written this biography at the prompting of my publisher Will Atkinson, who has been the book's friend and editor. Thanks, too, for James Nightingale's editorial help, and to Tamsin Shelton who has been, as with previous books, a punctilious copy editor. Clare Alexander was a kindly and judicious agent.

I could not have written this book without the friendship and help of Angelika Tasler, who escorted me from archive to archive in Germany, and who took the time to dictate many of the more illegible manuscripts. Our happy time together in Coburg and Gotha taught me more about Prince Albert and his brother than many solitary hours had done previously. Thanks too, in Germany, to Frau Rosemarie Barthel and Frau Cornelia Hopf in Gotha, and to Herr Dr Klaus Weschenfelder, Herr Rudolf Fuhrmann and Herr Dr Alexander Wolz in Coburg. I am especially grateful to the staff at the Kupferstichkammer who so kindly opened up the museum and

Kupferstichkammer specially for me when I chose a day to visit on which the public are not normally admitted.

At the Royal Commission for the Exhibition of 1851, Angela Kenny was endlessly helpful. It was my great good fortune, both at the Royal Archives and in the Royal Commission of 1851, to encounter fellow researcher Jana Riedel whose insights into the life of Prince Albert and the atmosphere of the nineteenth century have been invaluable. Christopher Guyver alerted me to the Seymour Diaries in the British Library.

The staff of the London Library, of the Bodleian Library, especially in the glorious new Weston Library, and in the British Library were, as ever, role models of everything librarians should be.

I have been working on the subject of the Victorians, and of their royal story in particular, for a long time, and feel gratitude to all the friends and colleagues whose expertise in the field have been exemplary.

While writing this book, I have picked the brains of Antonia Fraser, Daisy Goodwin, Ulrike Grunewald, Ruth Guilding (whom I first met at Osborne House while she was the curator and I was making a television film there about Prince Albert in 1988), Tanya Harrod, Tristram Hunt, Philip Mansel, John Martin Robinson, Alice McArdle, Anne Somerset, Margaret Stead, Roy Strong, Hugo Vickers and Lucy Worsley.

Karina Urbach and Jane Ridley both read the book at different stages of composition and were of invaluable help.

ARCHIVES AND MANUSCRIPTS CONSULTED

THE GREATER PART of this book derives from primary sources, above all from the Royal Archives at Windsor. The endnotes record which manuscripts have been consulted.

An invaluable source for the Great Exhibition is the Royal Commission for the Exhibition of 1851, housed in Imperial College London.

In addition, I have been lucky enough to consult manuscripts in the Forschungsbibliothek in Gotha, administered by the University of Erfurt, and at the Library of the Schloss Friedenstein. The archives came from the Staatsarchiv of Gotha and the Staatsarchiv Thuringen. In Coburg I consulted the Coburger Hausarchiv and the Staatsarchiv Coburg from which so much of my earlier book, *Victoria*, derives.

I have also consulted the wealth of manuscript material in the British Library, in particular the Peel Papers, the Aberdeen Papers, the Liverpool Papers, the Graham Papers and the Seymour Diaries.

ILLUSTRATIONS

13. Lord Palmerston. Imagno/Hulton Archive/Getty Images.
14. Lord Aberdeen. General Photographic Agency/Hulton Archive/Getty Images.
15. Etching showing Queen Victoria and Prince Albert's dogs, Eos and Cairnach, by Prince Albert, after Sir Edwin Landseer, 1844. Courtesy of Royal Collection Trust/© Her Majesty the Queen Elizabeth 2019.
16. *Five studies of the Royal children* by Queen Victoria, 1845. Courtesy of Royal Collection Trust/© Her Majesty the Queen Elizabeth 2019.
17. *Queen Victoria's Bedroom, Windsor Castle* by Joseph Nash, *c.* 1847. Courtesy of Royal Collection Trust/© Her Majesty the Queen Elizabeth 2019.
18. *The Children in the Pleasure Garden at Osborne* by Queen Victoria, 1850. Courtesy of Royal Collection Trust/© Her Majesty the Queen Elizabeth 2019.
19. *Fishing in Loch Muick* by Queen Victoria, 1851. Courtesy of Royal Collection Trust/© Her Majesty the Queen Elizabeth 2019.
20. *Balmoral Castle from the Approach (Abergeldie side)* by Queen Victoria, 1852. Courtesy of Royal Collection Trust/© Her Majesty the Queen Elizabeth 2019.
21. *The Queen and the Prince* by Roger Fenton, 1854. Courtesy of Royal Collection Trust/© Her Majesty the Queen Elizabeth 2019.
22. *The Queen, the Prince, Duchess of Kent and Royal Children at Osborne* by Dr Ernst Becker, 1854. Courtesy of Royal Collection Trust/© Her Majesty the Queen Elizabeth 2019.
23. Moving machinery at the Great Exhibition of 1851. Hulton Archive/Getty Images.
24. Henry Cole. Hulton Archive/Getty Images.
25. Joseph Paxton. Courtesy of Royal Collection Trust/© Her Majesty the Queen Elizabeth 2019.
26. Royal Patriotic Building, Wandsworth. Karl Dolenc/Alamy Stock Photo.

ONE

PRINCESS BEATRICE'S WAR WORK

In May 1943, Great Britain and Germany had been at war for over three and a half years. Churchill had secretly gone to Washington DC for the first wartime conference with Roosevelt – codenamed 'Trident'. In Berlin, Dr Goebbels proudly announced that after sixty days of 'work', the German capital was at last '*Judenfrei*' – free of Jews. In North Africa, the most brilliant of the German generals, Rommel, had suffered setbacks, with Tunis and Bizerte falling to the Allies.

At this point, the King of England heard from his great-aunt Beatrice: 'I congratulate you on the tremendous victory in Tunisia, which fills me with thankfulness and pride. It must be such a relief to you, and make you look with so much further confidence to the ultimate complete victory of our army'.[1]

With the world in flames, and the future of the free world in the balance, the King could have been forgiven for not devoting too much attention to letters from an old lady, concerning the papers of his Victorian ancestors. That was her reason for writing. 'Your daughters know that I have been engaged in translating my Father's correspondence with his stepmother'.[2] War fever is a strange group psychosis. This old lady was translating some harmless letters from German at a time when there were British people, whipped up by the war, who felt it was unpatriotic to read Goethe or listen to Beethoven; and when even a more reasonable majority, for years after the Second World War, associated the very word 'Germany' not with that nation's great poets, musicians and philosophers, but with the criminals of the Third Reich.

The old lady writing to George VI, a woman who was herself three quarters German, and who had been married to a German prince, was the King's Great-aunt Beatrice, the youngest daughter of Prince Albert and Queen Victoria. She was spending the war at Brantridge Park, Balcombe, Sussex. She had been born in 1856, before there even was a country called Germany. Her father, Prince Albert, had come to London from Coburg, the small town in Thuringia where his father was Duke of Saxe-Coburg-Gotha. His aunt Victoire had come to London in 1818 as the bride of the Duke of Kent. Their daughter, Victoria, became Queen of England in 1837 and Albert's bride in 1840.

Between them, Victoria and Albert had rescued the British monarchy from grave crisis. They had nine children, of whom Beatrice was the last. They established a dynasty. Almost from the beginning they were, as well as being real people in an actual historical context, semi-legendary figures. This was partly because they really did save the monarchy, and therefore established the kind of country Britain would become over the next century. Moreover, Albert was a person of prodigious gifts. Not only was he politically astute. He had administrative gifts which could have made him a great general. He was scientifically informed. He understood, and was enthused by, modern technology. He was a knowledgeable art collector. He was a musician – himself a composer. He designed his two houses, Balmoral in Aberdeenshire and Osborne House on the Isle of Wight. He was the father of a family. Unlike the royal personages and aristocrats they had known as they were growing up, Victoria and Albert were true to their marriage vows. So unusual was this, that they were considered 'bourgeois' or middle class for doing so. The solidity and moral rectitude of the Royal Family was something which greatly strengthened the standing of the monarchy in the Victorian age, as the middle classes – upper middle and lower middle – burgeoned in number and political strength.

It all came to an end, the idyll of Victoria and Albert's marriage, when they were just forty-two years old. For the next forty years, Victoria was a widow. She still exercised a busy role as a constitutional monarch behind the scenes. Public shows of majesty, however, were not emotionally possible for her without her 'Angel', as she called him, without any irony, at her side. She was lonely, and increasingly dependent on her companions. The burden fell heaviest on the youngest child, 'Baby', as Beatrice was known by her mother. The Queen regarded it as a betrayal when Baby announced her desire to marry Prince Heinrich von Battenberg, and made it a condition of the marriage that the bridal pair should remain at the mother's side. Baby was the Queen's dogsbody for the last decades, and Victoria made Baby her literary executor.

This was a formidable task, since Queen Victoria was a prolific diarist and letter-writer, leaving behind written records which, it has been estimated, would fill a library of 700 volumes if ever printed and bound in book form. Victoria had been in love with Albert. There was no doubt about the Queen's love for her Angel, and her love did not diminish with the years. She it was who had masterminded and, almost certainly, largely written the five-volume tribute *The Life of His Royal Highness the Prince Consort*, which was co-authored by a Germanist called Sir Theodore Martin.

Victoria had the intensely shy person's belief in, and use of, the written word as a means of self-expression. She was not someone whose instincts were to suppress or disguise the truth. From her girlhood onwards, she had poured out her impressions of life into copious journals and she was not one to hide her often changeable feelings. Beatrice, for reasons which do not have anything to do with us in this present study, was a very different person. She conceived it as her duty, when her mother died, to 'edit' Queen Victoria's journals. Admittedly one reason for this was that the Queen's handwriting was semi-legible, and Beatrice wrote a good clear hand. The chief reason, however, as she copied out the words her mother had written, was to discard passages likely to cause pain and embarrassment to herself and her siblings. Victoria had been candid in her often acerbic views of her children, and about the wider reaches of her and Albert's dynasty – nine children, forty-two grandchildren and innumerable cousinage. She somehow always knew when one of them was getting into a scrape or making a fool of themselves, either politically or in the area of private scandal. Down it all went into her letters and diaries. We know this because Beatrice was unable to censor, for example, the copious correspondence Victoria had with her eldest child Vicky – eventually the German Empress. And there are still notebooks and other remains in the Royal Archives which give us some clue as to the sort of material which has been lost from the official, censored, Princess Beatrice-version of the journals which we can now all read online.[3]

Princess Beatrice probably acted from motives which she considered good ones. She was, however, the archivist's dread. Those of us who take an interest in the past want the truth. Without all the material to hand – the damaging, as well as the adulatory, the good and the bad – the truth can never be told. Often it is painful and complicated. The art of biography, as demonstrated in its finest forms, is akin to that of Tragedy and the Novel. Writer and reader learn to take heed of the tragic flaws of our heroes and heroines. Queen Victoria is herself a case in point. She would have been the first to acknowledge her faults. Indeed, she was in fact the first, as is witnessed by a sad but revealing volume which escaped the attention of Princess Beatrice, which the Queen entitled *Remarks, Conversations, Reflections*,[4] in which she chronicled her tempestuous relationships with her mother, her husband and her children, all of whom she loved in different ways, but with none of whom did she enjoy a relationship of pure sunshine.

When I came to write her biography, I found my admiration for Queen Victoria deepen, because she was a woman who confronted her demons and, on the whole, who overcame them, without the help of therapy, or even of much advice from friends. The often only just legible scribble was her outlet, her psychiatrist's couch, her confessional, her secret beehive to which she disclosed her inmost heart. Not everyone liked her, and she was aware of that. Not everyone likes her to this day. It is hard, however, to dismiss her. She undertook a gigantic political and symbolic role when she was only a teenager – the Head of State of the most powerful economy in the world, the figurehead of what would become a global empire. She did her job throughout the next sixty years and she handed on the family business in good shape, when most of the European powers that seemed to be so powerful in 1901 (the year she died) were in fact great liners heading for the iceberg.

The success story of the British Royal Family was unquestionably to be laid at the door of Prince Albert, who, since his death in 1861, had been canonized. After he died, the Angel turned into a man who could do no wrong. His statues were to be seen all over the Empire. Albert Halls, Albert Squares, Albert Streets filled every English-speaking town, and many of the towns in India. He had, of course, been perfect. Baby really had been little more than a baby when he had died in 1861. She was just six. And the world in which she found herself during the Second World War was one which her father could not have envisaged in his most vivid nightmares. From his early years, as the young son of a German prince in a small Duchy in Thuringia, Albert and his brother had yearned for a united Germany. 'Deutschland, Deutschland, Über Alles!' was not a song of triumph when it was written. It was a Hymn to Liberty. It was an aspiration for the benign unity of German-speaking peoples which had been the wish, not only of German-speakers, but of all enlightened Europeans since the Middle Ages. How that benign union was to be achieved, that was the question. It was the hope that one day, the scattered princedoms, duchies, city-states and electorates of the Holy Roman Empire (which had been disbanded by Napoleon), together with the Kingdom of Prussia and the Empire of Austro-Hungary, could fashion some new political entity. The peace and prosperity of Europe depended upon it. Albert and Victoria's marriage plans for their children reflected their desire to infiltrate the autocracies of Russia and Prussia with their brand of political liberalism: not liberalism with a capital L, but a form of constitutional monarchy which allowed for parliaments to have their say, and for an enfranchised electorate to be represented in those parliaments. Neither Victoria nor Albert were democrats in the modern sense of the term, but they had the political nous to adapt the monarchy for a world where democracy in its varied forms would one day be adopted.

Albert would die in 1861. He would not live to see the decade in which all these dreams were ruined: Austria, which might have provided an umbrella of some sort for such states as Saxe-Coburg-Gotha, where Prince Albert was born, had been defeated in a painful, bloody war with Prussia. The north German kingdom had triumphed, and the new united German Empire, created after the defeat of France in the war of 1870, was a Germany dominated by a Prussia of militaristic ideology and autocratic political ideas. The journey from 1870, when the Germans triumphantly declared themselves an empire in the Hall of Mirrors at Versailles, had been escorted by the gods and Valkyries of war. It had seen the humiliation of the Germans in that same Hall of Mirrors, when France exacted its revenge in 1918–19. The reparations which the Germans were forced, in that mirrored hall, to pay to the French for the next decades led directly to the German desire for further reprisal, the rise of the extreme right and the tragedy of the Second World War, in which Britain was now engaged.

Beatrice's war work, translating the letters of Prince Albert's stepmother, opened a little window into a heart-rendingly different time, in which Germans and British intermarried, in which the Germans looked to Britain for political inspiration, and in which the British Royal Family – admittedly having to endure a certain degree of xenophobia from the press and the diehard politicians – was seen to be an entirely benign German import.

And the marriage of Victoria and Albert was, of course, seen as almost sacred; it was a picture of the ideal marriage, a pattern to be followed by all the good, decent families of the Empire. It would seem likely that Baby, who lived so close to her mother, and who had witnessed the canonization of her father, took this version of events entirely literally; all the more so because her own marriage had been far from idyllic, and her husband – Prince Henry of Battenberg as he was known in England – was not always faithful.

When she had finished her latest batch of translation work, she wrote to the librarian at Windsor asking if there was anything else which she might see.

Let the librarian at Windsor Castle of that time, O. F. Morshead, take up the tale, in a letter he dispatched, with some anxiety, to the King's Private Secretary, Tommy Lascelles. He told Lascelles that Beatrice had been translating the letters of Queen Adelaide to the Queen of Prussia. 'They were very dull, as it happens,' Morshead confided. Then she had turned her attention to the Prince Consort's letters to his stepmother. 'The contents were quite innocuous.' So far, so good. The aged Princess was kept busy, and no damage was being done to the Royal Archives. The librarian had then sent a box down to Surrey, labelled 'Diary &c. Prince Consort's notes on the birth of the Royal Children'.

> This seemed harmless enough, and likely to be of interest to the princess, & I accordingly sent it. I am really sorry to find that it contained inflammable material. I know that the prince and the Queen did not always agree during their early married years; but I suspected no revelations within these particular covers. I feel in view of what has happened that I ought to have been more on my guard, and I must apologise most sincerely for having inadvertently brought about a delicate situation.[5]

Morshead wrote that letter on 14 May 1943. The box which he had sent to Princess Beatrice, without opening it, contained not only reflections on Queen Victoria's accouchements. There was also a cache of letters in which the Angel constantly upbraided his wife for her displays of irrational ill temper. They are controlling letters. 'You ask me to promise "not to scold you again before your children". To that I willingly agree – what you call scolding [*schelten*] I would call simply the expression of a difference of opinion.' The coldness of the letters blows like a winter breeze from that box even at the distance of all the years since they were written. He records the misery of marital discord: 'You have again lost your self-control quite unnecessarily. I did not say a word which could wound you' – a statement which can only be contradicted by the mournful facts reflected in the letter – 'and I did not begin the conversation, but you have followed me about and continued it from room to room'.

The Prince's handwriting, which in youth had been so neat, and which still could be neat when writing to statesmen, has become, in these passionately angry, buttoned-up expressions of marital hate, *sprawling*. His hand was shaking as he wrote. Victoria had every opportunity in her lifetime to destroy these letters from her husband, which also contain, among their sentences of cold reproach, attempts at reconciliations, and expressions of joy when rapprochements are achieved. What she left behind to be read by posterity was a record of a marriage which was extremely difficult. Both she and her husband were strong characters, and they were often at odds, especially in the last decade of their lives together when she hated the loss of control and the crippling depression which the repeated experience of child-birth brought. Victoria was a sufficiently complicated, and sufficiently realistic, writer to want us to know what her marriage had actually been like.

Baby did not read these letters in this light. Not at all. Born a doormat, treated as a doormat, she was probably quite unaware of the psychological motives which prompted her to exact a doormat's revenge on her parents, by trying to make their colourful lives as dull as her own. Serenely happy marriages are not unknown to history, but they are the exception which proves a general rule, namely that men and women who undertake to spend the whole of their lives together are unlikely, at all times, to find this easy. Albert's parents divorced in a scandalous manner which kept the gossips' tongues wagging all over

Europe. He had been determined that his own arranged marriage to his cousin should not be a repetition of the parental disaster. Victoria, who had fallen head over heels in love with him, made him the recipient not only of her wildly affectionate needs and desires, but also the object of all her buried childhood nightmares, her uncontrollable angers and passions. Cold and stiff to a degree, he was unable to deal with this other than upping the ante of control, control and yet more control. Of himself, of the children, and of her.

Baby was so appalled by seeing the complicated truth revealed that she reacted as she had done to her mother's journals. She wanted to censor. Busy as King George was with the war, she wrote to him, asking his permission to burn these letters, which her mother had so carefully preserved. On 21 May, she wrote:

> My dear Bertie, I am most grateful to you for returning the box, & saying I may burn its contents. I felt sure you would agree that they ought not to be kept, particularly after having read 2 or 3 of the letters. It is very kind of you to reassure me about the contents of the Archives being kept entirely private and confidential. After that dreadful book of Arthur Ponsonby's, I have naturally got more apprehensive of any wrong use being made of what should be considered as strictly confidential.[6]

(She was referring to the recently published *Life* of Queen Victoria's Private Secretary, Sir Henry Ponsonby, by his son, the Labour MP and noted pacifist Arthur.)

The story, however, does not end there. Clearly, someone realized that to tamper with historical evidence in this way, when the characters were as central as Victoria and Albert, was a violation of truth. The original letters do not survive, but slipped inside the box are photographs of the originals, as well as tantalizing photographs of some pages from the Prince Consort's diaries. (It is known that he kept a diary, extracts of which appear in Sir Theodore Martin's *Life*, but no trace of it, apart from these photographs, has ever been found apart from a few leaves transcribed for Martin by Princess Christian of Schleswig-Holstein.) It is from the records of 'dreadful' books and 'private and confidential' archives that we are able to uncover the remarkable truth about remarkable lives. Albert was Victoria's 'Angel' but, as she alone fully knew, he was a fallen Angel. His daughter's wish for him to be a plaster emblem of perfection did a service neither to her, nor to him, nor to us. The truth is always more interesting than its opposite.

TWO

HIS MOTHER

QUEEN VICTORIA AND Prince Albert were cousins, and anyone who wanted to know what they held in common, at the deepest level, could learn much from intruding into their bedroom at Windsor Castle in the late 1840s. By 1847, the royal pair were the parents of five children, and Albert, with his zest for organization and interior design, felt the time had come completely to reorder their accommodation at Windsor. The nurseries were enlarged. The old school room was converted into a wardrobe for the Queen, and Albert's old sitting room was made into an enlarged school room, with the governess, Miss Hildyard, living in his former sitting room.[1] But it is the bedroom which concerns us.

Cynical old Bismarck would describe Coburg as the stud farm of Europe. Given the dynastic ambitions of the Coburg family, it was a fair enough description. The neo-Tudor bed at Windsor, with its emerald-green hangings, was the centrepiece of this dynastic ambition. As far as Albert's idea of Europe was concerned, the bed was central to all his plans, for from the fruit of his and Victoria's union could stem a new world of righteous children, taking on the roles of constitutional monarch, and spreading his political vision, of moderate conservatism, European federalism, free trade and Protestantism, from Madrid to Moscow, from Berlin to Budapest, from Sweden to Sicily. The bedroom, however, which was the actual, as well as the symbolic, seedbed of this dynasty, also took cognizance of the shared heritage of the married couple. Albert and Victoria hung the room with portraits of their immediate ancestors. They commissioned watercolours by Joseph Nash the Elder of all the refurbished rooms. That of the bedroom is particularly evocative of the married life of the pair, of all they shared, and above all their shared origins. Nash did many watercolours of the different rooms at Windsor and at Claremont House and Buckingham Palace. This bedroom picture, however, was of such significance that the Queen had it copied in oils by William Corden.[2]

On a sofa beside the fireplace is a garment thrown down as if Victoria had just momentarily left the room. And staring from the walls are the four parents. There is the Queen's mother and Albert's aunt, Victoire, Princess of Coburg-Saalfeld. There is Victoire's brother, Albert's father, Ernst. And then, two faces which stared at them out of their past but not out of their memory: Edward, Duke of Kent, who died when his daughter Victoria was just a baby. And the figure who, in some ways, for a biographer of Prince Albert, is the most fascinating figure in the entire drama: his mother.

Prince Albert was a puritan of almost obsessive rectitude. His brother Ernst would by contrast follow the primrose path of dalliance which, for many royal and noble personages in history, has been the norm. Even by the relaxed moral standards of their class, however, Ernst and Albert's parents laid a peculiar groundwork of emotional chaos which more than explains Albert's later horror of uncontrolled passion.

Prince Albert's letters, particularly those to his eldest daughter, show a sensitive awareness of the interior life. He wrote thoughtfully about the way in which we can compartmentalize our sorrows and memories. He lived before the days of psychoanalysis and never underwent therapy. Had such phenomena existed in his day, he might very well have had a robust attitude, and not wished to take all the emotional baggage out of the closet. His great-great-grandson, Prince Philip, Duke of Edinburgh, who had a comparably awful childhood, remarked to a royal biographer, 'suddenly my family was gone. I had to get on with it'.[3] Albert could easily have said the same. Nevertheless, a twenty-first-century observer surveying the bare facts would consider that Albert carried a heavy psychological burden.

Three days after Prince Albert's fifth birthday, in August 1824, his mother Luise[4] left Coburg. He would never see her again. His parents' divorce, never printed in the *Almanach de Gotha*,[5] the directory of the European High Nobility and Royalty, and never (of course) mentioned in either of the biographies authorized by Queen Victoria, formed the background of all his early psychological history. While enjoying a cordial relationship with his stepmother throughout his adult life (she was also his first cousin, his father having married a niece), Albert cherished the memory of his lost mother and retained an abiding dread of family discord, sexual scandal and emotional chaos.

Princess Dorothea Luise Pauline Charlotte Friederike Auguste of Saxe-Gotha-Altenburg – always known as Luise – herself had been a baby of only two weeks when her mother died as a result of a complicated delivery. When Luise was a child, her grief-stricken widowed father (who had been just twenty-eight when he lost his wife) wrote her a poem, addressed to her sleeping form:

> Sleep, little Princess, sleep…
> Guiltless and pure, sleep on…
> Far from the throng of the Court
> With its anxious muddles,
> Its stifling confusions,
> Its yawning boredom,
> Here in your cot, you can be happy,
> Sleep, little princess, sleep.[6]

Her father, Emil Leopold Augustus, Duke of Saxe-Gotha-Altenburg, was himself never happier than when asleep. A lazy, lackadaisical man, he would have loved to act upon his own advice, to retire to bed, and to indulge his passion for literature and whimsy.

Had he been born fifty years earlier, it might have been possible for the Duke to pursue a life of political disengagement, presiding over his tiny domains and ignoring the movement of events. Even had such a life been possible, however, he would not have escaped the one demand which determined the lives of all the minor German princes, dukes and electors: the dynastic requirement. The post-1870 state which we call Germany did not, at this date, exist. The Habsburgs ruled in Vienna as emperors. The multifarious little states which filled the landmass of central Europe, most of which had formed part of the Holy Roman Empire, had no political cohesion, other than the links of kinship. It was entirely appropriate that Gotha, the birthplace of Albert's mother, should best be known for its *Almanach*. For, whether the domains were small or great over which these families governed, it was deemed essential that they should all remain in the hands of the hereditary princes and aristocrats who had ruled them for a thousand years. Mesalliance was unthinkable. The German usage, among the nobility, is to refer to themselves as high-born – *hochgeboren*. Often, however, this is shortened simply to *geboren*. Is such and such a person born? As far as they were concerned, not to belong to their rarefied gene pool, not to be able to trace a pedigree back through the twigs and branches of the family trees of royalties and nobilities, all of whom were intermarried and related, was not completely to exist.

So it was that when Duke Emil Leopold Augustus of Saxe-Gotha-Altenburg married Luise Charlotte, a Princess of Mecklenburg-Schwerin, they were simply enacting a few inevitable steps in a genealogical dance which had been going on certainly since the Thirty Years War of the seventeenth century, and, in many cases, since the times of Charlemagne. Had they produced a male heir, the forested lands of Thuringia, a part of Saxony, would have been secure, and they could have looked forward, when that male grew to manhood, to the possibility of extending the dukedom by marriage to a neighbouring dukedom, such as that of Coburg.

As it was, Luise, being her father's only heir, carried the burden, from the moment she lost her mother, of having to be married off. With her the dukedom of Gotha came to an end, so it was inevitable that her father should be obliged to find a husband who could, as it were, marry it, and preserve it by filling Luise's womb with male offspring. However much her father might have loved her as a pure baby, and cherished her childish beauty in his sentimental verses, her only function in the political scheme of things was to grow to womanhood and marry another German prince. The show must go on. Breeding was all. It was fairly obvious, even as she lay in her cradle, that the neighbouring Duchy of Coburg would be the one into which she would be married off.

We have said that, had he lived in the peaceful days of the old regime, Duke Emil Leopold Augustus would only have had the dynasty to worry about. As it happened, he lived during a time of revolutionary turmoil and war. A glance at the map will tell any reader why Saxony could not escape the effects of the French revolutionary wars and the Napoleonic campaigns. Whether he was heading for Vienna, Austerlitz, Danzig or Moscow, Napoleon and his armies would pass this way. Unlike many of the German royalties and nobility, Duke Emil was a passionate admirer of Napoleon, a fact which would cost him, and even more his subjects, dear. When Prince Albert had finished building his beautiful marital home, Osborne House, one of the first pictures he hung in the billiard room was a portrait of Napoleon. Albert was fascinated by Napoleon, and had pictures of him in all his houses, an interest which perhaps made a nod to his maternal grandfather's hero-worship. Queen Victoria's interest in the Bonapartes was even more intimate. Her own mother, Victoire, Albert's aunt, had been on the shortlist of possible substitutes for the Empress Josephine when the Emperor's marriage was coming unstuck and he was looking for a fertile replacement who could provide him with an heir.

Napoleon's several visits to Gotha during Luise's girlhood had an almost emblematic significance. In October 1808, he stayed in the

Schloss Friedenstein on his way to the Congress of Erfurt: his power was perhaps at its zenith. The Duke pledged a regiment – the Duke of Saxony's – to the Rheinbund army that fought on Napoleon's side in the battles of Kolberg and Tyrol, and in Spain in 1811 and Danzig in 1813. The losses of men in each of these engagements were catastrophic. Few returned to Gotha. In December 1812, Napoleon stayed at Gotha, in a state of humiliation and depression, on his way back from the disastrous Russian campaign. He left behind his hat, which may still be seen in the Schloss Friedenstein. By 1813, the King of Prussia, Friedrich Wilhelm III, had called his people to arms, and the German peoples who sided with Prussia were enlisted to fight Napoleon. They would eventually play the decisive role in the victory of Waterloo. Albert's grandfather, Duke Emil, however, and his unfortunate subjects were still committed to fighting for the French. There were demonstrations against the Duke in Gotha when Luise was a child of thirteen, with crowds singing patriotic songs, and calling for them to unite for the German Fatherland. Perhaps it was in deference to popular feeling that Duke Emil did not have the Emperor to stay again. When Napoleon passed through Gotha for the last time, on 25 October 1813, he stayed in an hotel, at the Gasthof zum Mohren.[7]

Luise was fifteen years old when the inevitable alliance with the Duke of Coburg was agreed. Tiny, cherubic, slightly plump, with light brown hair and bright blue eyes, she still looked like a child, apart from her pronounced bosom. When she was confirmed, in the chapel of the Schloss Friedenstein, it was the first time she had taken part in any major public ceremony. As her sacred commitment to Christ was pronounced, she stood on the very spot where her mother had been buried, and burst into tears. The next year, in 1816, Ernst, Duke of Saxe-Coburg-Saalfeld, made his visit to the neighbouring dukedom, and his mother, the genius matchmaker Dowager Duchess Auguste, told her diary, 'I should not be surprised if I did not soon have a daughter-in-law from there'.[8]

From the Coburg point of view, the marriage could only be advantageous, since such an alliance would mean – upon the death of Luise's father – the Gotha Duchy would pass to her husband. All seemed as if it were going Coburg's way, when, in November 1816, there was a setback. Ernst's mother told her diary, 'This afternoon the Court Chamberlain, Count Salisch, arrived from Gotha to see Ernst, and to explain to him with many excuses that for the present the discussions about the betrothal could not proceed any further.' The reasons given were the precarious state of the Duchess of Gotha's health – that is, of Luise's stepmother. Auguste, however, who was a wise old thing, told her diary, 'I am somewhat upset and perturbed, and whenever I think of that marriage, I have a sort of secret dread that nothing may come of it.'[9]

Ernst was a notably handsome man of thirty-two years old. He had a long face, divided by a sharp nose and framed by whiskers. His head was crowned by a profusion of curling dark hair. He was aware of his good looks, and was unable to enter a room without looking for his image in any available glass. It would be unrealistic to suppose that a man of that age, and in that position, had not enjoyed a rich emotional life. Ernst's was more complicated than most, as his mother knew well – hence her misgivings about the prospects of his making Luise happy.

When he was twenty, Ernst had been betrothed to the sister of Tsar Alexander I. His sister Julie was married, very unhappily, to the Tsar's brother Constantine. Ernst, however, blew his chances with the Russians, and after four years, his betrothal to the Grand Duchess Anna Pavlovna was broken off. This was because of his scandalous liaison with a woman always known as La Belle Grecque.

In 1807, Napoleon had restored the Duchy of Coburg to Ernst and, the next year, accompanied by his youngest sibling, Leopold, the twenty-three-year-old Ernst had made the journey to Paris to pay homage to the Emperor. It was during this visit that he met and fell in love with Pauline-Adélaïde Alexandre Panam – at that date an animated fourteen-year-old. They started an affair almost at once, and she soon became pregnant. Ernst, unable to face up to his responsibilities, persuaded Pauline to go to Amorbach and to be looked after discreetly in her pregnancy by Victoire, his widowed sister (the future mother of Queen Victoria).

It was not a happy arrangement. Pauline complained that she was accommodated in the 'residence of the keeper of a coal-shed'. The person put in charge of feeding her was, she claimed, the chimney-sweep. She soon made her way to Coburg. Ernst had made a cowardly flight to make an (unsuccessful) attempt to patch up his relationship with the Romanovs. In Ernst's mother Auguste, Pauline met her match. 'The Duchess Mother began pouring forth a volley of invectives. In vain did I fly to avoid her – in vain did I conceal myself in the corners of the apartments whither she had pursued me. I incessantly heard the clacking of her enormous slippers, which echoed over the flooring and announced her coming, or rather, her Fury-like approach'.[10]

By this stage, Pauline claimed, she was being pursued not only by the furious Dowager, but also by Ernst's amorous brother Leopold. It was all too possible. Together with Ernst's Private Secretary and military adviser, Maximilian von Szymborski, the Dowager drew up a draconian agreement. Pauline would be paid an allowance of 3,000 francs per annum, and she would be accommodated in an apartment whose furnishings would be paid for by the Coburgs. This arrangement, however, would only be valid for so long as Pauline left the estates of His Supreme Highness the Duke of Saxe-Coburg. (We know little of this *éminence grise*, Szymborski, but he would play a sinister and cruel role in the destiny of Luise when she made her fateful marriage to Ernst.)

In the spring of 1809, in Frankfurt, 'on a solitary bed of straw, and without one farthing to prepare the requisite linen',[11] La Belle Grecque gave birth to a son. Pauline would eventually (1823) publish her scandalous memoirs. Ernst was by no means the only European prince to have had an illegitimate child, but the punctilious detail with which Pauline washed his dirty linen in public ensured her a wide readership. The English translation of her book would be a best-seller and would make it hard for anyone in Britain to take Ernst, Duke of Saxe-Coburg-Gotha, quite seriously. This would be a source of huge embarrassment to Albert when he arrived in London as Queen Victoria's future husband. The scandal of his parents' marriage went before him, and it was something which he would spend his entire adult life living down.

It is not clear, when Ernst of Coburg came to marry Luise of Gotha, how much her father knew of La Belle Grecque. If the Gothas feared that they were buying a pig in a poke, the Coburgs were evidently able to set their minds at rest, for, a few days before Luise's sixteenth birthday, the betrothal was announced, and a Court Ball was held in Ernst's honour. The wedding took place on 31 July 1817. There was yet another Grand Ball. A service of Thanksgiving was held in the chapel of the Schloss Friedenstein. Then the bridal pair, accompanied by crowds of happy peasants in traditional costumes, who had been well nourished with beer and sausage, made their way to the summer palace. The gardens were adorned with 15,000 twinkling oil-lights. There was dancing and music lasting five days.[12]

Throughout the jollity, Luise, as is made clear to her childhood friend and confidante, Auguste von Studnitz, was only half-aware of what was happening. She fancied herself in love with the handsome bridegroom. At the same time, after a highly circumscribed youth, she was still a child. She was quite unprepared for what would happen to her now that she had become the Duchess of Saxe-Coburg-Gotha. On 7 August, she left Gotha, and childhood, behind her, and, with Ernst at her side in the carriage, began her ceremonial journey to Coburg.

Gotha is a windblown, bleak town entirely dominated by the Schloss Friedenstein, which is in effect a gigantic Baroque barracks. Coburg is very different. The town, with its cluster of royal palaces, its half-timbered houses, its medieval churches and its well-proportioned market square, is an embodiment of what Germans mean by the word *gemütlich*. It feels intimate, innocent and friendly. It nestles among wooded hills. On one of these eminences stands the Veste, the medieval castle in which Prince Albert's ancestors ruled. If this sturdy castle calls to mind Luther's famous hymn 'Ein feste Burg', this would be appropriate, since the Augustinian friar turned Reformation firebrand took refuge in the Veste, fleeing the Imperial forces after the Diet of Worms. It is said of the Veste in Coburg (as it is said of a number of places in Germany) that this was where Luther worked on his translation of the Bible, a work which not only made possible his dream that every ploughboy could read the Scriptures for himself, but which also in effect invented the modern German language. In the pretty town itself, beneath the Veste, the Coburgs built for themselves a Baroque Schloss in the seventeenth century, known as Schloss Ehrenburg. Its gradual, sweeping staircase leads to a series of impressive salons and state rooms, none more fantastical than the Hall of Giants, in which huge white plaster of Paris giants appear to be holding up the ceiling. It was in the Schloss Ehrenburg that Luise, eleven months after her marriage, gave birth to her first son, Ernst.

The town, however, is not the only place which the Coburgs adorned with princely architecture. Some four miles from the market-place, in the wooded meadows which Prince Albert would always say were so like England, is their shooting lodge, the Schloss Rosenau. Yellow-gabled, Biedermeier-furnished, Gothick-casemented, cottage wallpapered, the Rosenau could plausibly claim to be the most beautiful house in the world. It was here that, on 26 August 1819, Prince Albert was born.

To be born the child of royal or upper class parents in the early nineteenth century was a risky undertaking. This was because, unlike the sensible poor who gave birth with the assistance of their mothers or sisters, or perhaps of a female midwife, the rich were assisted by doctors. Princess Charlotte is only one dire example of many who died in childbirth. The only daughter of George IV of England, she had been married to the younger brother of our Duke Ernst of Saxe-Coburg – Leopold.

The Duchess Auguste, Albert's Coburg grandmother, was a matchmaker extraordinaire who had managed to marry her children to great advantage. Ferdinand George married the heiress of the Prince Koháry in Hungary. Their son became King Consort of Portugal, by his marriage to Queen Donna Maria II. Sophia married Count Mensdorff-Pouilly, a French aristocrat who, emigrating because of the Revolution, became an eminent diplomat in Austria. Her sons were close friends of Albert, and her daughter Marie became Albert's stepmother. Antoinette married the Duke of Württemberg, brother of the Empress of Russia. Julie married Grand Duke Constantine of Russia, the younger brother of the Emperor. Victoire, pretty and vain, married the Prince of Leiningen, and her son Charles became one of the more plausible claimants to the dukedom of Schleswig-Holstein – the disputes over which played so crucial a role in later nineteenth-century European history.

Duchess Auguste, in other words, by the dynastic marriage of her children, had a finger in many a pie, and, when you consider how tiny Coburg is, there is something all the more impressive in the fact that Auguste's grandchildren and great-grandchildren occupied thrones in Sweden, Belgium, Bulgaria, Russia and Portugal.

Her youngest son, Leopold, looked likely to have possessed himself of the richest plum. Married to Princess Charlotte, he was poised to become, in effect, the King of England when her highly unpopular father, George IV, came to die. As things turned out, however, Charlotte died before her father.

As arranged marriages went, it would seem that Charlotte and Leopold were well matched and happy together. Charlotte spoke of him as '*mon Leopold qui fait mon unique Bonheur*' (which she spelt 'bonnheur').[13] Moreover, she got on well with her in-laws, and befriended Victoire, Leopold's sister, who had been left a widow with two young children in Amorbach. All the books repeat the story that, after Princess Charlotte's death, her uncles, in a faintly absurd race to produce an heir to the throne of England, abandoned their mistresses and collected German brides for the purpose. In the archives at Windsor, however, is a volume of the Duchess of Kent's papers put together by her son-in-law and nephew Prince Albert. It is clear that Charlotte was very fond of Victoire, to whom she wrote as her '*adorable soeur!*' Moreover, she urged her uncle Edward, Duke of Kent, to visit Victoire in Amorbach, and it is obvious from the letters that he went there in 1816, a whole year before Charlotte died, with a view to marriage. When he came away, he wrote to Victoire saying that he was going to consult his brother the King, asking permission to propose to her. The letters he wrote to her, before and after their marriage, are love letters. At the end of December 1818, he wrote, 'God bless you. Love moi as I love you, ton fidèle Edward'. He had clearly already taken leave, in his heart if not in his bed, from his long-standing mistress Madame de Saint-Laurent. This bundle of letters is labelled by Victoire herself, '*Briefe meines geliebten Mannes als Er um meine Hand anhielt*' – 'My beloved husband's letters when he proposed to me'.[14]

In other words, before she died, Princess Charlotte had made the links between the Coburgs and the British Royal Family even stronger, not merely with her own marriage to Leopold, but also by ensuring Victoire's marriage to Edward, Duke of Kent. She had thereby prepared the path down which Queen Victoria would tread into the world.

Charlotte's death at Claremont House on 16 November 1817 ruined Leopold's hopes of becoming King of England in all but name. It also put a potentially catastrophic check on the ambitions of a figure who will play a large part in our story, and whom we should get to know here – even though he would not enter Albert's life until much later.

Leopold's mentor and backer, one might almost say agent, was the tiny and manipulative figure of Christian Friedrich von Stockmar, a figure whom Elizabeth Longford brilliantly described as half Merlin, half Puck. He was born in Coburg in 1787, the son of a successful lawyer. After attending school at the Coburg Gymnasium, one of the best schools in Germany (alma mater of Goethe's father), he studied medicine at Würzburg, Erlangen and Jena. He founded a military hospital in Coburg in 1812 to cope with the flood of French, Allied and Russian soldiers who were hobbling through the town after the disastrous Napoleonic campaign. Typhus was rife and he himself caught the disease. In 1814, he moved to Worms where he established another military hospital, and was commanded by the German officer in charge to treat the Germans first and the other nationalities afterwards. Stockmar won universal renown for saying that he would treat all the wounded in strict priority of need, regardless of their nationality. Prince Leopold was so impressed when he heard this story that he eventually asked Stockmar to become his personal physician, and the young German doctor came to London. 'The country, the houses, their arrangement, everything, especially in the neighbourhood of London, delighted me, and so raised my spirits, that I kept saying to myself, "Here you must be happy, here you cannot be ill".'[15]

Stockmar fell in love with England, and was enraptured by its political system. He soon became much, much more to Leopold than a mere doctor. 'Stocky', as Charlotte nicknamed him, was now a power near the throne, and since George IV, obese and wheezing, was clearly unlikely to live much longer, it looked as if Stocky was going to become a person of real influence in English political life. When Charlotte's accouchement drew near, Stockmar, who was staying with her at Claremont House, near Esher in Surrey, carefully distanced himself from the scene. Had she given birth to a live child, and had she lived to become the Queen of England, Stocky would have been smilingly standing behind her throne. In the hour of her distress, however, he needed to distance himself as far as possible from any suggestion of blame for a bungled birth.

Charlotte was in labour for fifty-two long hours, as her mother-in-law Auguste pityingly recorded.[16] She was delivered of a large dead boy, and died herself, some hours afterwards. The incompetent trio of English doctors who were attending her went through the routine of bleeding her and giving her brandy. When, in panic, the English doctors summoned Stockmar from his room, well past midnight, she squeezed his hand and said, 'They have made me tipsy.' The death rattle was heard in her throat, and she called out, 'Stocky! Stocky!' in a loud voice. They were her last words.

We shall return, later in this story, to the influence of Stockmar upon the destinies of Victoria, Albert, and indeed the whole of the monarchy. For the time being, the crucial fact was that he had escaped any taint of association with poor Princess Charlotte's death. One of the doctors, Sir Richard Croft, felt such a burden of guilt that he shot himself. The other two, the King's surgeons David Dundas and Everard Home, performed the autopsy. They found that 'The child was well-formed and weighed nine pounds, every part of its internal structure was quite sound'. As for poor Princess Charlotte, 'The uterus contained a considerable quantity of coagulated Blood and extended as high as the navel, and the Hourglass contraction was still very apparent.'[17]

To Stockmar, the Duchess Auguste poured out her grief. 'My whole being is torn apart with sorrow and pain,' she told Stockmar. 'I can only trust that God and Time will pour soothing Balsam in the burning wounds'.[18] Leopold was, for the time being, out of the running. Clearly it was necessary, as God and Time healed the burning wounds, that they also provided other Coburgs who could step up to the mark and replace Leopold as a person of influence in Britain. Stockmar's beady eyes were now fixed upon Victoire as the bearer of hope for Coburg influence in London.

When the time came for Victoire, as Duchess of Kent, to be delivered of a child, in Kensington Palace on 24 May 1819, they were determined that there should not be a repetition of the Charlotte tragedy. The English doctors were banished. Dr Marianne Heidenreich von Siebold, the first woman to qualify as a doctor in Germany and herself a distinguished accoucheuse, delivered the baby who would become Queen Victoria. She then travelled back to Germany, and was in Coburg in time for the Duchess Luise's second confinement in August.

Ernst, as we have said, had been born in the Schloss Ehrenburg; and there is something more than appropriate about this. Not only would he, as Ernst II, rule as Duke of Saxe-Coburg; but he was *mondain* and he aptly entered the world in a grand palace plum in the middle of the town. Albert, the intellectual, the scholar, the dreamer, the younger son, the Angel, the aesthete, was to be born in the beautiful little Schloss Rosenau, surrounded by woods and meadows, gardens and birdsong. The same Madame von Siebold, who had delivered his cousin Victoria less than three months earlier, also brought Albert into the world.

Since Duke Ernst would divorce Albert's mother for adultery only a few years after he was born, it is inevitable that questions should have been posed about the young Prince's paternity. Such rumours, interestingly, were not put into print until the twentieth century. An anonymous writer in 1915 published a book called *A German Prince and His Victim*, which contained the unsubstantiated claim that a Court Chamberlain, Baron von Meyern, 'a charming, handsome cultivated man of Jewish extraction, much her senior, was... supposed to have taken Ernst's place in the fair Luise's heart'.[19] This suspicion was adopted by Lytton Strachey in his *Queen Victoria*, and by Laurence Housman in his play *A State Secret* – one of the series which made up his charming *Victoria Regina*. Given the fact that Duke Ernst I and Maximilian von Szymborski went out of their way to vilify Luise and to depict her as a serial adulteress, it is interesting that they did not name the mysterious Baron von Meyern as one of her lovers.

Nowadays, if you are shown round the Schloss Ehrenburg, you will probably be told by the guide that the likeliest candidate for Albert's father is his uncle Leopold, who came back to Coburg during the autumn of 1818, when Luise was confined to bed with chickenpox.

In 2008, Richard Sotnick published *The Coburg Conspiracy*, in which he put forward another possibility, one Friedrich Blum, 'a working member of the Court at Coburg'. It is not entirely clear who this Friedrich Blum was. A man named David Bloom, in 2004, claimed that his grandfather, Sam Blum, brought up in Tukums, now in Latvia, was the grandson of Friedrich Blum. It was alleged that Sam had claimed that Friedrich had an affair with 'the Princess';

that a baby had been born, 'who had been accepted as a prince of the family, and whose name had been Albrecht'. A photograph of Nathan Bloom, some sort of relative, taken around 1930, is one of the illustrations in Sotnick's book, bearing the caption, 'The likeness to Prince Albert in his later years is striking'. Nathan Bloom had a high, bald forehead, it is true, but I have shown the picture to a number of people, randomly chosen, not one of whom could see a resemblance which could dispassionately be called 'striking'. Nathan, who was a bootmaker in Stepney, in the East End of London, was the elder brother of Sam. The family, we are told, 'have never wavered in their belief that Friedrich Blum was both their direct antecedent and the father of Albert, the Prince Consort'.[20]

The trouble with the theory, apart from the lack of resemblance between Nathan Bloom and the Prince Consort, is that the Bloom family have not been able to establish that they do in fact descend from the supposed Friedrich Blum. And, rather strangely, the author of *The Coburg Conspiracy*, having apparently supported the plausibility of the claim, added a short chapter at the end of his book, in which he appeared to think Uncle Leopold a more plausible candidate. In substantiation, he calls as a witness Karoline Bauer, one of Leopold's many mistresses, who recalled Luise as saying, 'I do not know whom I love more, Ernst or Leopold.'

This, really, is the point where claims about Albert's paternity sink into the quicksands. Karoline Bauer's memoirs were published in 1884. She was a cousin of our friend Christian von Stockmar. She had considerable success as an actress, and was playing at the Royal Theatre in Potsdam when Leopold, then aged thirty-eight, was in the audience. She was said to have borne a very strong resemblance to Leopold's English wife, Princess Charlotte, whose death eleven years earlier he always mourned. The following year, in July 1829, in a villa which Leopold had hired for her in Regent's Park, the pair would undergo a sort of 'marriage' ceremony, performed by her strange little cousin Stockmar. There she resided with her mother, true to the title of her English biography, 'A Prisoner in Regent's Park'.[21]

Fairly unsurprisingly, Leopold did not honour this 'marriage'. The year after he became king of the newly founded realm of Belgium in 1831, he married Louise-Marie of Orléans, the daughter of King Louis-Philippe. Bauer's posthumous memoirs were an act of revenge upon Leopold and all the Coburgs. D. A. Ponsonby, the biographer of both 'Lina' Bauer and of the Duchess Luise, surely made a very strong point when she wrote:

> Caroline Bauer hated the Coburgs all her life, because of her bitter resentment of the way Leopold had treated her. She was a German, writing in German: she had neither the need nor duty not to publish what she would of the Prince Consort's parents… Anyone who has read the whole of Caroline Bauer's long memoirs will agree that no consideration for others' feelings would have prevented her from smacking her lips over the gossip that the second-born son of the Coburgs was fathered by another than the Duke – if such gossip had existed.[22]

Pass another half century and more after Ponsonby's books, and Ulrike Grunewald, in 2013, wrote the most thoroughly researched biography of Duchess Luise yet to appear: *Luise von Sachsen-Coburg-Saalfeld*. This sympathetic account, by a young, modern German researcher and author (and television documentary-maker), who plainly has no axe to grind, and is not some monarchist lickspittle, finds no evidence at all to support the notion that Prince Albert was not the son of Duke Ernst I, though she now believes that if anyone other than Ernst was Albert's father, it could, at a pinch, have been his uncle Leopold. There is no evidence for this, however.

THREE

THE WETTINS

ERNST WAS A cruel man, who enjoyed teasing his young wife, and playing with her feelings. An example occurred early on in the marriage, back in 1817, when Coburg was preparing for the three hundredth anniversary of the beginning of the Reformation – a story in which Ernst's ancestors had played so crucial a role. Luise, his very recent bride, was nervous about the public role she would be playing in the day's ceremonies. Instead of attempting to put her at her ease, he told her the order of the day's events – a church service in Coburg, and then a procession to the Rosenau, where there were to be medieval jousting, dancing and other amusements. Ernst sternly told her that he had decided she was not to accompany him to the Rosenau after all. He said this merely to torment her. She wept for a whole day before he told her he had only been teasing.

From the very first, he made no attempt to provide her with companionship while he indulged one of his keenest passions, hunting. In very early days, she would accompany him onto the hunting field while he rode out. Once she had become pregnant, however, this was deemed unwise, and she spent many days cooped up, entirely alone, while Ernst went in pursuit of deer and wild boar.

One of Luise's most dangerous enemies at Court – though she was far too naïve to realize this at first – was Ernst's Chamberlain, a homosexual misogynist called Maximilian von Szymborski – as his name implies a Pole, which made him especially unpopular in Coburg.

He did everything he could to pour poison into Ernst's ears about Luise, and this would eventually have its disastrous consequences.

In the latter stages of Luise's second pregnancy, Ernst went to stay in Karlsbad with his sister Antoinette and her husband Alexander, the Duke of Württemberg. Even before he had set out, Luise was complaining to her penfriend in Gotha, Auguste von Studnitz, 'You can readily imagine how unhappy I am made in advance by embarrassing feelings of jealousy for all the beauties there in Karlsbad! Luckily he will only be gone for eight days'. It was no secret that Ernst fancied the Württemberg daughter – his niece Marie, who would one day become his wife. She was the same age as Luise, just sixteen at the time of this visit. 'I am very alone, and very sad and, for the first time in my life, I wish that Time would fly'.[1]

As her time approached, Luise awaited the birth in Schloss Rosenau, with Dr Siebold in attendance. The presence of this first-rate midwife-obstetrician, who had delivered Victoria in Kensington three months earlier, gave the state prayers, read out in all the churches in Coburg, a stronger chance of being answered favourably. 'We believe in Thee, O Lord our God! Thou hast not left Thy people unheard. Bless and protect our Duchess, His Highness Our Duke and all the Ducal Family! That Thou shouldst vouchsafe Joy to Thy people, Amen! Amen!' The Duchess was 'happily delivered of a healthy and well-formed Prince' at 6 a.m. on 26 August 1819.[2]

The birth of Albert provided consolation of a sort. He was baptized by Professor Genzler, a distinguished old Lutheran pastor who, the previous year, had presided over the wedding, in the Hall of Giants, between Albert's aunt Victoire and the Duke of Kent. The baptism took place in the hall of the Rosenau. 'The good wishes with which we welcome this infant as a Christian,' said the preacher, 'as one destined to be great on the earth, and as a future heir to everlasting life, are the more earnest when we consider the high position in life in which he may one day be placed.'[3] It was clear from the outset that the Coburgs had positioned Albert as the potential future companion of the Queen of England. If all went according to plan, Albert would be able to succeed where the death of Princess Charlotte had caused Leopold to fail.

He was baptized in the Lutheran faith which he would never, in his heart, forsake. The Church of England, to which he would officially belong after his marriage, meant nothing to him. His baptismal names were Franz August Carl Albrecht Emanuel[4] but he was always known as Albert, not Albrecht. They were preparing him for his English destiny. 'My children are the pride and joy of their grandparents,' Luise wrote dotingly, in French, 'the more they grow, the more amusing they become. The elder, particularly, has great spirit, and the little one captures every heart with his beauty and his gentillesse.'[5] By the time he was nearly two, he had begun to resemble his mother. She, in turn, had begun, in her intense boredom and growing marital dissatisfaction, to realize her own capacity to capture hearts.

It was after the birth of Prince Albert that the marriage of his parents began to unravel. Luise's most recent biographer, Ulrike Grunewald, makes a convincing case, when she emphasizes that the first scandal caused by Luise in Coburg was not herself committing adultery, but complaining about the adulteries of her husband. It simply was not *done* for the wives of nobility or royalty to raise objections to the behaviour of their males, behaviour which might have been scandalous for a member of the middle classes, but for a prince was considered perfectly normal.

In the first year of Albert's life, Luise began to recover from her low spirits. She doted on the new baby, who always seemed to be smiling, and whose blue eyes were angelic. She also made no secret of doting on her brother-in-law, the widower Leopold. Even before Albert was born, she was writing to Auguste von Studnitz (in December 1818), 'Yesterday evening my brother-in-law Leopold came here and spoke to me a lot about Gotha. He had seen you and praised your beauty, though he thought you looked a bit pale. O, dear Auguste, how happily would I have swapped places with him, and been able to see you again. You'll have found him very beautiful and loveable, no? Tell me honestly, which of the two do you find the best looking, him or Ernst?'[6]

The exuberant teenager's love of handsome men, however, was to go too far for her husband's pleasure. Like Browning's 'Last Duchess', 'she liked whate'er she looked on, and her looks went everywhere'. In the April after Prince Albert was born, she and her husband reached a fateful turning point in the marriage. Ernst was invited on an official visit to Vienna, and Luise accompanied him. The glitter, the size, the energy of the Imperial Court overwhelmed her. They were there over Easter. Apart from over thirteen dinners, concerts at Court, visits to the opera (which she found unsatisfactory, 'they have no tenors'[7]), she was also, as a Protestant with no previous experience of Catholic liturgy, deeply impressed by the church ceremonies: the Emperor and Empress kneeling, on Maundy Thursday, to wash the feet of twelve men and women in commemoration of Christ washing the feet of the disciples at the Last Supper; the lighting of the Easter Fire on Holy Saturday; the processions of priests in their exotic vestments, followed, on their way to St Stephen's Church on Easter Monday, in carriages which went back to the time of the Emperor Charles V.

But more impressive, even, than any of this was their 'Premier Ministre, Prince Metternich. Usually I have thought Ministers very tedious, but this one has changed my opinion. He is so kindly, cheerful also – when he wants to be – sentimental, that he must captivate the people. He is also very good looking'.[8]

Almost certainly, at this stage, Luise knew nothing about La Belle Grecque, the subject uppermost in Ernst's mind. His mistress's threat to go public with their affair, and with her ill-treatment at his hands, and the birth of their son, was a source of perpetual anxiety, which he shared with Metternich; this was, indeed, his chief reason for wishing to visit Vienna – to plan 'damage limitation', were Pauline to carry out her threat of publishing memoirs. Luise's artless, prattling crush on Metternich must have greatly irritated her older husband.

It seems as if Luise first heard about La Belle Grecque sometime in 1822. By this point, the little Court at Coburg was hissing with rumour and gossip, some courtiers being especially vindictive towards Luise. In the early summer of that year, Luise's father, Duke August, made his last visit to Coburg. He was only fifty, but seemed much older, and spent most of the visit lying in bed, in a dream world of his own. Karoline Bauer, future lover of Leopold and chronicler of royal scandals, claimed that already the tongues were wagging about Luise's infidelities. Bauer, Stocky's cousin, was at this stage a young actress in the Coburg theatre. Perhaps Duke August kept to his bed because he could not bear to hear his daughter vilified. Whatever his reason for keeping to his bed, he was not well. He returned to Gotha and on 17 May 1822 he died.

Luise confided in one of her girlfriends, Julie von Zerzog, that she was completely shattered by the loss of her father. She had lost her last feeling of security. Julie did not touch, in her *Reminiscences*, upon the reasons why relations between Luise and her husband descended so catastrophically into hatred from this point on. One reason could have been money. Ernst had colossal debts – over 541,031 Reichsthaler.[9] She inherited some money from her father, but not enough to cover liabilities on this scale; and it would have been strange if she had not resented her husband purloining her assets at this grief-stricken moment of her life. One thing is for certain. At whatever point Ernst decided to get rid of his wife, he was determined not to do so until he could add the Gotha dukedom to his own titles and assets.

Ernst and Luise paid a visit to Gotha in January 1823. Her uncle Friedrich IV – her father's younger brother – had inherited the dukedom. Though less than fifty, he had led an invalid's life since being badly wounded in the Napoleonic Wars. Ernst knew that he would not have to wait long (in fact, until 1825) before Friedrich died, and he could inherit at least the bulk of his estates, and the Gotha title, changing his style to Duke of Saxe-Coburg-Gotha.

During their stay in the Schloss Friedenstein, Ernst spent most of his days out on the freezing, dank hunting field. Luise, back in the gigantic house where she had spent her girlhood, had little to do, except to walk its endless corridors and salons, look at the portraits of her ancestors staring down from the walls, and cheer herself up as best she may by flirting – in particular with a court chamberlain called Baron von Münchhausen.

It was one of several flirtations which made her husband angry. Perhaps, in 1823, his accusatory bullying of Luise intensified because he knew that the scandal of La Belle Grecque was about to break. The long-dreaded memoirs were published in Paris. German Law enabled translations of the book to be banned in Ernst's own dominions, but these domains were not large. There was little to stop copies of the book coming over from France. The book caused a European sensation. It was especially popular in London, which, naturally enough, caused profound embarrassment to those two defenders of Coburg honour who remained in England, Prince Leopold, and his sister Victoire, now widowed and living with her little daughter Victoria in Kensington Palace.

Given that Leopold himself would become, much later in the century, the butt of a woman scorned, when his mistress Karoline Bauer's Proust-length exposé of her racy career was posthumously published in 1884, there is some amusement to be derived from reading his letters to Ernst about La Belle Grecque. Leopold initially excited Ernst's ire by offering to act as a go-between, possibly to adopt and educate her son by Ernst. Ernst told Leopold to mind his own business, and Leopold lobbed a grenade back, from which their fraternal relations never fully recovered: 'You write as head of the family but you forget that in that capacity you have a duty towards the Honour and Wellbeing of Your Own.' He went on to warn his brother that he had a duty to avert shame from members of their family, that he did not seem to realize how embarrassing La Belle Grecque's memoirs made Leopold's own position in England. Maybe to Ernst it did not matter what people were saying about him in London. 'To us [i.e. to Victoire and Leopold] it is important that we depend on our good name, and a reputation for decent behaviour'.[10]

Leopold was especially worried, because Ernst appeared to be on the verge of making a public reply to La Belle Grecque, in a periodical called *Journal of Luxury and Fashion*. To allow the German public to infer/gather the whole truth from documentary evidence appeared to Leopold an inappropriate defence. On the contrary, he advised an Iron Silence.

> Speak with almost no one about the matter and remember that you and yours have Enemies. You threaten to contradict her by public statements, and believe me, you will be doing yourself some really rather strange favours. I hope that you will first wait for everything to come together before you proceed. For my part, I do not threaten, but I warn you, to do nothing which could bring about an unavoidable breach between us; you would thereby lose friends you can rely upon; whether you have an abundance of those, I do not know.[11]

The public scandal did bring about a rift between the brothers in which Leopold clearly called into question Ernst's capacity to be leader of the Saxe-Coburg family. He went so far as to threaten the elder brother with the withdrawal of his support. Europe's 'late Dynasty' depended upon the split.

Leopold believed himself to have public opinion on his side by making a very public breach with his brother. 'In the worst case,' he wrote, 'I believe that the Public would forgive me, if I took enough trouble for the honour of the Meiningens, at the cost of undertaking the education of my brother's child and to avoid a situation where Madame A. will otherwise begin the shabbiest kind of show.'

He ended his letter to Ernst, 'Many greetings to Luise whose letter I've received. Adieu.'[12]

Luise's closeness to Leopold, whether platonic or not, was a further reason for Ernst's anger – anger with both brother and wife. Certainly, by now, Luise was playing the dangerous game of making her husband jealous by flirting, as her revenge for Ernst's unfaithfulness. On a visit to the Duchess of Gotha, her stepmother, one of the Ladies-in-Waiting, Countess Charlotte von Bock, accused Count Solms of being in love with Luise. He roared with laughter, but was also quite flattered, and tactlessly told the Duke. 'Had he been sensible, he would have laughed too, but he took it seriously and was very angry with me, until we made it up, and the whole thing ended in tears... I can laugh now that my anger is over,' she told her friend Auguste.[13]

These spats continued throughout the tense period when La Belle Grecque was threatening Ernst, and then acting upon her threats by publishing her memoirs. The Coburg Court was divided between those who sided with the Duke and fed the rumour-machine about Luise's supposed unfaithfulness, and those who hated Szymborski and blamed the troubles on him. She remained extremely popular with the people. Karoline Bauer, many years later, was to write in her memoirs, 'To avenge herself for many acts of infidelity on the part of her husband, and to distract herself in the loneliness of her heart, the Duchess had now likewise begun to spread love-threads on her own account. In Coburg the very sparrows on the roof twittered stories of the amours of both Duke and Duchess.'[14]

The sparrows were not necessarily telling the truth. By January 1823, however, when the pair were in Gotha for Luise's first visit to her childhood home since the death of her father, Ernst's agent Javon made his final offer to Madame Pauline – an income of about £200 per annum if she would be silent. It was not enough. A few weeks later, a Parisian publisher came out with *Mémoires d'une Jeune Grecque. Madame Pauline-Adélaïde Alexandre Panam, contre S.A.S., le Prince-Régnant de Saxe-Cobourg*. When the English translation appeared – which, to the embarrassment of Ernst's siblings in London, it inevitably did – it carried a quotation from the Psalms on the flyleaf: 'The mighty have stained my forehead with blood. From the depth of my misery, O Eternal! I have invoked thy hand against their injustice.'[15]

Nothing was going to be quite the same again in Ernst and Luise's marriage. Yet, as Karoline Bauer cynically and accurately noted years later, he was not willing to risk losing the title of the Duchy of Gotha being added to his Saxe-Coburg-Saalfeld; therefore he was unwilling to divorce her.

There was also the simple fact that Luise was extremely popular among the people of Coburg, and Ernst (and perhaps even more his evil counsellor Szymborski) was hated. Her final departure from the Rosenau on 28 August 1824 – shortly after Prince Albert's fifth birthday – was accompanied by a big public demonstration.

'The people loved me to the point of worship,' she wrote in a letter to her friend Auguste, 'and they considered my departure not as a reasonable agreement but as a betrayal and an act of violence. They got together secretly, and by word of mouth agreement, they gathered in their thousands at the Rosenau. Everything was peaceful and still'.[16]

The press were strictly forbidden to report the incident. When Luise came out and got into her carriage, the people unharnessed the horses and they themselves pulled the carriage along the five-mile journey back to Coburg. When she appeared on the balcony of the Ehrenburg to thank them for their love, the huge crowd simultaneously burst out with the hymn 'Nun danket alle Gott'. Luise recalled how moving it was to hear the thousands of voices singing. Doris Ponsonby, Luise's English biographer, speculated that the crowd believed Szymborski to have been more than Ernst's counsellor; she believed that the anger of the crowd, and Luise's slightly mysterious willingness to leave her sons and get out of Coburg so fast, stemmed from the belief that Ernst had stepped over the boundary of 'normal' heterosexual impropriety. It is unlikely we shall ever know the truth of that. For whatever reason, Luise, despite the adulation of the crowds, took her leave. She headed for the town of Sankt Wendel, and lived in the little Schloss which she had inherited from one of her family. From now on she went under the name of the Countess Pölzig and Bayersdorf.

Five months later, her uncle Friedrich, Duke of Gotha, died, and the title passed to her, and through her to her (still legally) husband Ernst. His Duchy now became the Duchy of Saxe-Coburg-Gotha.

Ernst had cynically waited before he divorced the wife who had brought the title to him. Once he was the Duke of Saxe-Coburg-Gotha, she could be discarded. In March 1826, Luise married her lover Alexander Elisäus von Hanstein, a twenty-three-year-old officer in the Prussian army who took her title – he was styled Count Pölzig. It seems to have been a happy marriage, though it was a short one. In 1831, she and her husband went to Paris to consult a celebrated gynaecologist called Dr Antoine Dubois. They resided at the Hôtel Mirabeau in the rue de la Paix. After a visit to the opera, Luise suffered a severe haemorrhage and was carried out unconscious. She lived another five months.

Her stepmother wrote that during this last illness, her thoughts were constantly with her sons. 'The thought that her children may have forgotten her entirely greatly distresses her. She wanted to know if they ever talked of her. I told her,' reported her stepmother, 'yes, they are far too good to forget their mother, but they are not being told of her suffering, as that would make them too unhappy.'[17]

She died on 30 August 1831, aged thirty. Albert always hated Paris. When one reads the story of his parents – an older royal prince, marrying a beautiful, inexperienced, lively girl; his intense unpopularity with his own people; her capacity to wow the crowds; his refusal to give up a former mistress; her taking lovers; her death in Paris – it is hard not to sense history's repetitions.

FOUR

CHILDHOOD

'TOUT RAPPELLE L'HOMME À SES DEVOIRS'

ERNST I WAS a deplorable husband, but he was an intellectually enlightened man, determined that his sons should be heirs not merely to titles, but to the riches of European culture. He took pride in his extensive library in Coburg, and in the art collection and vast collection of prints in the Veste. When he finally became the Duke of Gotha in 1825, he was an appreciative curator of his immense art collection in Schloss Friedenstein. In Coburg he founded a teachers' training college, and in Gotha an excellent grammar school (the Ernestinum). He also spent money which he did not possess restoring two theatres, one in Coburg and one in Gotha. (The theatre in the castle in Gotha is the earliest theatre in Germany, a Baroque gem.)

Johann Wolfgang von Goethe, between 1775 and 1832, lived not far away in Weimar. He remarked in his conversations that Germany's lack of unity or political cohesion was one of the reasons for its cultural richness. Whereas the most talented French actors, musicians, artists were ineluctably drawn to Paris, there was no place in Germany which was comparable. This meant that small German towns and courts could multiply the numbers of good opera houses, good universities – there were over twenty – good theatres and schools, and over a hundred public libraries. There was nothing like this in England, France or Russia.[1]

The part of Germany where Prince Albert and his brother Ernst grew up was especially rich in culture. Not only did Duke Karl August of Weimar attract Goethe to be his Privy Counsellor and companion, nearby Jena had Fichte, Schelling, Hegel and Schiller walking its streets and attending its salons and seminars. Gotha, Coburg, Meiningen and Erfurt, nearby, benefited from their proximity to all this richness. Ernst was determined that his sons should benefit from it too.[2]

He was an attentive father, not in the sense of spending a great deal of time with his sons when they were young, but as one who took them, and their education, very seriously. Two things stand out immediately. One is Ernst I's determination that the two sons should not be cut off from 'ordinary' people. At this date in England, royal princes might have received military or naval training (Queen Victoria's father was trained as a soldier in Germany, her uncle William IV, as an officer in the Royal Navy) but no suggestion ever arose that they should have been sent to school, or that they should have been taught their lessons alongside non-royal children. Ernst I decreed that throughout the boyhood and youth of Albert and Ernst II, they should be taught alongside their contemporaries – in the Rosenau, with local Thuringian boys; and, later, at a German university, they would not only share lectures but also lodgings with non-royal youths. Every Sunday they were in the Rosenau, a dozen or so local boys came up to play with the Princes; for their birthdays, there would be a huge crowd of local children – as many as 1,300 – playing and picnicking in the meadows near the Rosenau.[3] Queen Victoria never conquered her shyness, and when she was alone, after Albert's death, her inability to mix with people, or to show herself in public, became a major political handicap. Many English people found Albert irritating when he arrived as the husband of their sovereign – because he was a know-all who found it difficult to suffer fools gladly. Those who met him on any professional level, however, whether sitting on committees or discussing one of his pet subjects, always found him natural, unaffected and without 'side'.

The second thing which Ernst I laid down for his sons was that they should be given the widest possible education: English, French, Latin, Greek, Mathematics, Science should all be taught; attention should also be given to the arts: painting, music and dance. When they were very young, there was plenty of time assigned for play. Their toy soldiers, in the uniform of the Saxe-Coburg army, are preserved in the art collection in the castle at Coburg.

Perhaps because relations with his wife were so disastrous, Ernst took the strange decision to remove them from any female care at the age of four. The nursery maids were replaced not, as would have been normal, by a female teacher or governess, but by a male tutor.

It would seem that the man selected to supervise their education could not have been better chosen. He was Johann Christoph Florschütz, the son of a teacher at the famous Coburg Gymnasium. With a high brow, a shock of dark brown hair, bright eyes looking out from a sharp, bony face, Florschütz was a kindly, imaginative teacher to whom Albert remained devoted. His official title was 'Ducal Counsellor and Princely Instructor' (*Herzoglicher Rat und Prinzen-Instructor*). The boys called him 'Herr Counsellor'.

The mother was now gone forever. The father was as often as not four miles away in Schloss Ehrenburg in the centre of Coburg. From 1832, he lived there with his second wife, Duchess Marie von Württemberg – the teenaged daughter of his sister Antoinette. There is no record at all of how Albert and his brother responded initially to their father's second marriage. Nor do they appear to have had anything to do with their father's three illegitimate children – Berta Ernestine, who was the same age as Ernst, and twins Ernst Bruno and Robert Ferdinand von Bruneck who were younger than Albert. The two little motherless boys in the Rosenau, deprived of emotional compensations, grew very close.

Florschütz noted of the elder brother, the Crown Prince, that he 'possessed an extraordinarily kind heart' that made him very popular. 'For his brother he entertained the tenderest love. In any of those critical moments or doubtful situations which might crop up anywhere, he would rather give way and put himself at a disadvantage, even though complete self-control could never be said to be one of his most outstanding characteristics'[4] – so the rather convoluted Florschütz observed. The tutor appeared to be saying that, in those conflicts which always arise between siblings, Albert, the younger, more 'delicate' one, mysteriously controlled the more obviously open-hearted elder brother; that Ernst felt protective towards Albert, and at the same time a little deferential towards him.

Queen Victoria cannily noticed, both in her private *Recollections*, and in the published *Early Years of the Prince Consort* compiled at her behest and dictation by her secretary, General Grey, that Albert, from the age of five until he met his future wife, was focused chiefly on males. She observed that it was normal for a child to show distress, when transferred from the care of a female nurse to a male tutor. Albert, by contrast, 'showed a great dislike to being in the charge of women, and rejoiced instead of sorrowing over the contemplated change' from nurse to tutor.[5]

One might think that this was wishful thinking on his widow's part – the wish being to exclude all female rivals for his heart until, during his teenage years, he met her. Though she idealized Albert, however, Victoria did know him better than anyone, and her judgement is sound. Albert's only female playmate as a young child was his cousin Victoire – the future Duchess of Nemours, who was destined to die in childbirth at Claremont, like Princess Charlotte. Victoire, three years younger than Albert, was the closest he had to a sister. Though he remembered her fondly, and would be desolated by her early death, they could never be as close as siblings, and the age difference of three years is huge in early childhood. He was fond of both grandmothers. The very fact that he adapted with apparent ease to his father's remarriage, and the fact that he appeared to have an easy and happy relationship with his stepmother, suggests that with his mother's departure he sealed off a part of himself. He was not entirely unusual, as a nineteenth-century male, in being brought up in an all-male world. It did mean that when he encountered the Feminine in the full-blast personality of his cousin Victoria it would be a powerful experience – not one with which he ever entirely came to terms.

Albert was a clever boy, who responded to the Herr Counsellor's exacting standards with eagerness to succeed, and with grief when he failed. Florschütz made the boys keep journals. When he was only six years old, we find Albert writing, 'At home, I wrote a letter. But because I made so many mistakes in it, Herr Counsellor tore it up and threw it in the stove: at which I wept'.

There is a lot of crying in the journal. On 23 January 1825, when he was six, we find, 'When I woke up, I was ill. I've got the cough again, only worse. It was so fearful, I wept.' For 26 January, 'We recited, only I cried, because I could not recite because I had not paid enough attention. After the meal, I was not allowed to play because I had cried during the recitation.' On 28 February, 'At the lesson, I cried, because I could not find the verb, and Herr Counsellor pinched me, to show me what a verb is, and that was why I cried.'[6]

Queen Victoria would maintain that Albert often spoke of his childhood as the happiest time of his life. His daughter Vicky, however, remembered, 'Papa always said, he could not bear the memory of his childhood, he was so unhappy and wretched, and he had continually wished himself out of this world'.[7] As anyone who can remember a childhood, especially a 'sensitive' childhood, would testify, *both* could be true.

When Queen Victoria visited the Rosenau in 1845, she stayed overnight; after breakfast with Ernst and his wife Alexandrine, 'we went upstairs (2 flights) to where Albert and Ernest used to live when they were in the Nursery, & to the last Moment. The view is beautiful from these small rooms, under the roof. The paper on the walls is still full of holes made by them when fencing'.[8]

Her description of the experience is comparable, in vividness, to Yeats, in his poem 'Among School Children', suddenly imagining his beloved Maud Gonne in her infancy – 'And thereupon my heart is driven wild:/She stands before me as a living child.' Looking round at the apple-green walls, and the pretty, Biedermeier furniture, the imaginative Victoria was accompanied by Ernst and Albert as two young men; but in that attic retreat, she momentarily resurrected the Lost Boys and their vanished childhoods.

Most of their boyhood was spent in the Rosenau, whose elegant rooms were the backdrop of their earliest recollections. The well-planted garden, with its pergolas and shrubberies, was where they played, and which they drew and painted over and over again. Its surrounding parks and meadowland were the scenes of their boyhood games and their early sports. Like all Thuringians, they were, from their earliest years, enthusiastic for hunting and shooting. (Ernst used to say that he was able to stave off a revolution in Coburg in 1848 by, very reluctantly, allowing the general public to hunt on his land.[9])

Florschütz was the two brothers' tutor for the next fifteen years. Given their prodigious range of skills – mathematical, musical, linguistic, aesthetic – it is clear that Florschütz had an enormous influence. 'Ancient languages have a more theoretical application for children, modern languages more practical,' he noted.[10] He was determined that they should have plenty of both, as is made clear by the timetables he drew up for them, preserved today in the Kunstsammlung at the Veste in Coburg. By the time he was ten, Albert had a schedule each day which lasted eleven hours. The first lesson would be at 8 a.m., the last finished at 7 p.m. The subjects ranged from World History and Geography to Mathematics and Chemistry, from Latin and Music to English and French.

His school exercise books are preserved at Windsor. Latin, painstakingly copied in a fair hand when he was eleven: 'Mens tranquilla omnia optime facere potest'; or 'Parentes rectissime liberas suoas hominibus sapientibus comittunt' [*sic*].[11] In his French exercise book, the little boy inadvertently wrote the whole story of his life: 'Tout rappelle l'homme à ses devoirs' ('Everything recalls a man to his work'); the sentence prompting the thought in any reader – whence comes the need to be a workaholic? Presumably the narcotic of constant work blots out the painfulness of acknowledged emotions. As he himself put it, in the English exercise book which showed, aged ten, that he still had a long way to go before mastering the language, 'Amidst all his sufferings, he hade this consol~~gon~~tion' [*sic*]. When Florschütz set them to read *Hamlet*, he glossed the words 'to befit', 'self-slaughter', 'jelly', 'fowl', 'melt', 'thaw' and many others. One wonders what children so young, with so poor a grasp of the language, made of Shakespeare.

By the time he was twelve, he had begun the serious study of German history, beginning with Charlemagne, and, within two years, working his way forward to the Thirty Years War. It is refreshing, in all these relentlessly serious and demanding tasks, to find that he also liked doodling, especially drawings of dachshunds. As he put it in his English book, 'One must try to unite the useful with the agreeable'.[12]

Albert's fondness for drawing is evidenced by the drawings and paintings by the two brothers so lovingly preserved in Coburg. Albert's best efforts in this area are drawings of dogs. His passion for dogs never diminished. But both boys also loved to draw uniforms and warriors, both modern and historical. One of Albert's most detailed pencil drawings, done when he was twelve or thirteen, depicts a military commander in sixteenth-century armour, proudly displaying his coat of arms.

Heraldry was of huge importance to both children. In Windsor[13] he kept his childish notes of the Chivalric Orders of Europe, from 'Der St Annenordnen' in Russia to the Order of the Thistle in Scotland, the Order of St Elizabeth of Hungary, and the heraldic elephants of Denmark, which are so much in evidence at Gotha – a forebear having married into the Danish Royal House.

It was by studying heraldry, orders and the ritualized outward trappings of royalty and chivalry that they both learned and displayed the complexity and nobility of their lineage. The last Duke of Gotha, Friedrich, died in 1825, and it took several months for the Saxon Commissioners to decide how the surviving lands and titles should be distributed. The old Duke of Meiningen was pressing for himself to take over the Coburg dukedom, giving the lesser one of Gotha to Ernst and his heirs. It was with some triumph that Ernst in fact managed to retain and acquire both, in exchange for ceding the dukedom of Saalfeld to Meiningen.[14] Such matters will probably seem esoteric and boring to the majority of modern readers, but to Albert and his family they were central. The Saxon Princes, in their reading of history, had played a vital role in checking the unbridled powers of the Holy Roman Emperors. The division of his princely family, the Wettins, into two branches at the time of the Reformation was also a source of interest and pride for religious reasons. The elder, senior branch of the family, known as the Ernestine, had been among the first to embrace the doctrines of Luther in the early decades of the sixteenth century. In the barons' wars which broke out after the Reformation took hold in Germany, John Frederick the Magnanimous, the direct ancestor of both Prince Albert and Queen Victoria, was defeated at Mühlberg by the Emperor Charles V. The Ernestine, senior branch of the Wettins, was deprived by the Catholic Emperor of some of their richer spoils. The younger branch, the Albertines, became the Kings of Saxony. The name of Ernst was carried proudly by the Dukes of Coburg through all the generations down to that of the two boys in the Rosenau. Their family, after all, through three centuries, had been living proof of the Importance of Being Ernest.

In the summer of 1831, Duke Ernst visited his sister in London. George IV had died the previous year, to be succeeded by his brother, the Duke of Clarence, who reigned as William IV. Now all eyes were on the Coburg Duchess of Kent in Kensington Palace, and her little daughter. Their old mother Duchess Auguste, back in Coburg, wanted Ernst to remind Victoire that the little Flower of May (as she called Victoria) was destined for yet another arranged Coburg marriage – namely to Albert. No doubt Ernst was designated to remind the new King, too. It was not enough to have Coburg females installed in the centre of British power. It was essential, for Auguste's dynastic ambition to be fulfilled, to reverse the calamity of poor Leopold's loss of Princess Charlotte.

In their father's absence, the two boys visited their step-grandmother in Gotha.

On 13 September 1831 – less than a month after Albert's twelfth birthday – the people of Gotha were informed by a notice in the local newspaper that their former Duchess, Luise, Albert's mother, had died in Paris. Luise and her second husband Alexander stayed at first in the Hôtel Mirabeau in the rue de la Paix, but evidently (see below) after a while, moved to cheaper lodgings. When they had been in the city a couple of weeks, they decided to go to watch the famous ballet dancer Marie Taglioni. It was during this performance that Luise suffered a severe haemorrhage and was carried out of the theatre unconscious. Four doctors were consulted, and she was pronounced to have advanced cancer of the womb. She lived for five months, but her husband, though a devoted companion, was not actually with her when she died. The only person in the room was her companion, Anna Metz. 'Your highness recognizes me? You know who I am?' Luise smiled, and died.[15]

Her will,[16] which Prince Albert preserved at Windsor, is a Balzacian document, written in French, and describing her as '*Madame Dorothée Louise Pauline Charlotte Frédérique Auguste Duchesse de Saxe, contesse de Polzig et de Baiersdorf actuellement épouse de Mr Maximilien Elysée Alexandre comte de Polzig demeurant ordinairement dans la ville de Saint Vendel, principauté de Lichtenberg de présent à Paris logé avenue des champs Elysées: près la rue d'Angoulême, trouvée dans une pièce donnant sur l'avenue au premier étage d'une maison dont est propriétaire M Chernay architecte*.' Oddly, the will is dated almost one month *after* her death, 26 September 1831. It names her two sons as her '*heritiers universels*'. She did, however, leave her husband, the handsome officer for whom she had wangled the title of Count Pölzig, the 25,000 Thaler which was her legacy from her uncle Friedrich – last Duke of Gotha. This sum, Ernst doggedly refused to pay.[17] The wranglings after her death, between her two husbands, were all but interminable. In the state archives at Coburg there are 630 pages of legal documents and letters, all to do with her estate and her expressed wishes about her burial.

Count Pölzig was not merely impoverished by her death. He was burdened with the question of what to do with her body, which, upon her instructions, was embalmed. She wanted to be buried with him, in the crypt of the church of Sulzbach. For some reason, permission was not given by the relevant authority. He paid 40,000 francs for the embalming, and took the corpse back to Sankt Wendel, where it was temporarily laid to rest. The ever-gossipy Karoline Bauer recorded in her memoirs that Luise had left instructions that he must never be parted from her body, on pain of losing his inheritance. 'For years the luckless Count Pölzig dragged about with him the embalmed corpse of his spouse from place to place, but one morning to his terror he found that the precious coffin had vanished. He had, in the meantime, married a Fräulein von Karlowitz. When he found that his allowance continued to be paid to him, he soon got reconciled to his loss.'[18] This wildly inaccurate anecdote gives a flavour of Bauer's style, and explains why Albert's brother, when the book appeared, did his best to have it suppressed.

Nevertheless, like much wild gossip, it was not pure invention. Luise really did leave instructions to be embalmed, and, thanks to the recalcitrance of her first husband, she was not able to be buried where she wished. It was years later that Ernst and Albert arranged to have her body moved to a mausoleum where she was laid beside her hated ex-husband. The gesture is surely an eloquent expression of Albert and Ernst's wish, deep into their grown-up lives, that their parents had never been separated.

Three months after the death of their mother, the boys also lost their Coburg grandmother, Auguste. She was seventy-four. One of the most formidable networkers and matchmakers in royal history, she had lived long enough to see her son Leopold chosen as the first King of the newly created Belgium. She had made the journey to

Brussels to see him. It was not perhaps as distinguished as being Princess Charlotte's consort, but a kingdom is a kingdom. (He was previously offered Greece, and, after much dithering, turned it down; a decision he always regretted.) She had high hopes, as she died, that Albert would make a great dynastic match with the Flower of May. Leopold did not come back to Coburg for the death-bed, but she died in the arms of her two other sons, Ferdinand and Ernst I.

FIVE

'THESE DEAREST BELOVED COUSINS'

ALBERT'S FIRST VISIT TO ENGLAND

As the two children became teenagers, the programme of teaching became more intensive. The court painter, Sebastian Eckhardt, who had captured the likeness of the two Princes, was officially engaged to teach them to draw. At about the same time, their cousin Victoria, in Kensington Palace, was being given drawing lessons for two hours a day.[1] It was Victoria's mother Victoire, incidentally, who had first engaged Eckhardt as teacher and court painter in Meiningen, back in 1818, the year of her marriage.[2] Biographers of Victoria have tended to emphasize the role played in her education by Victoire's evil genius, Sir John Conroy. And undoubtedly, with his notorious 'Kensington System' – a scheme designed to keep the young Princess at a distance from the Court and the King – Conroy played a part. He was, apart from anything else, always on the spot. It would be wrong, however, to underestimate the influence of Victoire's brother Leopold on the turn of events. It was he who helped to draw up the timetable for Victoria's lessons, which involved two hours every morning of the serious subjects, such as Latin, Geography,

History and modern languages – English, French and German, in all of which Victoria was fluent – with music and drawing in the afternoons. Uncle Leopold, determined that she and Albert should be principal characters in History, emphasized to Kensington and to Coburg the vital importance of the History lessons.[3]

The boys in Coburg were being given an even more rigorous educational force-feed, with additional tutors being brought in at Florschütz's discretion to help in specialist areas. Herr Ruth, for example, taught them Latin to a very high level. In 1832, when Albert turned thirteen, they were translating most of Ovid's *Metamorphoses* into German.[4] By 1835 they had read through Sallust's *De Conjuratione Catalinae*, and much of Cicero. In History, as they advanced through their teens, they had progressed from the Thirty Years War to the French Revolution, the implications of which haunted Albert for his entire career. How to preserve monarchy while accepting the principles of the Enlightenment? Leopold I, as the King of Belgium, believed that he had cracked the secret of how to do this, and we watch Leopold in turn being guided by Stockmar.

Interestingly enough, Stocky held back from open interference at Coburg during the boys' early teens. Albert scarcely knew him at this point, whereas Victoria, and the English politicians, saw him all the time. The file of Letters and Papers of Baron Christian Stockmar,[5] preserved at Windsor Castle, shows that Stocky was in constant touch with the Duchess of Kent (Victoria's mother), with Conroy, and with the brother of the late Prime Minister, the 3rd Lord Liverpool, with whom the Duchess of Kent and Conroy, with the little Princess, often stayed, with Stocky in attendance.[6]

Stocky left nothing to chance. 1832, the year in which Albert first started to contemplate the French Revolution and to mull over its implications for monarchies of the nineteenth century, was also the year in which Britain saw the passing of the Reform Bill. There was no question that the Whigs, who – against the fiercest opposition by the King and by the Tories – had brought in this really very modest reform of the electoral system and very minor extension of the franchise, viewed it as a modification of royal power. One Tory MP remarked that William IV must either accept the Reforms of Earl Grey 'or start for Hanover' – i.e. be dismissed, and simply be the King of Hanover, a title he held in conjunction with the crown of England.[7]

Meditating upon the American Revolution in one of his private political essays, Stockmar saw that the problem facing the Founders of the Republic had been 'the need to satisfy the will of the people, whilst limiting their influence; to give weight to the will of the people, whilst holding in balance the Constitution. And they sought this balance by the establishment of the Senate'.[8]

Something very similar, Stocky saw, was going on in England. The Whigs did not want democracy in any modern sense. They wanted to retain power for themselves. They had limited the power of the monarch, but it was necessary for them to give at least the appearance of offering more power to 'the people'. Stockmar saw the position of any future monarch – preferably a Coburg-controlled monarch – as potentially crucial here. Rather than giving real power to the people, or leaving it all in the hands of the Whig aristocracy, a reformed representational system could make use of a politically astute monarch. 'The state of the English Constitution at the present day takes its origin from the former unbridled power of the monarch and the feudal arrangement of boundless power of the aristocracy. Only very slowly and gradually did the people, through a process of enlightenment, redress the balance'. For the 'balance' to be maintained, it was essential to keep the monarchy.[9] As Leopold settled in to the newly founded Kingdom of Belgium (he accepted the role on 4 June 1831), Stockmar saw his reign as a sort of 'dummy run' for a future Coburg Queen – or King – of England, where the sphere of influence was so much broader, the power base so infinitely stronger. It was in readiness for all this that Florschütz and Duke Ernst I were preparing the two young Princes during their daily lessons at the Rosenau.

In 1832, Albert and Ernst made their first visit to Brussels to see their uncle Leopold. They remained in constant touch with him, and in 1836, when Albert was approaching his seventeenth birthday, it was agreed that he and Ernst should spend a year in the Belgian capital.

When one says 'agreed', it was not accepted by all Duke Ernst I's advisers that Brussels was a healthy environment in which the youths should be let loose. On 20 February 1836, Ernst had decided that they had reached a new threshold. In addition to Florschütz, who remained their Director of Studies, he appointed Oberstleutnant von Wichmann as their governor. Baron von Wichmann was a soldier of the old school, who had fought at Waterloo, described by one biographer as a 'stiff introvert'.[10] He was by no means enamoured of what he heard of Brussels, which he feared was a nest of modern ideas. He favoured a period in Geneva, but he was overruled and they set out to spend ten months in the Belgian capital. It was a carefully budgeted excursion. Wichmann himself was given a salary of 2,500 francs; Florschütz 1,500. Four indoor servants accompanied them, on pay which varied between 228 francs and 180 francs, and three grooms, who were paid 72 francs for the year, with 20 francs to spend on their livery.[11]

Before the Brussels period began, the Princes had a whistle-stop tour of German-speaking cities – Dresden, Prague, Vienna, Pest and Buda; then Leopold took the Princes to make their first visit to London – in June 1836. It was the first time they had been apart from Florschütz in a decade, which was – as Albert admitted in a letter to his stepmother – 'the only thing which spoils our pleasure'.[12]

The London visit was not a casual thing. It was a deliberate statement of intent by the Coburgers. They had heard that William IV had invited the Prince of Orange and his two sons to attend the summer parties to commemorate his seventy-first birthday and the seventeenth birthday of his niece Princess Victoria. William made no secret of the fact that he wanted Victoria to marry one of the Dutch Princes, and to bring the Coburg influence in England to an end. Stocky and Leopold had to act fast, and, heedless of protocol, to push in regardless of the Orange faction. Lord Palmerston, the Foreign Secretary, ever destined to be Prince Albert's enemy, urged Sir John Conroy not to allow the Coburg party to stay at Kensington

Palace during their time in London. The King considered it inappropriate that Victoria and Albert should sleep under the same roof, and Palmerston suggested that it would be better if Albert, Ernst, their father and entourage were accommodated in an hotel.

'I am really *astonished* at the conduct of your old Uncle the King,' Leopold wrote to Victoria. 'This invitation of the Prince of Orange and his sons, thus forcing him upon others, is very extraordinary... really and truly I never heard anything like it, and I hope it will *a little rouse your spirit* now that slavery is even abolished in the British Colonies, I do not comprehend *why your lot alone should be to be kept a white little slavey* in England, for the pleasure of a Court, who never bought you'.[13]

The Orange Princes got to London before the Coburg ones, but they were so charmless that Victoria never felt tempted to marry either of them, whatever her old uncle desired. Albert, aged seventeen, had seen Coburg and Gotha, tiny little cities, little more than country towns with palaces. True, they had just been to Vienna and Prague, but nothing could have prepared them for London, the Great Wen, a huge cauldron of humanity sprawling along the banks of the Thames and rapidly expanding in population by the decade. Vienna at this date, capital of the Austro-Hungarian Empire, had a population of 270,000. London was approaching two million, and by 1860 would be over three million. No city in history had ever been of such monstrous size.

Nothing could have prepared them, either, for the bustle and crowds at Court. Nor for the filthy water. Naturally, Albert fell ill; but even the robust Ernst and other members of the German party found themselves with 'gippy tummy'. Somehow, they managed to stagger through, though one brother or the other found themselves too ill to get up on at least a few of the days. Nine days after Cousin Victoria's seventeenth birthday, Albert wrote back to Florschütz:

> On the 24th, my dear cousin's birthday, the King gave a great ball at St James's Palace, the one at which my partner and I fell, & I was taken ill & had to go home during the first dance.

> The first day I came out, after a most unpleasant illness, was the 27th, to the King's levee at St James's Palace. We went at 2 o'clock & stood for two and a half hours in a dense crowd of people. The ambassadors of all Europe, from the Russia to Hundustan [*sic*]... passed by the King and exchanged a few words with him; then came the whole of London society, famous and tedious people. ['*kam die ganze Londoner, berühmten und langweiligen Männer*'.] To me the most interesting were the Duke of Wellington, General Hill, General Beresford, Lord Grey, Sir Robert Peel, Sir John Russell [*sic*], Lord Palmerston, the Speaker, the Lord Chancellor, the Lord Mayor & the Archbishop of Canterbury... I was terribly tired, for we had to stand from nine o'clock till 2 in the morning & I dared not go to sleep.[14]

Four days later, it was the King's birthday. William IV for much of his life had been a good-natured, sybaritic naval officer who 'affected the language and manners of the rough and hearty tar'.[15] He was not (quite) as much of a buffoon as his Whiggish opponents believed. Having lived with one of the greatest actresses of the age (Mrs Jordan) for twenty years, and fathered ten children by her, he had chosen wisely in his Queen, Adelaide of Saxe-Coburg-Meiningen, an astute, generous-hearted woman who did much for the popularity of the monarchy, devoting most of her money to charity, and trying (unsuccessfully) to make peace between the King and Princess Victoria's mother.

It was William's keenest ambition to live long enough to allow Victoria to reach her majority, thereby thwarting the ambitions of Conroy to become the power behind her throne. In this ambition, William would succeed. In his attempts to thwart the Whigs, and the progress of Reform, he did not always show much political nous. In 1835, he had lost his temper with the liberal reformists and exclaimed, 'Mind me... the cabinet is my cabinet; they had better take care, or, by God, I will have them impeached.' Such sayings were widely derided. Stockmar could see that, if the monarchy was to have a hope of surviving in a Europe where republicanism was strong, it would need more sophisticated defenders than King William. The young David, snatched from the sheepfold for the purpose, was now being conducted into the incomprehensible crush of William IV's seventy-first birthday.

> That day 3,600 ladies and gentlemen filed before the King and Queen in the grandest apparel... & so on – more balls. On Monday evening my aunt gave a very brilliant ball here at Kensington. I wiped out my disgrace at the last by dancing without a break until 4 in the morning, when the sunlight took the place of the wax candles...
>
> Today we are to see a new English opera & after that I go with Papa to a Ball at the Duke of Devonshire's. Ernst, unfortunately, cannot accompany us, as he is in bed; he is suffering from an attack similar to the one I had last week. None of us can get accustomed to the climate here, the way of life, or the food... You can imagine that all this makes it very difficult for us to see the wonders of London.[16]

Albert was no doubt agog to see the Parthenon Marbles in the British Museum, the Italian Old Masters at the National Gallery and the splendour of George III's library. The wonder of London, however, which was of most concern to his minders was, naturally, his animated, plain little Coburg cousin Victoria.

Of course, Victoire defied the King, and the Coburgers all stayed together in Kensington Palace. The pattern of Victoria's future life with Albert was foreshadowed by a letter the Princess wrote to her uncle Leopold. 'I am sorry to say that we have an invalid in the house in the person of Albert who, though much better today, had had a smart bilious attack. He was not allowed to leave his room all day yesterday, but by dint of starvation, he is again restored to society, but looks pale and delicate.'[17]

There is a bouncing exuberance about Victoria, as she vividly continues, to this day, to live on the written page. Her energy, her appetites, her passion for dancing, appeared boundless. At her seventeenth birthday party, she danced five times – with the Duke of Brunswick, with both the Dutch Princes – William and Alexander – with Ernst and with Albert. In the country dancing after supper, she danced with Albert again, but he was compelled to retire to bed, whereas she stayed up until half past three and felt 'all the better for it next day'.[18]

She was not attracted to Albert. He was less tall than Ernst, and at this stage he was rather plump – she went so far as to say 'very stout'.[19] He was much too young to marry, and so was she. Her companion, Baroness Lehzen – who had been governess in turn to her Leiningen half-siblings and then to her, and who was her first really close companion and confidante – urged her not to take her loose agreement with Uncle Leopold as a firm commitment to marry Albert, and she agreed. Indeed, she later said that she did not consider herself 'bound to marry Albert at all'. Blood is thicker than water, however, and, if, on her first meeting with Albert, she felt no sexual attraction, she had become aware of the powerful ties of family. She always felt closer to her mother (irritating as she was) and to Uncle Leopold than she did to William IV. She was always more a Coburger than she felt herself to be a Hanoverian.

On their last day, she told her journal, 'At 9 we all breakfasted for the last time together! It was our last happy happy breakfast with this dear Uncle and these dearest beloved Cousins, whom I do love so very, very dearly, much more dearly than any other Cousins in the world. Dearly as I love Ferdinand, and also good Augustus, I love Ernest and Albert more than them, oh yes, MUCH more.'

Stockmar, who did not really know Albert at this stage, had been highly impressed by what he saw of the young man in London. Even his youth did not count against him in Stocky's beady little eyes. 'His youth was the guarantee of his pure unspoilt nature.' His good education would 'enable him to give the Princess the political support she will one day so badly need... In the whole *Almanach de Gotha* there is not a single Prince of riper years to whom we could entrust the dear child, without incurring the gravest risk.'[20] Luckily for Stockmar, this broad claim would not be put to the test. The Princess was not attracted to either of the Dutch Princes. No one else in the *Almanach* would be considered. Before Albert and Victoria were put together in the royal cage, however, Albert's pure unspoilt nature required a little more polish. Every cultivated German boy at this period would need, before he began his adult or professional life, to study at a university, and to travel. Two or three years at university and a year of travel would have been normal. But Stocky and Leopold were middle-aged men in a hurry. Albert would have a Grand Tour in miniature, and a speeded-up, truncated university career.

SIX

EUROPEAN JOURNEYS

THE TEN MONTHS in Brussels were a vital part of the plan, consolidating their closeness to their uncle Leopold. Here was Albert's first practical political lesson. Not only could he watch a constitutional monarchy in action. Leopold, from the neutral position of the newest European nation (and one of the smallest), could also open Albert's eyes to the international political picture. He could explain to Albert the actual situation in Great Britain – a complicated political scene in which the Whig complacencies and their recent Reform Act could go nowhere near to solving: unrest in Ireland, on the mainland, the vast growth of the urban poor, the development of a real socialist movement which showed every sign of gaining ground, the instability of agricultural economy, the potential calamity of the big landowners holding on to tariffs to keep up the price of corn while the population were hungry. Leopold, married since 1832 to an Orléanist Princess, Louise-Marie (daughter of Louis-Philippe), acquainted Albert with the factions in France that would keep that country unstable for the next two decades. In Spain there was a civil war. In northern Italy, Italian nationalists were poised to break away from the Austro-Hungarian Empire under whose dominion they had lived for centuries. Poland, similarly, was hoping for independence from Russia. Russia, like Austria, was dealing with the movements for change by an ever more draconian conservatism. Metternich in

Vienna was urging his Emperor to be equally reactionary. This in turn was having its effect on the future of the other German-speaking lands. Only by reconciling the military ambitions of the Protestant northern Kingdom of Prussia, and the Catholic absolutism of the Habsburgs could the future of Germany be envisaged. And how many of these European countries could see, and understand, why Britain, for all its social problems, overcrowded cities and political unrest, was becoming prodigiously richer than anywhere else: because of trade and the advance of technology?

Albert learned these lessons quickly, and would waste no time, when he eventually married his cousin, in trying to make the British monarchy follow the pattern and examples of Uncle Leopold's in Belgium. (Leopold eagerly encouraged the building of the Belgian railways, for example, against the fierce opposition of the Catholic Church and the forces of political reaction.)

Nor need Albert's governor, Baron von Wichmann, have worried about Albert becoming infected by the spirit of left-wing reform. Although his uncle Leopold, by Wichmann's standards, was a radical, the truth is that they were all, in their modified ways, advocates of absolutism. Albert was an economic liberal, but he never truly became one in politics, and his university education prepared him, in his own eyes, to become an enlightened despot.

The choice of a university at all for a royal personage was, by British standards, eccentric. Ernst I, however, had no doubt that it was the right course for his sons, and it simply came down to the choice, not whether they should have a university education, but which one they should select. Berlin was unsuitable on several grounds. Prussia was seen by the Coburgers as a 'Parvenu' state. Berlin students had a reputation for being dissolute. Stockmar believed it suffered from 'an epidemic of dissolute behaviour, like catarrh'.[1] There was also the sense that in Berlin they would learn little except practical studies – military and administrative. Vienna was considered unsuitable because it was Catholic, and because the young men would be studying there very much under the shadow of the Emperor. One can imagine some princes deriving benefit from the deep classical learning on offer at Göttingen, the science at Berlin or Cassel, the metaphysics at Jena and Heidelberg. Wichmann would no doubt have considered these latter places to contain radical elements with whom it would be dangerous for the young Coburgers to mix. The truth is Albert, though a brilliant man in so many ways, was not an abstract thinker, and the choice they made for him and his brother – the University of Bonn – could not have been better. It combined practical and intellectual subjects. Nor could it have been lost on Ernst – even more musical than Albert – that it was the city of Beethoven's birth. They had also, though they were not to know this, arrived just one year after the departure of one of Bonn's most celebrated students – Karl Marx.[2]

They arrived in Bonn in April 1837, with their greyhounds – Albert's was named Eos – and their valet, Cart – Isaac François Daniel Cart, a French-speaking Swiss who was also fluent in German, and who had been their personal servant since they were little boys of seven and eight.[3] They would spend three semesters there as students. As was usual for sons of the nobility or royalty, they did not lodge in university halls, but found a house in the town, where their dogs and personal servants could try to make them feel at home. Their friends were to be the Duke of Mecklenburg-Strelitz, Prince William of Löwenstein-Wertheim and Count Erbach. Prince William was an accomplished musician, something which provided a bond with the Coburgs – both of whom had already tried their hand at composing. William recalled, after Albert's death, that they had 'passed many an evening in song... to the despair of Colonel von Wichmann'. He also remembered amateur dramatics. He is one of the few to remember Albert for his sense of the ludicrous, but perhaps this is slightly different from having a sense of humour? He remembered Albert doing caricatures of the lecturers and imitating fellow students, but 'the great business of his later life, the many important duties he had to fulfil, soon drove into the background the humorous part of his character'.[4]

Although Baron von Wichmann remained their companion, it would seem as if Florschütz was sent back to Coburg, for we find

Albert writing to him, rather pathetically, a month after their arrival in Bonn, 'I long so for a word or two from you in our loneliness'.[5]

In addition to homesickness, Albert also had begun to be worried, young as he was, by the prospect of baldness. The baldness gene, which we see so clearly in the present British Royal Family – with two of Queen Elizabeth II's sons being bald, and Prince William also – was a strong one. 'In order to save my hair from coming out entirely,' he wrote desperately, 'I have embarked on a radical treatment at the hands of Herr Smakur, who rubs my head every evening with a very greasy oil. He promises the best results'.[6]

The two Princes were not expected, however, to live for pleasure, and Albert soon found himself with a heavy programme of study. The lectures on which he was urged to concentrate were those on Jurisprudence, Politics and Philosophy. When he had a son of his own about to start at university, Albert told him that Law was 'a subject in which I took the greatest interest'.[7] He kept very full notes on the lectures of Professor Moritz August Bethmann-Hollweg, grandfather of the Chancellor of Germany at the outbreak of the First World War. Bethmann-Hollweg was a follower of Kant, and taught the students that Roman Law was a reflection of Natural Law. He placed enormous emphasis on the study of Ethics. States were, in Bethmann-Hollweg's view, therefore not purely human or utilitarian inventions. They existed as it were by divine decree. And although princes were not there by Divine Right, exactly speaking, the monarch was a symbolic embodiment of the unity of the state in one person.[8]

While for most of Bethmann-Hollweg's audience these were purely theoretical considerations, for Albert they were not. As Bethmann-Hollweg intoned that 'The whole of the Law concentrates itself in the personality of the Prince', Albert contemplated his destiny. He had only been at the university for a few weeks when the news came that old King William IV had died in England, and Cousin Victoria was now the Queen. Although he was only eighteen, Albert must have been made aware, by his father and by his uncle Leopold, that he should be ready at any time to secure her as his bride. They were on edge about this. Already, the young Emperor of Russia had proposed to her, and although she had given the Coburgs the understanding that she would marry Albert, nothing had been settled.

To make quite sure that they were not overlooked in this matter, Duke Ernst I accompanied his brother Leopold to London for the coronation to show solidarity with their sister Victoire. The young Queen, in banishing Duchess Victoire's villainous companion, Sir John Conroy, had also distanced herself from her mother, and this was not perhaps the best moment for the Coburgs. All sorts of other influences could now be brought to bear on Victoria.

What was happening? Stockmar's closest contact with Victoria's Household was with the late Prime Minister's brother, Lord Liverpool. Conroy, Liverpool told Stocky, was lying. His children were going about telling the world even in 'pastry cooks' shops' that 'Sir John is enchanted with the Queen'.[9] (They in fact loathed one another.) The Queen held her first Drawing Room and 'enchanted everyone'.[10] Liverpool asked Stocky to use his influence, and that of Leopold, to prise the Duchess of Kent away from Conroy's dire influence. 'Our dinner on Monday went off very well. I handed the Duchess of Kent into dinner & altho under restraint [*sic*]... I could not help thinking she was altogether better dressed and more at her ease than I should have expected. In fact I really think that if King Leopold comes over soon and talks firmly but kindly towards his sister, & that [*sic*] Melbourne supports him, is the absolute necessity of Conroy's removal'.[11] When he did come, Leopold discovered that relations between Victoria and her mother had broken down, and wrote desperately to the Duke of Wellington about 'the unfortunate disagreements which have occurred between the Queen and my sister respecting Sir John Conroy'.[12]

Absorbed as he was in his university studies, Albert was homesick. They had some fun – they took part in a stag hunt; they attended amateur dramatics. But his stomach pined for home food. 'The music and fried sausage at the fortress must be wonderful'.[13]

He was painfully aware that if Uncle Leopold, and Stocky, and his father, were successful in their attempts to ingratiate themselves with the Queen and the British politicians, his own future would lie not in beloved Coburg but in unknown Windsor and London, surrounded by politicians who had no sympathy with him. Already, the Queen had formed a close bond with her Whiggish Prime Minister, Lord Melbourne.

Sensing the diminution of Albert's days of freedom, when he could travel, study, please himself, Uncle Leopold suggested that the two boys, at the end of their semester, should do a tour through Southern Germany, Switzerland, and, if they had time, take in Venice. Uncle Leopold was also anxious that the Princes should be out of the public gaze, for rumours were circulating that his engagement to Victoria would be announced, and in Bonn, all eyes were upon him.[14]

It was a two-month vacation, during September and October 1837. He sent two souvenirs of the Swiss journey to his cousin in Kensington, which she treasured for the rest of her life. One was a pressed flower – 'Rose des Alpes' – and the other was a scrap of Voltaire's handwriting, given him by an old servant at Ferney, who had worked for the satirical playwright and sower of discord. Voltaire, who had befriended Frederick the Great and Catherine the Great, was the inspiration not only for the Revolutionaries, but also for the monarchs of the Enlightenment. The scrap of paper, and the philosopher's pencilled hand, signalled Albert's wish to be an enlightened despot in the mould of Frederick the Great.

They had thick rain when they came into Switzerland, but the weather cleared after their first week. By the time their diligence had penetrated as far as Brunnen, they abandoned the carriage in favour of ponies; and when they had climbed further – Altdorf, Amsteg, the Devil's Bridge, Furka – winds, snow and storms were blowing about them. Between the squalls and flurries of snow, there were glorious moments of clear sunshine, when the depths of the valleys and the soaring snow-peaked heights above them filled both boys with awe.

When his firstborn son was in Chamonix in 1858, Albert wrote to him:

> I was certain that the Alps would make a deep impression on you. No one can behold them without being struck by the Majesty of God's Creation and the insignificance of Man! One feels oneself almost morally improved by the contemplation of such grandeur and beauty. The 'Met glace' is a wonderful sight, & I well remember the pleasure which the tour to the Jardin gave me exactly 20 years ago... I thought of anything rather at that moment (just of Vicky's age) than that my son should be there 20 years later, on the same spot, & coming from England.[15]

It was a glorious tour, all too short. From the frisson of meeting an old man who had served Voltaire, to the sublime Romantic experience of the mountains; from the peaceful anonymity of Swiss hotels, to the sight of Venice, where they had only two October days, it was a thrilling holiday. 'What thanks I owe you, dear Papa, for having allowed us to make such a beautiful tour! I am still quite intoxicated by all I have seen in so short a time.'[16]

The Princes returned to the Rosenau for a few days, but by the beginning of November, they had to return to Bonn and resume their studies. They worked there over Christmas, but at the beginning of the New Year, 1838, Leopold summoned them to Brussels. He had been over to London to consult with the Queen about the coming nuptials. Victoria's reaction was that, at eighteen, she was too young, and that Albert – in appearance alone – was far too young. By March 1838, Leopold was writing to Stockmar in England:

> I have had a long conversation with Albert, and have put the whole case honestly and kindly before him. He looks at the question from its most elevated and honourable point of view. He considers that troubles are inseparable from all human

> positions and that therefore if one must be subject to plagues and annoyances, it is better to be so for some great or worthy object than for trifles and miseries. I have told him that his great youth makes it necessary to postpone his marriage for a few years.[17]

Albert's response reveals the high sense he had of his own dignity. He would be prepared to wait, he said, as Victoria suggested, 'But if after waiting, perhaps for three years, I should find that the Queen no longer desired the marriage, it would place me in a very difficult position and would, to a certain extent, ruin all the prospects of my future life.'[18]

It is clear that Albert had entered wholeheartedly into the idea that he was being groomed and prepared to become Britain's Enlightened Despot. Only the Little Woman on the throne was thwarting his ambition. He agreed to a plan which Victoria and Leopold had concocted together. Albert should complete two more semesters at Bonn, and then go on a tour of Italy. This would take them into the middle to late months of 1839, when the whole question could be revisited. To spare Albert's dignity, and to avoid what he considered the danger of ruining 'all the prospects of my future life', the conversation was supposedly a secret, even though all Europe now knew that Leopold and Stockmar were working busily to strengthen the Coburg connection. Little as he could afford it, Duke Ernst I hurried to witness the coronation, in the summer of 1838, while Albert and Ernst toiled at their lectures.

'It must,' mused Albert, 'have been a sight never seen before. The Queen is said to have behaved with complete sangfroid and calm. She was the only person who was not in the least tired by the ceremony. Papa found her somewhat grown, but otherwise not in the least changed.' With Ernst returned to Germany, Leopold prepared to go to London in July. 'I hope much from his presence,' Albert said. 'Although he is said to be no longer in high favour, his mere appearance will restore him to it.'[19]

It would be a mistake, though, to think of the remaining two semesters at Bonn as a mere marking time before Victoria was ready to marry her prince. The basis of their study was, as has already been seen, the Law. At Bethmann-Hollweg's lectures, they had considered the philosophical basis of Jurisprudence. From Perthes, they learned about German Law and German Constitutional Law in particular. Here, anyone approaching Germany in the late 1830s from the perspective of the twenty-first century needs to be careful how they use the term 'liberal'. Clemens Theodor Perthes was regarded by his contemporaries as a 'liberal' – far too liberal for such as Baron von Wichmann. While being a professor at Bonn he also served as a Conservative Councillor, later becoming a member of the Second House in Berlin, during Bismarckian times. He was on the 'liberal' wing of the German Conservatives, which meant that by modern standards, he was not truly a liberal at all. When Albert's brother, in his memoirs, looked back on their university days, he saw Perthes as an ultra-reactionary, a supporter of enlightened despotism. A more nuanced twenty-first-century perspective[20] can see that Perthes was a liberal by comparison with his reactionary contemporaries. Those who follow Albert's political career can perhaps, in part, trace some of their puzzlements to the influence of Perthes. Perthes did not regard the monarch as the all-powerful representative of God on earth, but he did think that monarchies were more orderly than republics. Ernst II wanted to represent him as a defender of absolutism, but this is not true.[21] Of the three types of governance on offer in Europe – between oligarchy, republicanism and monarchy – he favoured the latter because it maintained and consolidated the unity of the state in a single person. Probably no one in England, since the onset of George III's madness, quite wanted to contemplate what the philosophical basis of their Constitution might be. The thought of British Unity (if it existed at all) being embodied by the corpulent George IV or the jolly Jack Tar ultra-reactionary William IV was not really sustainable. Perthes's idea of state unity and state stability focused in a person remains, however, to this day part of the British monarchical idea.

The British press, always, by and large, hostile to Albert, were not slow to blame his political outlook on the dire influence of German university life. In November 1854, for example, the *Daily News* would denounce Albert as 'a Prince who has breathed from childhood the air of Courts tainted by the imaginative servility of Goethe, who has been indoctrinated in early manhood in the stationary or retrograde principles of the school of Niebuhr and Savigny'.[22]

Even Sir Theodore Martin, who wrote his five-volume hagiography of the Prince in collaboration with Queen Victoria, would write, 'The Universities? Their training was too one-sided and theoretical for one whose vocation would be to deal practically with men and things on a grand scale.'[23] Since Martin was a Germanist with a deep reverence for German universities, we can only conclude that this contempt for the 'theoretical' comes from that essentially grounded, common-sense, *bodenständig* personage, Albert's widow.

As well as studying Constitutional Law and the Philosophy of Law, Albert attended pure philosophy lectures by Immanuel Hermann Fichte, making hundreds of pages of closely written notes. This Fichte, son of the famous Johann Gottlieb Fichte, was named after Immanuel Kant, his father's hero. Whereas his father Johann is seen, by historians of philosophy, to pave the way for Hegelian Idealism, Immanuel Hermann, who had attended Hegel's lectures, was shocked by what he deemed their pantheistic tendencies. He wanted to fashion a philosophy which gave, if not proof, then justification for religion. He saw religion less as a matter of personal salvation, than as a useful social glue, a system which held societies together: 'Theology as a branch of Anthropology',[24] as a modern commentator has called it. Albert was never a particularly religious person, but he would seem to have absorbed the younger Fichte's essentially utilitarian explanation of the religious phenomenon. Albert accepted theism as a reinforcement of Ethics, Kant's 'categorical imperative'.

Fichte's course began with the philosophy of Kant, then went back to the Greeks and came to consider the entire history of philosophy, from Plato onwards. 'You can well imagine that our studies are very interesting, but none the less difficult,' Albert wrote to Florschütz on 18 May 1838.[25] Compared with the demands of philosophy, Albert was relieved to be able to study the history of art, anticipating the longed-for Grand Tour in Italy, his truncated *Kunstreise*. The Professor of Art History was Johann Wilhelm Eduard d'Alton, best known as an anatomical and zoological engraver. D'Alton had lived in Italy for many years, and now, in his late sixties, was in an ideal place to advise the Prince about his forthcoming trip. (He had taught the young Marx only a couple of years before and knew him well: it would be wonderful to think that he had told Albert about the future author of the *Communist Manifesto*. It is difficult to think of two figures more different from one another, in the entire nineteenth century, than D'Alton's two famous pupils.)

D'Alton taught Albert in the late afternoons, four days a week. The Prince was also adding Botany and fencing several times a week to the packed timetable. 'The army of lectures increases day by day... I am at my writing-table, therefore, punctually at five, as in the winter, in order to get through the work. Physical exercises are now to be started, for the development of the body. Today a gymnasium is to be erected in front of the glass door, to my design, how far it will penetrate the hard head of the Bonn carpets, I have as yet no means of judging'.[26]

As we contemplate Albert's time at Bonn University, in which he and Ernst had managed to squeeze three years' worth of work into three semesters, it is difficult not to feel admiration. At the same time, impressive as it all is, can we find here a misfortune? Albert was a clever man, but he was not content with this, and he would have liked to be a universal genius. Frederick the Great, at his Potsdam palace of Sanssouci, had a Music Room where he played the flute with Bach, a salon where he held philosophical conversations with Voltaire, as well as parade grounds where he inspected his well-disciplined troops and planned his political domination of Eastern Europe. Albert would have liked to be a monarch in this mould, though without Frederick's military ambition. Britain, however, neither wanted nor needed a Frederick the Great, even though Frederick's most voluminous biographer, and ardent hero-worshipper, Thomas Carlyle, lived in Chelsea. Albert's multi-talents would be put to the service of the state when he married Victoria, but he would be doomed to an everlasting sense of frustration.

One element which made Albert dread the coming Italian tour, however, was the fact that he would make the journey without Ernst. While Albert was to acquaint himself with the masterpieces of Renaissance art, his brother was to be sent to Dresden for military training – or, as Albert himself put it, 'to sacrifice himself to Mars… The separation will be frightfully painful for us. Up to this moment we have never, as long as we can recollect, been a single day away from each other! I cannot bear to think of that moment!'[27]

The separation, the severance of their daily companionship, was to be permanent, and they both knew this. The Italian tour was part of the preparation for Albert's English life. Lest there could be any doubt about this, he was accompanied on the journey, not by one of his father's courtiers, but by Baron Stockmar, who came specially from London. Stocky had made just such a tour with Leopold when he was on the verge of manhood. The journey, as Sir Theodore Martin said, was 'familiar ground'.[28] It was familiar in more than a geographical sense. This was the journey Stocky had made when he was training up a young Coburg Prince to be the consort of a Queen of England. Second time lucky, provided Victoria did not, like Princess Charlotte, die in childbed.

They set off in December, passing through Munich and staying with the King and Queen of Bavaria. Albert was placed next to the King ('unfortunately terribly deaf'[29]) at dinner. Next day he had tea with the Queen Mother and met Lady Hester Stanhope, the Levantine traveller, whom he described as 'the famous beauty Lady Stanhope'. He still had not mastered the niceties of English titles and modes of address.

Albert was never one to talk down his own discomfort. His upper lip was seldom stiff. He wrote desperately to Florschütz, back in Coburg, after four days of being pent up in a freezing coach, and suffering from toothache, 'I flung myself from side to side, but nowhere could I find my beloved Thus [i.e. Florschütz] to sob out my pain in his arms; but I must admit that Herr von Stockmar was very sympathetic and really did everything he could to relieve me.'[30]

It was indeed an adventurous time of year in which to be travelling. From Munich to Verona, they were journeying in deep snow and temperatures of minus 12 degrees Celsius (10–12 degrees Fahrenheit). Damp hotel rooms in the Apennines were not borne with stoicism. He patiently recounted to Florschütz 'another attack of throat trouble'.

By the time they reached Pisa, he was feeling a little better and would have liked time to explore the Campo Santo and the other stupendous works of art and architecture. Royal duty, however, compelled him to attend 'an invitation from the Duchess of Nemours to attend the funeral of poor Marie Württemberg who was relieved of her sufferings on the 2nd [of January]'.[31] Eventually, they reached Florence, where they were to spend three months – January till March 1839.

They took rooms in the Casa Cerrini, the palazzo of the Marquis Cerrini. They had a beautiful salon, upholstered in red silk, with Louis XV furniture and ceilings painted with allegorical scenes. The walls were covered with pictures. Albert was given a little study, where he worked, as if he were still a student swotting for exams. He rose at six, worked till noon, dined at two, drinking only water, did sight-seeing in the afternoons, and went to bed at nine.[32] Is it possible that Stocky, for all his desire to mould his young charge into a model Prince Consort, actually found Albert's company a little intense? The Prince noted with puzzlement that 'although he is not ill, he has not been able to leave his room since we arrived here'. For the first two weeks, Albert took all his walks alone 'so that I often feel very lonely'. The wildest dissipation which he allowed himself in the evenings were games of dominoes with Stockmar, followed by 'political or philosophical discussions which are often very lively. You know I have always enjoyed them, and love defending my views to the last ditch.' Stocky, disappointingly for the Prince, does not appear to have shared this need for disputation, and 'rarely emerges from his fortress to launch an attack himself'.[33]

Apart from being lonely, and failing to get a rise out of his older companion, Albert was nursing another disappointment. The famed Italian art did not live up to his expectations. As he paced alone round the Uffizi and the Pitti Palace, he was bound to admit, 'I can't attain to the state of ecstatic enthusiasm in which so-called connoisseurs envelop themselves as in a mantle in order to hide their real failure to understand the nature of the art. I have been clever enough not to let anyone notice my coolness, or even to confess it, otherwise I should be regarded as a frozen Vandal from the ice-bound regions of the North.'[34]

Although in later life he prided himself on his knowledge of the visual arts, music was really always the art which meant most to him. He played the piano; he installed a small chamber organ in their apartments; and he obtained permission to slip into the Church of the Badia when it was closed to the public, and to play the organ there. He had begun music lessons with 'a very muddle-headed teacher of counter-point'.[35] He also took singing lessons three evenings a week from a Signor Cecherini, worked at composition, and tried to make progress in Italian, feeling humiliated by how often he was obliged to say '*non intendo*' ('I don't understand') when he was addressed.

Stockmar, meanwhile, to obviate the Prince's need for a quiet evening of dominoes and philosophical speculation, began to take Albert into society, escorting him to two balls at the British Embassy, a ball by the Duchess Courant and another by the Duc de Montfort (Jerome Bonaparte). He felt 'knocked up' by the late nights. 'You know how much I depend on getting to sleep early.'[36]

Florence was full of European aristocrats or royalty with time hanging heavy on their hands. 'Wherever you go here you are sure to tread on the toes of some Italian, Polish or French prince or duke.'[37]

Desperate to find a young companion for Albert, Stockmar had enlisted the help of King Leopold. He had English friends in

Brussels called Henry and Margaret Seymour, whose son Francis, six years older than Albert, was an ensign officer in the 19th Regiment of Foot. They wangled Francis leave from his regiment for a few months to come to Italy where he acted as Albert's groom-in-waiting. It was Seymour's task to accompany Albert to evening parties at the Embassy (where the Minister, the Hon. Henry Fox, with his wife Augusta presided over a glittering salon – they were the future Lord and Lady Holland). Albert found it intensely boring. 'I have danced, dined, supped, paid compliments, have been introduced to people, and had people introduced to me: have spoken French and English – exhausted all remarks about the weather – have played the amiable – and in short have made "*bonne mine à mauvais jeu*".'[38]

When spring began, it was a relief to accompany Seymour on a walking tour in the Apennines. They covered about '26 Italian miles a day. It caused a remarkable sensation here – the lazy Italians, who, like the foreigners, never leave their carpeted salons or the walks just outside the town, would not believe it.'[39]

In March, they left Florence, spent the night in Arezzo, passed on to Perugia and reached Rome, putting up at the Hotel Europa in the Piazza di Spagna. He immediately set off for a walk into the city, accompanied by Seymour. Once again, as in the Florentine picture galleries, he was not enraptured. 'I felt no enthusiasm. I struck my forehead, called myself an unfeeling block, incapable of any lofty sentiment, but all in vain.' The next day, they went to see the Basilica of St Peter's. 'I have long yearned to look upon something great and uplifting: here I found it, but it oppressed me to the depths. Here I found the standard by which men's achievements must be measured. I felt as though I were looking at the Alps from the summit of the Faulhorn, to which they seemed so near that one could stretch out to touch them, and then again, so immense that one could imagine them one with the heavens. So it is here. Everything is in such perfect harmony'.[40]

The Papal Mass, however, was less impressive than the architecture which provided its stage setting. 'What a collection of absurd ceremonies! The Pope [Gregory XVI] looked to me like a pagoda. Then he was undressed again and fumigated and again dressed up; then his hand was kissed by the Cardinals and later his foot; all this fumigation, dressing and undressing, putting on and taking off his cap went on endlessly.' Moreover, the music was indifferent – '*mittelmässig*'.[41]

They managed to obtain an audience with the Pope through the influence of the Saxon Minister (i.e. Ambassador), Herr Plattner, a convert to the Catholic faith. Plattner led Albert, Seymour and Stockmar into the Pope's 'little room'.

Gregory XVI was in part a namesake of the Prince, who, before his elevation, had been an aristocratic monk named Bartolomeo Alberto Cappellari. A professor of science and philosophy at the University of Venice, he was an arch-reactionary, who condemned most manifestations of modern life – calling railways '*les Chemins d'enfer*'. He was well informed about art and antiquities, encouraged the excavation of the catacombs, and founded the Egyptian and Etruscan Museums in the Vatican.[42] For nearly half an hour, Albert gave the Pope the benefit of his thoughts, expressed in faltering Italian, on subjects of which he possessed at best a schoolboy knowledge. 'The Pope asserted that the Greeks had taken their models from the Etruscans. In spite of his infallibility, I ventured to assert that they had derived their lessons in art from the Egyptians.'[43] The Pope, being, as the Prince conceded, 'kind and civil', did not point out that he had spent many years studying his Egyptian and Etruscan collections.

There followed a rather heavy piece of slapstick. Herr Plattner fell to his knees to kiss the Pope's foot at the very moment when Gregory took a step backwards to ring the bell, for his guests to be accompanied down.

> Thereupon Plattner lost his balance & fell on his face, nevertheless, he crawled after His Holiness on his stomach & grasped his raised foot; the Pope, a very stout, heavy man whom one foot only could not support, began to stagger,

> and made vigorous efforts to free himself, which made him kick Plattner's mouth, outstretched to kiss, ten times at least. Seymour was on the point of bursting out laughing at this extremely comic scene when Herr von Stockmar pushed him out through the door.[44]

Hitherto, Albert – despite claiming in a fanciful letter to his Bonn fellow student Löwenstein that he had been 'intoxicated with delight' by the Florentine galleries – had not got his eye in. He had not learned how to look at paintings. In Rome, despite his very mixed feelings about the churches and their architecture, this was to change. He learned to view Italian art through educated German eyes. This was because, while he and Seymour were staying at the Hotel Europa, Albert encountered the community of German artists, antiquaries and archaeologists who were living there. His guide to the city was Dr Emil August Braun, ten years his senior, secretary of the Prussian Archaeological Institute and a native of Gotha. (Both his father and uncle had been in ducal service; his father as a forester, his uncle, Wilhelm Ernst Braun, the Keeper of Ducal Collections to Albert's grandfather.) Emil Braun, having trained in medicine at Göttingen and studied archaeology at Munich, had joined the staff of the Berlin Museum. Albert had an immediately receptive mind. There was nothing original about him: he was in this sense docile, and Braun almost literally opened his eyes to Italian antiquity and art. Albert described him as 'an excellent young man, but with one of the most devilish tongues I have encountered in a long time'.[45]

As well as showing the Prince antiquities and paintings, Braun introduced Albert to his friends – artists Friedrich Overbeck and Peter von Cornelius, the latter trying to revive the medieval art of fresco painting. August Kestner, Barthold Georg Niebuhr and Christian Karl von Bunsen were in Rome as diplomats, but they were all learned archaeologists. From Rome, Albert's party went down to Naples, visited Pompeii, slept at the top of Mount Vesuvius, saw the sun rise over Salerno, visited Paestum, Amalfi, Castellammare,

Sorrento and 'the lovely island of Capri'. In May, Seymour was still with him, and Albert wrote to Uncle Leopold, begging him to get Seymour's leave extended. He was so very 'pleasant and exactly fitted for this post'. The post being, someone who was paid to be Albert's friend. Apart from his brother, Albert never had a friend in his entire life. When he went to Britain, many politicians, courtiers, inventors, academics and artists became his eager collaborators in his many admirable enterprises. There were no friends. Here was one of the huge differences between himself and his future wife. Victoria made friends, often very eccentric ones, and they were a vital emotional need for her: Melbourne, Disraeli, Augusta Stanley, John Brown, Abdul Karim, and others. Albert had only his brother.

Stockmar was impressed by Albert's cleverness and application, but he had begun to see his limitations as well. He was socially clumsy and tongue-tied with women. 'On the whole he will always have more success with men than with women, in whose society he shows too little *empressement* [enthusiasm]'.[46]

In May, they made their way north. Albert's father came to join them in Milan, which allowed Stockmar the chance to rejoin his family in Coburg, and Seymour, who accompanied the Coburgs to Lake Geneva, to go back to England. In blood-relation Switzerland, Albert was once again safe with his family. They were joined by his aunt Julie, the wife (long estranged) of Grand Duke Constantine of Russia, and his cousin Count Hugo Mensdorff-Pouilly. He was really happiest with family.

His reunion with Ernst was glad; and after accompanying their father to Karlsbad, where the older Ernst was taking the waters, they all returned cheerfully to Coburg, where they found their dear old tutor, the 'Rath', had got married.

The summer at the Rosenau was Albert's last. For August and September, he and Ernst could be together again, and sickly, studious Albert could 'enjoy some days of quiet and regular occupation'.[47] It was all soon to change. In October 1839, Ernst and Albert were scheduled to revisit England.

SEVEN

A SOMEWHAT ROUGH EXPERIENCE

IT HAD BEEN an open secret since 1836 that the Coburgs – Duke Ernst, Uncle Leopold, Stockmar – intended Albert to marry Victoria, as the Duchess Auguste had planned since their birth. It was a more or less open secret, too, that the Queen had asked for time. If Albert displayed impatience about this, it was less because he was ardently in love with her and more because he feared that the longer she was allowed to lead a single life, the harder it would be, when she eventually married, to control her.

'Victoria is said to be incredibly stubborn, and her extreme obstinacy to be constantly at war with her good nature'.

Who 'said' these things about Victoria? Since Albert was expressing these thoughts in Germany, in 1838, it seems fair to conclude that he derived his information from Stockmar. The assessment of his future bride is chilling in its passionless absence of even the pretence of affection.

'She delights in court ceremonies, etiquette and trivial formalities. These are gloomy prospects: but they would not be so bad if there were not such a long interval for them to harden, so that it will be impossible to modify them. She is said to take not the slightest pleasure in nature, to enjoy sitting up at night and sleeping late into the day.'[1]

He was viewing his coming marriage as a task – the Taming of the Shrew, or of the Minx – rather than a pleasure or privilege. It may be that she sensed some of this, for as the summer of 1839 wore on, she told Uncle Leopold that she wished to consider the 'affair' to be broken off, and the matter of her marriage not to be considered for four years.[2]

Evidently, however, Leopold and Stockmar believed it was worth testing the water, and a visit to England by Ernst and Albert was arranged to take place in October. Albert claimed that he went 'with the quiet but firm resolution to declare, on my part, that I also, tired of the delay, withdrew entirely from the affair'.[3] Maybe this was the case. Maybe he was telling himself this story, as a hedge against humiliation.

The two young men left Brussels on 8 October and arrived at Windsor Castle on the 10th, after a stormy journey. Their luggage was delayed in following them, so the young Princes had only the clothes they stood up in as they arrived at the Castle. They stepped into what was, in effect, a Whig country house party, in which all the Queen's friends, principally Lord Melbourne, the Prime Minister, were gossiping and joking.

It was only six months since the Bedchamber Crisis, an event whose political significance was clear to the politicians, but not to the Queen. Lord Melbourne, the Whig Prime Minister who had become her instant and close friend upon her accession, had been defeated in a vote in the Commons. (It was actually a matter which did the Whigs some credit – they were attempting to reform the appalling prison conditions of plantation workers in Jamaica and to impose direct rule from London until the owners saw sense.) Sir Robert Peel, leader of the Tories, was poised to become Prime Minister. The Queen, who was platonically in love with Melbourne, was heartbroken, and found Peel cold and unattractive. Peel believed that it was his duty, as an incoming Prime Minister, to ask the sovereign to make some changes in her Household, and to replace at least some of her Whig Ladies-in-Waiting (or Ladies of the Bedchamber, hence the nickname for the 'crisis') with Tories.

Victoria, as well as her passion for Melbourne ('Lord M.'), had a rooted loathing of the Tories because they were the party favoured by Sir John Conroy. So, she refused to have any Tory Ladies in her Household. Peel riposted that, in that event, he refused to form an administration.

Victoria did not realize what had happened. It was the last time in history, really, that a British sovereign exercised her or his will in defiance of the advice of a Prime Minister. She thought she had won her spat with Peel. In fact, he had demonstrated to the world that the sovereign was in place only with the acceptance of the Parliament and its elected representatives. Because Peel refused to serve her on the terms she demanded, he had in fact demonstrated her impotence. She did not understand this. Melbourne returned as a lame-duck Prime Minister, but his days in that role, when the Princes made their visit from Brussels on that October day, were numbered. What Albert and Ernst stepped into was, in effect, a party not unlike Chekhov's *The Cherry Orchard*, in which the principal players were facing The End. They were doing so, of course, with the exquisite *élan* and jokiness of that brilliant circle, who had known the Holland House set, and the Smith of Smiths, and even (disastrously as far as Lord M. was concerned, for his wife had her very public affair with the man) Lord Byron. Anything further removed from the experience of Prince Albert, with his early bedtimes and his sheafs of Fichte's philosophy notes, would have been hard to construct.

Victoria's magnificent journals, which she kept throughout her adult life, form, for the next week, what is in effect a short romantic novella about the duration of the Coburg Princes' stay. The entry on 10 October starts with a mysterious episode in which she discovered that someone had thrown stones through her dressing-room window, breaking the glass. The police were called. Through the golden autumn day, she signed her state papers, and then, as usual, gossiped and laughed with Lord M. and his friends. She kept saying that she was sure that the Princes would arrive, and Lord M. said he did not think they would.

'At half past seven, I went to the top of the staircase and received my two dear cousins Ernest and Albert, whom I found grown and changed and embellished. It was with some emotion that I beheld Albert who is beautiful.' Because they were in their '*negligé*', the boys had to eat in their own apartment. Lord M. thought that convention should have been overruled and that they should have been allowed into dinner in their day clothes. '"I don't know what's the dress I would appear in, if I was allowed", said Lord M., which made us laugh so.'[4] Prompted by Lord M., they asked the boys into the drawing room for an hour or so. Later, as the Queen sat up with Lord M. chatting, 'I asked him if he thought Albert like me, which he is thought (and which is an immense compliment to me). "Oh! yes, he is," said Lord M., "it struck me at once".' The two sat up together until half past eleven.[5]

The next day, she lolled around and breakfasted at ten. Albert brought her letters afterwards, from Leopold's wife Louise, Uncle Ernest and Uncle Ferdinand. 'Albert really is quite charming, and so excessively handsome,' she wrote afterwards. Lord M. advised her later in the day to determine who her partners should be in a dance which had been arranged for after dinner. She danced two quadrilles with Albert and stayed up until ten minutes past one, 'a charming evening'.[6]

The next day, Saturday, 12 October, she recorded in her journal:

> Lord M. had met my cousins in the passage with their dog (whom I said I hadn't yet seen); of his brother having some greyhounds, Spanish greyhounds being the best; of my trying to ride in spite of my head[ache] &c. At a little after 2 I lunched with M[a]., and my 2 dear Cousins; Albert's dog Eos was there; a charming black greyhound with part of her face, her 4 feet, the tip of her tail, and a spot on her neck, white; she is so gentle, and so clever; gives her paw; jumps an immense height, eats off a fork, etc.; and loves Albert so dearly – which is natural. I love and admire him more and more; those eyes of his are bewitching and so is the whole face.

On Sunday, the Duchess of Kent and the Queen took the boys to the service in St George's Chapel. They were enchanted by the music. In the afternoon, the boys walked down to Frogmore, where the Duchess had a house. Victoria stayed in the castle with Lord M. 'I said seeing them had a good deal changed my opinion (as to marrying), and that I must decide soon, which was a difficult thing. "You would take another week," said Lord M.' In the evening, Victoria played '2 games at Tactics with dear Albert, and 2 at Fox and Geese.'[7]

The next day, Victoria asked Lord M. whether the time had not come when she ought to tell Albert. When she said she would find it difficult, Lord M. laughed and asked, '"Has nothing ever passed... not even in writing; never, I said'.[8]

On Tuesday, 15 October, Victoria sent for Albert and received him in a little closet. 'I said to him, that I thought he must be aware why I wished him to come here,– and that it would make me too happy if he would agree to what I wished (to marry me); we embraced each other over and over again, and he was so kind, so affectionate; oh! to feel I was, and am, loved by such an Angel as Albert, was too great delight to describe! he is perfection in every way,– in beauty – in everything!'[9] When she informed her Prime Minister of what had happened, there were tears in Lord M.'s eyes.

After dinner, Albert gravely inquired of the Whig statesmen how they would explain the word 'peasant' in English. 'Common people', 'common fellows', said Lord M. Lord Palmerston added, 'Clods'. Albert went to bed early with a nosebleed.[10] On Wednesday, 16 October, Lord M. explained to his monarch how she was to announce the news to the Privy Council.

To his Gotha grandmother, Albert wrote:

> Certainly, dear Grandmama, my cherished home, my beloved country, will always be dear to me, and in my heart will find a friend who will frequently remind me of her. To live and sacrifice myself for the benefit of my new country, does not prevent my doing good to that country from which I have received so many benefits. While I will be untiring in my efforts and labours for the country to which I shall in future

> belong, and where I am called to so high a position, I shall never cease to be a true *German*, a true *Coburg* and *Gotha* man. Still the separation will be very painful to me.[11]

Albert was in a daze, unable completely to absorb the magnitude of what had happened. He was more than a little troubled by the intensity of Victoria's new passion for him. ('I am often puzzled to believe that I should be the object of so much affection'[12]) and this would be a difficulty forever, until his death. He never came to terms with her passionate intensity. Moreover, the visit to England had made him aware that he would be going there as a complete stranger, who understood nothing of its ways, customs or politics. 'My position will have its dark sides, and the sky will not always be blue and unclouded,' he wrote to his stepmother Marie.[13]

The wedding was fixed for February 1840. The Queen informed the Privy Council of her intention to marry on 23 November 1839. It gave them just a couple of months, either side of Christmas, to make the necessary preparations for Albert's arrival. When she dictated to Colonel Grey, her Private Secretary, her version of *The Early Years of the Prince Consort*, published in 1867, Victoria was intent upon emphasizing how popular her choice of husband had been. She based this claim on the fact that 'People… hailed with satisfaction the prospect of a final separation between England and Hanover – the union with which, no less than the Monarch who now occupied the Hanoverian throne (and who failing the Queen, would have ascended that of England) was in the highest degree unpopular.'[14]

This was true. Had Victoria died without issue, the English throne would have been inherited by her detested uncle Ernest, Duke of Cumberland and King of Hanover, a brutal (it was said he had murdered his valet) and ugly man whose political views were violently to the right of the most diehard English Tories. It would have been difficult to match the King of Hanover's unpopularity, but this did not mean that Albert – of whom the public knew nothing, and even the political class knew little – would have been an immediately popular choice. When, ten years after Grey's *Early Years of the Prince Consort* had been published, Sir Theodore Martin's *The Life of His Royal Highness the Prince Consort* hit the presses, his royal collaborator admitted that events in England at the time of his betrothal 'were calculated to give the Prince an impression that a somewhat rough experience awaited him in his future home'.[15] This is the truth.

Leopold's behaviour, since the death of Princess Charlotte twenty-two years earlier, had to some extent poisoned the well. He continued to own, and some of the time to occupy, Claremont House in Surrey where the Princess had died, even though he was the King of the Belgians. He continued to draw his Civil List payments, a whopping £50,000 per annum, long after he had ceased to be Charlotte's consort. And his interference in the Queen's life, aided and abetted by his diminutive counsellor Stocky, was duly noted both by the Court and by the Parliamentarians.

When Victoria's father, the Duke of Kent, had died, leaving a little baby in the charge of his widow, who could scarcely speak English, and her villainous companion Conroy, Lady Jersey, one of George IV's mistresses when he had been the Regent, exclaimed, 'Why should we have Germans to rule over us?'[16]

Twenty years later, it would still, to many British ears, sound a perfectly reasonable question. The Queen could have married an English aristocrat. If she had to marry from the royal breeding-pool, she could have married her cousin George, future Duke of Cambridge, with whom she always got along very well and who, though, like all Hanoverians, of German origin, was to all outward appearances a bluff English cavalry officer. Had it not been for Stockmar and Uncle Leopold, she probably would have made some such choice. But they were not the only ones responsible: that she had fallen in love, totally, was a key factor in the story. She was always a woman of intense passion, and when her passions were engaged, she was not easily to be denied.

This did not prevent the press, and the politicians, from wondering, 'why should we have Germans to rule over us?'

Perhaps the most notorious exponent of this viewpoint was that dyed-in-the-wool backwoodsman Colonel Charles de Laet Waldo Sibthorp, a splendid figure on the back benches of the Commons, immediately recognizable because he had not altered his manner of dress since the beginning of the century. He still wore white nankeen trousers, clanking top boots and a huge wideawake hat. Sibthorp combined extreme xenophobia with a hostility to Catholicism which bordered on mania. (His brother, a clergyman, became a Roman Catholic not once, but twice, which Sibthorp regarded as a tragedy.) When on form, Sibthorp could deliver very memorable, if not always effective, parliamentary speeches. He had, naturally, opposed Catholic Emancipation in 1829, and been violent in his denunciations of the Reform Act three years later. The prospect, when it was put before the Commons, of awarding Prince Albert a grant of £50,000 was more than Sibthorp could endure.

He put it about that Prince Albert was himself a Roman Catholic. Although Melbourne did his best to reassure, there were many, including the Duke of Wellington, who thought that this idea might as well be given an airing, if it upset the Whigs and kept the Queen on her toes. Far from the Tories loving the Royal Family at this date, they found it difficult to accept the Queen's dotage on Melbourne, her tactlessness over the Bedchamber Ladies, and, most damaging of all, her monstrous claim that Lady Flora Hastings, one of the Tory Ladies who was allowed into Court, a childhood friend of hers and, unluckily for her, a friend of the Conroys, was with child by the villainous Sir John. Victoria had insisted that Lady Flora underwent medical examination to prove her innocence. The doctors revealed that she died shortly afterwards, a post-mortem showing the swelling to be advanced cancer. Flora Hastings was a virgin.

Against this background, neither Parliament nor people felt much inclined to heap rewards on a Coburg prince.

Lord M. told the Queen, "'They [the Tories] are very angry with you,"... I said (as I do God knows!) I hated them beyond everything and never would forgive them... I said I paid £34,000 every year in pensions, charities &c. "Good God!" said Lord M.'[17] Robert Peel, who was poised to form a Conservative Government at the first wobble of the Whigs, seized on Sibthorp's speech and gave it his enthusiastic backing. Melbourne had assured Victoria that Parliament would grant Albert £50,000 per annum, that it would go through on the nod. Much as he wanted to please the Queen, however, Lord M. was unable, from the House of Lords, to prevent the Commons from reducing Albert's money to £30,000. Peel reminded the House of Commons that when Leopold had been granted *his* £50,000, it had been regarded as much too much.

When the Queen heard of this speech, she exploded. Peel was 'a nasty wretch', and Sir James Graham too. 'As long as I live I'll never forgive these infernal scoundrels, with Peel at their head; as long as I live for this act of personal spite!!'[18] She was particularly angry with the Duke of Wellington. Even more so when they came to discuss the question of precedence – where Albert came in the royal pecking order. At a grand dinner, for example, should he go in before the King of Hanover? Before the Queen Dowager?

The Queen assumed, very naïvely, that Albert would, as her husband, take precedence over all. Wellington and the Tories were determined to remind her that this was not the case. Wellington insisted that Albert should have the same precedence as Prince George of Denmark when he married Queen (then Princess) Anne and Leopold when he married Princess Charlotte. Though the Queen had another outburst, and said she would never eat dinner with the Duke of Wellington again, Lord M. calmed her down, and made her realize that there was nothing, at this stage, which she could do to alter the time-honoured protocols.

The next few weeks before the wedding, which took place on 10 February 1840, were in some ways a playing-out, or a foreshadowing, of the following twenty-one years, with Victoria in a state of passionate fury – against her politicians, and against those who checked her will – and Stockmar rather ominously 'vexed at Albert's misapprehensions about the various things'.[19]

One of his misapprehensions was that he would be allowed to bring with him his own German entourage, whereas, naturally enough, Lord Melbourne and the other elder statesmen saw it as essential that Albert should learn the ropes and have an English Private Secretary, to help initiate him into the mysteries of English life – social, political and courtly. He was allowed a German secretary, Herr Doktor Schenk, but otherwise, his Secretary was to be George Anson, hitherto Melbourne's own Secretary.

After the exchange of love letters which had become their habit over the previous month, it was a shock for Albert to receive a very firm negative, not from officialdom, but from his own beloved: 'As to your wish about your Gentlemen, my dear Albert, I must tell you quite honestly that it will not do.'

He replied, 'I am very sorry that you have not been able to grant my first request… Think of my position, dear Victoria; I am leaving my home, with all its associations, all my bosom friends, and going to a country in which everything is new and strange to me – men, language, customs, modes of life, position. Except yourself I have no one to confide in.'[20]

She refused. As Lady Lyttelton, who became the governess to the Queen's children, would remark, 'a vein of iron runs through her most extraordinary character'.[21]

On 14 January, Lord Torrington and General Sir Charles Grey (later the Queen's long-suffering Private Secretary and author of *The Early Years of the Prince Consort*) left England, with the band of the Coldstream Guards, to accompany the Prince back to his destiny.

Festivities in the streets of Coburg had gone on for nights and days. By the time Grey, Torrington and the Coldstreamers reached Thuringia, the Court had moved in progress back to the Schloss Friedenstein at Gotha, the childhood home of Albert's mother. They were received with a great dinner at four o'clock in the afternoon, followed by a masked ball in the delightful Baroque theatre. The next day, 23 January, Duke Ernst, himself a Knight of the Garter, sat on the throne. Albert was invested by Prince Charles of Leiningen, his brother-in-law, another KG, into that most noble Order of Chivalry. It was done with tremendous pomp in the glorious throne room. The corridor was lined with German soldiers, and the band of the Coldstream Guards played. Dinner was served for 180 people, and they then went to the theatre to see a production of Weber's *Der Freischütz*, with its rousing chorus of '*Viktoria! der Meister soll Leben*', and its ironical aria by the peasant who beats his master at a shooting competition, '*Schaut der Herr mich an also König!*' ('Let the Gentleman gaze on me as King!') Albert would be singing this song in his head for the rest of his life, unheard and unheeded by the British who, however much he was admired, would never regard him as their King.

On the next morning, 28 January, the journey back to England began. The twelve carriages which passed through crowded streets paused, so that the two young Princes could embrace their Gotha grandmother for one last time. Then, on they went, on a slow progress through Germany, Duke Ernst, his two sons, the three Englishmen – Lord Torrington, General Grey and Captain Seymour – and a galaxy of German royalties and nobility, most of them relations. They paused at the Elector of Hesse, they visited the hilltop Schloss of Cassel – later Kaiser Wilhelm II's favourite residence – Arnsberg, Deutz, Aachen. It was in Charlemagne's ancient capital that Albert heard the news of his £50,000 grant being reduced by Parliament, and some of the political realities of his future life began to dawn on him. Colonel Sibthorp and Sir Robert Peel and the Duke of Wellington were not to gaze on him as King. George Hamilton Seymour, the British Minister in Belgium, was unimpressed by the splendour of the German entourage as it passed through Belgium en route for the Channel.

> It appears that the Queen pays the whole of the expence [*sic*] not only of Pce Albert's but of the Duke of S. Coburg's journey. The consequence is that these Princes travel like

> Princes. On the road they gave 10 & 12 Louis to the different bands of musicians. They have directed 100L. to be distributed to the Palace servants – and on the road a claim was put in by the Duke's private servant for band wages! To this Lord J. demurred. A[lbert] himself is a plunderer, that is clear – He talked with glistening eyes of the presents in diamonds which the Q. intended for him.[22]

They crossed the Channel in a Dover packet, the *Ariel*, in choppy waters. It took five and a half hours. Albert and Ernst were by now 'in an almost helpless state'.[23]

They spent the night at the York Hotel in Dover, and, because they were not due in London until Saturday, 8 February, they killed time in Canterbury, where the Mayor offered them a Loyal Address, and a great crowd assembled to cheer the German Princes. Eos, in the company of the valet, was sent on ahead, much to the delight and excitement of the Queen.

Eos, Ernst and Albert had been spared the more dramatic quarrels which had been raging for the previous two weeks about the guest list for the wedding. Charles Greville MP, Clerk in Ordinary to the Privy Council, and the best political diarist of the era, said:

> She has been as wilful, obstinate and wrongheaded as usual about her invitations and some of her foolish and mischievous courtiers were boasting that out of above 300 people in the Chapel there would be only five Tories. Of these 5, two were the Joint Great Chamberlains, Willoughby and Cholmondeley, whom they could hardly omit, and one Ashley, the husband of Melbourne's niece; the other two were Ld. Liverpool, her own old friend, and the Duke [of Wellington] but there was a hesitation about inviting them.[24]

They were married in the Chapel of St James's on 10 February. She was given in marriage by her strange old uncle the Duke of

Sussex, in his black skull-cap, and she sat between her husband and the Duke at the wedding breakfast afterwards in Buckingham Palace. The weather, which had promised to be foggy and wet, as it had been on the previous day, turned to sunshine. A huge crowd witnessed them leaving the Palace, and lined the route much of the way to Windsor. They were greeted by a chorus of the schoolboys of Eton, and two days later, they were joined at the Castle by Victoria's mother, by Duke Ernst, by brother Ernst, and the whole Court, with dancing until the small hours. On the 14th, they returned to London. There was not much time for them to be 'initiated in the sacred mysteries of creation, which ought to be shrouded in holy awe till touched by pure and undefiled hands' (Albert's description of losing his virginity).[25]

Victoria's more candidly sensual delight in the whole procedure is justly celebrated.

> He took me on his knee, and kissed me and was so dear and kind. We had our dinner in our sitting room; but I had such a sick headache that I could eat nothing, and was obliged to lie down in the middle blue room for the remainder of the evening, on the sofa; but, ill or not, I never, never spent such an evening!!… feelings of heavenly love and happiness, I never could have hoped to have felt before! He clasped me in his arms, and we kissed each other again and again! His beauty, his sweetness and gentleness,– really how can I ever be thankful enough to have such a Husband! – At ½ p.10 I went and undressed and was very sick, and at 20 m. p.10 we both went to bed; (of course in one bed), to lie by his side, and in his arms, and on his dear bosom, and be called by names of tenderness, I have never yet heard used to me before – was bliss beyond belief! Oh! this was the happiest day of my life! May God help me to do my duty as I ought and be worthy of such blessings![26]

The next entry is even more expressive:

> When day dawned (for we did not sleep much) and I beheld that beautiful angelic face by my side, it was more than I can express! He does look so beautiful in his shirt only, with his beautiful throat seen. We got up at a ¼ p.8. When I had laced I went to dearest Albert's room, and we breakfasted together. He had a black velvet jacket on, without any neckcloth on, and looked more beautiful than it is possible for me to say; we sat talking together till 20 m. p.10. Wrote to Lord Melbourne, and M[a]., from whom I had a letter. At 12 I walked out with my precious Angel, all alone – so delightful, on the Terrace and new Walk, arm in arm! Eos our only companion. We talked a great deal together. We came home at one, and had luncheon soon after. Poor dear Albert felt sick and uncomfortable, and lay down in my room, while I wrote to Uncle Leopold. He looked so dear, lying there and dosing [*sic*].[27]

Again, just as the scraps and misunderstandings before the wedding foreshadowed everything which happened in the next twenty years, this intimate vignette, so vividly captured by that superb diarist, is, in a way, the story of their lives, her rapturous love for the sickly figure who lies, looking so dear, 'dosing', and eventually, all too soon, dead.

At the same time, while she drooled over him, the Prince's bafflement that their honeymoon should be so short was met with monarchical rebuff. 'It is usual, is it not, for newly married people to stay up to four to six weeks away from the town and society... It might perhaps be a good and delicate action not to depart from this custom altogether'. So he had written from Gotha when she had explained to him how short the honeymoon would be. Her brisk reply was, 'My dear Albert, you have not at all understood the matter. You forget, my dearest Love, that I am the Sovereign, and that business can stop and wait for nothing. Parliament is sitting and something occurs almost every day for which I am required and it is quite impossible for me to be absent from London.'[28]

She would not have cause in the future to be the one who had to remind him that business can stop and wait for nothing. The only way in which she could be prevented from her persistent continuance in her monarchical duty would be to ensure that she was laid low, on a regular basis, in childbed. It would not take long. Meanwhile, he could content himself, in a letter to Florschütz, that 'I have made a very favourable impression on the little Sovereign'.[29]

He established early in the marriage, during the summer of 1840, that he wanted to be regarded as something more than a figurehead. He chose his area of public discourse well: namely, on 1 June 1840, he presided over a public meeting at Exeter Hall in the Strand, to promote the Abolition of the Slave Trade.

Much is often made of Queen Victoria's error of judgement, when the Whigs left office in 1839 and she refused to appoint Tory Ladies to her household, leading to the Bedchamber Crisis. The matter over which the Whig Government resigned in April 1839 is less often remembered. It was the condition of prisons in Jamaica where recalcitrant slaves were sent by their 'owners' and kept in conditions of appalling cruelty. Lord M.'s Government had done little enough for the slaves, but at least the Whigs felt that something should be done to force the slave-owners of Jamaica to behave with more decency, eventually with the view of abolishing their right to own slaves at all. (It was only the trade, among British subjects, which had been abolished in 1807, not the ownership of slaves. The Whig Government abolished slavery in the British Empire in 1833.) The Whigs in 1839 had proposed that direct rule from Westminster be imposed upon Jamaica until the slave-owners be forced to a more humane administration of the prisons. Because many of the Whigs had friends or family who owned sugar plantations, they only won their motion by a narrow majority (of five) and this was why Lord M. resigned. Among the Tories who voted against direct rule of Jamaica was William Ewart Gladstone, whose maiden speech in the House of Commons in 1833 – in old age he bitterly repented! – had been a defence of his father John Gladstone's ownership of slaves. In that year, John Gladstone is thought to have received over £75,000 compensation from the Government for releasing 1,609 slaves.[30] By 1840, all the slaves were meant to have been released, but many were not; and those slaves who lived in French, Spanish or Portuguese colonies, although their governments or 'owners' had been paid colossal sums in compensation from British taxpayers, were still being trafficked from West Africa across the Atlantic, and they were still being sold in Brazil and elsewhere.

Albert's espousal of the Cause was therefore completely laudable. It was not an anodyne piece of public do-gooding. There were still rich British families, both aristocratic and nouveaux riches like the Gladstones, who were tainted by involvement with the evil trade. But, as the career of W. E. Gladstone shows, all decent people wished to put the old ways behind them. And all liberal (small L) interventionists wanted Britain to lead the world, not merely by example, but if necessary by gunboat and force, in extirpating the trade wherever it was found. His speech was a brave thing. His English was still uncertain, and he was extremely nervous. Victoria made him repeat it to her by heart. He had an audience of many thousands, however, and it had two effects. Firstly, it helped the Cause. Secondly, it established his position in the country as an independent being. He had staked out his place. The Queen might be the Head of State but he was to be much more than a secretary, and a fertilizer of her womb. He told his father, 'My speech was received with great applause, and seems to have produced a good effect in the country.'[31]

EIGHT

UNBOUNDED INFLUENCE

THEY WERE MARRIED at the beginning of one of the most *mouve-menté* decades in British, and European, history. Turbulence in the economy was reflected in a complete disruption of the political order. The political quarrels about free trade generally, and the lifting of tariffs on cheap imported corn in particular, led to a complete realignment of British politics. While the economy grew and boomed, many of its victims went to the wall. This was the decade of the Irish famines. It was the decade which saw the flowering of Dickens's genius, as exemplified in *A Christmas Carol* (1843) and *David Copperfield* (1849–50). The debtors' gaols he immortalized in *Little Dorrit* (1857–8) saw an unprecedented rise in new inmates – 13,586 in 1843 alone.[1] (The entire population of England and Wales was 15.9 million.)

It was the decade of the railway, when Time itself was regularized: Greenwich Mean Time was imposed on the country at large to allow the new rail networks to function.[2] The Duke of Wellington complained that railways would encourage the lower classes to travel about;[3] but most of the lower classes were too busy, too hungry, too miserably housed, to allow for any such adventures. It was the era when modern Conservatism, as a political party, really began. It was also the era of Chartism, and, beyond the enthusiastic crowds of the Chartists, a much wider demand for an extension of the franchise. By the end of the decade, after the Year of Revolutions in Europe, 1848, Karl Marx and his friend Friedrich Engels would have penned *The Communist Manifesto* and set up residence in England.

That pair claimed that the spectre haunting Europe was Communism. It was not the only spectre. Bourbon absolutism in Naples – according to W. E. Gladstone, a 'rising political star' who was still a stern unbending Tory – was 'the negation of God erected into a system of Government'. From Portugal, Spain, the Bourbon kingdoms and Papal States of Italy, as from the vast regions of the Habsburg and Russian Empires, Europe felt the tension between liberalism and reaction, between crowned heads who hoped to rule as if the French Revolution had never happened, and the forces of change, the emergent power of the *bourgeoisie*.

It was an era, too, of seismic intellectual and spiritual revolution. 1844 would see the publication of *Vestiges of the Natural History of Creation*, an anonymous work of popular science which summarized the theories and discoveries of evolutionary biologists and geologists. It appeared no longer possible to believe the Creation Myths of the Bible. Where did this leave the Church of England by law established, whose Supreme Governor was Queen Victoria? And how would that Church withstand the defection to Rome in 1845 of its outstanding intellectual, John Henry Newman?

These convulsions, of minds and nations, Church and State, were the turbulent accompaniment of Victoria and Albert's marriage. Deeply engaged as he was, and would be, in European and British political issues, knowledgeable as he was, and would be, about scientific and technological advance, widely read in contemporary literature, Prince Albert would be no mere observer of these developments. He was determined, from the moment he left Germany to marry Victoria, that he should find a role for himself personally, and a role for the monarchy, in the new world which was dawning. Like the poet Tennyson, who would become his near neighbour on the Isle of Wight, he was acutely conscious of how this generation was to find its future 'In the steamship, in the railway, in the thoughts that shake mankind'.[4]

The royal pair were emblems of a new world. Given their extreme youth, they could not fail to be. Born post-war, they were going to become the images of what the new world itself would become. Would there be a return to the *ancien régime* in Britain, to the aristocratic oligarchy, and in Europe, as the Emperor of Austria and the Emperor of Russia both wanted? A return to monarchy, autocracy, the suppression of the new thinking in science, philosophy, religion? Or would the world move leftwards? Would it be impossible to ignore the aspirations of the middle classes, the pitiful condition of the working classes? Would the world discard both Throne and Altar which had, surely implausibly, been 'restored' post-1815 in some European countries? Or would neither of these things happen? Would Christianity survive, but in a radically evolved pattern, as Mr Newman predicted in his groundbreaking book of 1845, *The Development of Christian Doctrine*? Would technology and science be advanced alongside economic growth? As the world – at any rate, Britain – became less rural, more bourgeois, less aristocratic, would it inevitably embrace free trade? Would the shy German Prince, seen as so alien by old stagers in the press and in Parliament, actually turn out to be a harbinger and even an *instrument* of change?

On Sunday, 23 February,[5] the Queen recorded:

> Wrote letters and my Journal, & at 12 went alone with dear Albert (my Attendants following) to the Chapel Royal. There were numbers of people out, as we came along, who cheered very much. We got back after 2. Walked in the afternoon with Albert. Wrote and read, & Albert helped me with the blotting paper, where I signed. – Ld Melbourne dined. Sat between Albert & his father. Afterwards, in the Drawingroom [*sic*; of Buckingham Palace], I had some conversation with Ld Melbourne, who was very amusing & in high spirits. Talked of the last Indian victory, which Ld M. said was most splendid.[6]

Victoria plainly saw this as a charming scene, but Albert had been prepared, by Stockmar's unrealistic conversations, for a more dynamic role than that of the blotter of his wife's signatures, the sitter-in on her high-spirited conversations with Lord M. about the success of her Empire. (And, incidentally, their belief that all was 'most splendid' at this date in India was wide of the mark.) By May of that year, having had three months of marriage, Albert's frustration had erupted in more than one direction. To his Bonn contemporary Prince von Löwenstein, he complained, 'In my home life I am very happy and contented; but the difficulty in filling my place with the proper dignity is that I am only the husband, and not the master in the house'.[7]

At the end of May, Baron Stockmar insisted upon broaching the matter with Lord M. The Queen had told her favourite that 'Albert complains of a want of confidence in trivial matters, and on all matters connected with politics of this country'. Melbourne told Stockmar that there was a perfectly simple explanation: that is, the Queen simply enjoyed talking of trivial subjects with her husband, and that her political caginess with Albert was not attributable, as he furiously supposed, to the influence of Baroness Lehzen. From their very first meeting, there had been a fizzing chemical hostility between Victoria's old governess and Albert, and neither Stockmar nor Albert were convinced by Melbourne's explanation. Stockmar told Melbourne, 'The Queen is influenced more than she is aware of by the Baroness'. He openly told Melbourne, 'The Queen has not started upon a right principle. She should by degrees impart everything [to Albert]'[8] – by which Stocky of course meant she should effectively impart everything to him.

The simplest method of securing a bigger political role for Albert, however, was for Nature to take its course. Victoria became pregnant within weeks of her wedding, and on 22 November 1840, at about 2 p.m., 'a perfect little child was born, but alas a girl and not a boy, as we both had so hoped and wished for. We were, I am afraid, sadly disappointed.' When Dr Locock, the only medic in attendance, said,

'Oh, Madam, it is a Princess', the Queen replied, 'Never mind, the next will be a Prince.'[9]

Albert, in his still very uncertain English, had been busy drafting pencil notes about the baby's upbringing. Nothing was to be left to chance. Some of these notes related to security. One member of the nursery staff remembered that Albert had devised 'intricate turns and locks and guardrooms... and intense precautions, suggesting the most hideous dangers'. As this observer added, the dangers were 'not altogether imaginary'.[10] On 2 December, Mrs Lilly, the monthly nurse, heard a stealthy noise in the Queen's sitting room at one in the morning, and summoned a page, who found a boy rolled up under the sofa. This was the Boy Jones. (An earlier intruder was named the Boy Cotton.) Jones was seventeen, but of stunted growth, and looked younger. He was mentally ill. The son of a Westminster tailor, he admitted to freely wandering around the Palace, sitting on the throne, helping himself to food in the servants' hall. He had crept in through one of the windows opening on to Constitution Hill. He broke into the Palace three times, on each occasion being apprehended. The third time, he was sent to a House of Correction, put on a treadmill for three months and then sent to sea. The feeble lack of security had demonstrated to Albert's satisfaction that the Palace, under the supervision of Baroness Lehzen, was appallingly badly run.

The safety of the baby, who was named Victoria (Vicky) after her mother but was known by the parents as Fat Vic, Pussy or Pussette, was only the beginning of Albert's concerns for her. His memos outlined the duties of a huge nursery staff – a superintendent of the nursery, a monthly nurse, a nursery maid, an assistant nursery maid, a housemaid, a wet nurse and a 'footman-in-waiting downstairs to take the various messages'. The superintendent of the nursery was enjoined not to act on the physicians' orders without first consulting the parents of the child, and to insist that the wet nurse stood to feed the child. The logic behind the standing seems to have been that it ensured the wet nurse was young and strong.[11]

The child was christened Victoria Adelaide Mary Louisa – being named, respectively, after her mother (and Coburg grandmother), after the Queen Dowager, after Albert's stepmother and Albert's mother. Her name was in a sense a little biography of Albert's early life. The day before the christening, 12 February 1841, Albert had been skating on the thin ice on the lake in the gardens of Buckingham Palace, and went through into the water. 'The shock from the cold was extremely painful and I cannot thank Heaven enough, that I escaped with nothing more than a severe cold.'[12] The incident did not prevent the christening from taking place in the drawing room of Buckingham Palace. Lord M., charming as usual, said of the baby, 'How she looked about her, quite conscious that the stir was all about herself', prompting the Queen to think of Stockmar's saying, beloved of Albert, that a man's education begins on the first day of his life.

In her literary Albert Memorial, the five-volume *Life of His Royal Highness the Prince Consort* attributed to Sir Theodore Martin, Queen Victoria was at pains to emphasize the honourable and progressive motives which guided Stockmar's thinking. She/Martin quotes a letter which Stockmar wrote to the Prince in 1854, as an encapsulation of Stocky's credo:

> I love and honour the English Constitution from conviction, for I think that, under judicious handling, it is capable of realising a degree of legal civic liberty which leaves a man free to think and act as a man. Out of its bosom singly and solely has sprung America's free Constitution, in all its present power and importance, in its incalculable influence upon the social condition of the whole human race; and in my eyes the English Constitution is the foundation-corner and cope-stone of the entire political civilisation of the human race, present and to come.[13]

(It was common in the nineteenth century, as it still is in the German language, to use the word 'English' to mean 'British', even when speaking of the Scottish or Welsh.) It was essential, if this programme of civilizing the human race was to begin, that the process should begin at the top – with Albert convincing the Queen that the monarch's political role was to eschew party politics.

It was a summer in which they began to acquaint themselves, as a pair, with the aristocracy, with the class who still actually ran Britain, who owned most of its land and controlled its Parliament.

In June 1841, the University of Oxford held Encaenia, its annual speech day or prize-giving day, when the Chancellor bestows honorary degrees. Albert was made a Doctor of Civil Law in Wren's stupendous Sheldonian Theatre. It was a colourful day, the Doctors of the university processing through Oxford in their scarlet gowns to hear Latin speeches by the Public Orator, and recitations in the ancient tongues from undergraduate prize-winners. The Chancellor was none other than the Duke of Wellington himself. He was a national hero, but this had not prevented the Queen, at the time of her wedding, saying that she would not even look at him again. In Oxford, a Tory stronghold, the Duke was an idol, except among those High Church diehards who disapproved of the Duke's support, in 1829, for Catholic Emancipation. Albert was dismayed by the loud behaviour of the Tory undergraduates. They cheered the name of the Queen, but as Greville recorded, 'Her Ministers individually and collectively were hissed and hooted with all the vehemence of Oxford Toryism'.[14]

Victoria felt it was 'something which they ought not have done in my husband's presence'.[15] A revealing comment. It is as if she were the protective male, wishing a feminine partner to be spared male raucousness. At the party after the degree ceremony, known as Lord Crewe's Benefaction, the Duke invited Albert to attend his annual big dinner in London the following week. Shocked by the hoots and hisses of the Tory students, Albert snubbed the Tory, and national, hero, by making a feeble excuse. Greville thought the refusal very foolish, 'for the invitation was a great compliment, and this is a sort of national commemoration at which he ought to have felt a pride at being present'.[16]

The Duke was their fellow house guest at Nuneham Courtenay, the village near Oxford where the Archbishop of York kept his palatial manner of life. When the royal party arrived from Windsor, changing horses at Slough, Hurley Bottom, Nettlebed and Benson, they were met by a contingent of the Archbishop's own horses, and an escort of the Oxfordshire Yeomanry.[17] Walking about the grounds at Nuneham, Victoria and Albert reflected wistfully that it was 'much in the style of Claremont, but on a much larger scale'. Albert enjoyed the walks by the riverside and the distant views of Oxford's spires. Victoria felt 'a little low'. She was pregnant, and it was the first time since she was four years old that she had been separated from 'my dear Lehzen'.[18] At dinner, the guests included the Duke of Marlborough and the eccentric Dean of Christ Church, Dr Buckland, a distinguished though highly conservative geologist, famed for his omnivorous zoophagy. He claimed that moles and bluebottles were the most 'disagreeable' things he had eaten. It was at Nuneham that he devoured his most famous item. The Harcourts possessed the heart of Louis XIV, preserved in a silver casket. When Buckland saw it, he said, 'I have eaten many strange things, but have never eaten the heart of a king', and 'before anyone could hinder him, he had gobbled it up'.[19]

Archbishop Harcourt, the royal pair's host, was now well past eighty. When Victoria was a girl of sixteen, he had entertained her, with her mother, at York. Albert, as a Lutheran, had seen little of Prince Bishops in Germany. Nothing could have prepared him for Edward Venables Vernon-Harcourt, who had been Archbishop of York since 1808. He was immensely rich, partly because of the gigantic emoluments of the northern province, partly because his wife, Lady Anne Leveson-Gower (pronounced Looson-Gore), was rich, partly because he had inherited the estates of his cousin, the 3rd Earl Harcourt, in 1830. He was to live another six years, before, at the age of ninety, accidentally falling into a pond at Bishopsthorpe, his northern palace, and dying of a subsequent chill on a frozen Guy Fawkes' Day, 1847.[20]

There were no people quite like this in Germany. Apart from the fact that Albert was still far from fluent in the English language, his much greater difficulty was understanding the land in which he had arrived. He was completely the fish out of water.

Towards the end of July, the royal couple were entertained at Woburn by the Duke of Bedford. Here was even more splendour than they had encountered at Nuneham Courtenay. Here too – there seemed to be no getting away from the terrifying old hero – was the Duke of Wellington. His extreme deafness gave him the excuse to keep 'very much in the background'. Their host the Duke of Bedford was pleased that the Queen was 'received everywhere with the greatest enthusiasm, and an extraordinary curiosity to see her was manifested by the people, which proves that the Sovereign as such (and certainly she had no great share of personal popularity) is revered by the people'.[21]

While Victoria enjoyed the limelight, Albert was introspective, wistful. He took Anson, his new Private Secretary, with him to Woburn, and used the opportunity to take stock. In eighteen months of being married to the Queen of England, had he found a role?

Anson assured him that 'those who intended to keep him from being useful to the Queen, from the fear that he might ambitiously touch upon her prerogatives, have been completely foiled'. Parliament had voted, nem. con., for a Regency Bill, which made Albert regent, were Victoria to die leaving an infant heir. The Queen found her husband 'an active right hand and able head'. The good and wise looked up to him as 'leading a virtuous and religious life'. Albert could not find this entirely satisfactory. Virtuous he certainly was; religious, not especially. What was he meant to *do*? Anson assured him, 'Arts and science look up to him as their especial patron'.[22] That was not going to be enough for Albert. Years later, Greville remembered, 'Melbourne told me long ago that the Prince would acquire a quite unbounded influence'.[23]

In 1840, the Queen had still believed 'The Whigs are the only safe and loyal people, and the Radicals will also rally round their Queen to protect her from the Tories'.[24]

All British monarchs in the last century had sided with one political grouping or another. George III sided with Pitt; George IV with Charles James Fox; William IV with the Tories. It was, however, only a matter of time before Melbourne's administration fizzled out. The Bedchamber Crisis of 1839, when Victoria childishly refused to accept even minimal changes in her Household, had led to Peel's withdrawing from the Ministry. Thereafter, having defeated Melbourne's Government in Parliament, the Tories would go on to beat him at the hustings. Melbourne would finally resign in 1841. It was the end of the Whigs. When Lord John Russell formed his administration in 1846, it was a Whig–Peelite alliance, what might properly be called a Liberal Alliance, just as, five years earlier, in September 1841, the Cabinet formed by Peel was a Conservative Cabinet. This was the Tory Party which would split irrevocably over the repeal of the Corn Laws. Thereafter, there would be a coalition government for twenty years until the emergence of the modern bi-party system of Conservative versus Liberal. In such a volatile political climate, the attraction of Stockmar's political vision was all the clearer. In a Parliament in which Whigs, Radicals, diehard Tories, Peelite Conservatives and Irish nationalists were all in contention, Victoria's belief that the Whigs were the only safe people would have seen the monarchy tying itself not to the dominant governing party, but to an obsolescent class and sect. By following Stockmar's idea of being 'above' politics, Victoria and Albert could put monarchy at the heart of politics during years of uncertainty and change.

Not that Baroness Lehzen thought so. She spent £15,000 of the Queen's money bribing the electors to vote for the Whigs. Albert, when he discovered this, remonstrated with Melbourne, but Lord M. was impenitent, replying that £15,000 was nothing in comparison with what George III used to spend on bribes during elections. When the Queen and Prince Albert attended Ascot races, they were met with loud cheers and cries of 'Melbourne forever; no Peel!'[25] The Queen's guests were mainly Tory on this occasion, and were not best pleased.[26]

The General Election of 1841 saw a decisive Conservative victory, but it was very different from a modern election which, by virtue of instant electronic communication systems, can take place in a single day. Voting in the 1841 election took place over the entire summer. Parliament was dissolved on 23 June. The Reform Act of eight years earlier had widened the franchise a little and got rid of Rotten Boroughs, but the system was not what a modern voter would recognize as democratic. Only one in five adult males had a vote. (Women, of course, had no vote.) The larger boroughs had around a thousand voters, rural seats far fewer. The Conservatives fielded 500 candidates and ended the election with a Commons majority of seventy-six.[27] Parliament reconvened on 19 August, and the Whig Government was then formally voted out of office, defeated by ninety-one votes in the Commons and overwhelmingly in the Lords. Wellington's brother, Lord Wellesley, aged eighty, had written to Lord Liverpool on 21 July, 'The Conservative Cause is now (after so triumphant an Election) the declared cause of Parliament, & of the People, & according to our happy Constitution, is therefore the Cause of the Crown.'[28]

On the morning of 30 August, Peel went down to Windsor to kiss hands and take office as Prime Minister. A few days later, on 4 December, there was a meeting of the Privy Council. Old Ministers took leave of their offices, which were bestowed on their Conservative counterparts. The Queen 'looked very much flushed, and her heart was evidently brim full, but she was composed, and throughout the whole of the proceedings, when her emotion might very well have overpowered her, she preserved complete self-possession, composure and dignity. This struck me as a great effort of self-control and remarkable in so young a woman' – so wrote Greville.[29] She was also – which he did not write – well advanced (seven months) in her second pregnancy.

Composed, dignified, as she and the world took leave of the Whigs, the Queen may have been. She and Albert, however, were still very young and still had much to learn about the ways of the world. The tetchiness and coldness of Peel's initial exchanges with the Queen (she tetchy, he cold) were very marked. Victoria never liked ceding a point to one of her ministers but she could now at last see that what she had imagined was a victory over the Bedchamber Crisis had in fact been a defeat. She now accepted some of Peel's suggestions for changes to the Household; some, but by no means all. Over the next few years, if a Peel-appointed Lord-in-Waiting, Maid of Honour or Equerry left the Household, Victoria always made sure they were replaced by a person of her choosing. Peel's power, and that of all future Prime Ministers, had been firmly asserted after the Bedchamber Crisis. He could now afford, politically, to let her get away with trivial Household appointments. In the olive-branch letter extended to her, he could write graciously, 'My prevailing wish throughout the whole of the extensive and complicated arrangements for the New Constitution of the Household has been that the Change of Persons about Her Majesty's Court should be as agreeable to Her Majesty's feelings as it is possible to make it.'[30]

As the Queen's pregnancy advanced, it was more and more via Anson or Albert himself that the new Prime Minister communicated with his monarch. Sometimes, clearly, Anson was echoing the unmistakable tones of the Queen. For example, on 9 September 1841, it is clear she was still smarting from the emotional Council meeting in which she took a painful farewell of the Whigs. From Claremont, Anson was obliged to write, 'the first intimation which Her Majesty received of your intention to adjourn Parliament was either through the newspapers or by common report. It was a subject of interest which Her Majesty would like to have been informed of through you.'[31]

Albert's communications with Peel were, by contrast, and from the very first, more cordial. Even before Peel became Prime Minister, Albert wrote to him, begging him to sit for the young historical painter

Feodor Dietz, and to give him 'two sittings of one hour each'.[32] When approached by the painter himself, Peel had initially refused but when asked in person by the Prince, he relented. From now onwards, the communications between Peel and Albert were of a range and a seriousness to convince the Prime Minister that the Prince was a man of competence and judgement. In September, Anson told Peel that the Prince had taken over the running of two large farms within Windsor Park, 'taking the stock and crops from the Commissioner of Woods', and that if the Commissioner of Woods were agreeable, he would take over the running of a third farm, Home Farm.

On 3 October, Albert wrote to Peel from Windsor, 'when you were last here, our conversation turned upon the Nibelungen Lied [*sic*]'.

No previous Minister of the Crown, dining and sleeping at the Castle, had found himself discussing Middle High German poetic mythology. Peel, an intelligent, wide-ranging book-collector and connoisseur of paintings, warmed to this. Albert's letters accompanied 'a very fine edition of the work, which has lately appeared, and I therefore send it to you to look at'.[33]

The letter was written in huge, clear, copy-book handwriting. It looks as if a clever German schoolboy, accustomed to the *Alte Schrift*, has been practising his English hand. It is not an entirely naïve letter, however, because the gift of the *Nibelungenlied*, and the recollection of a pleasant evening at Windsor, pressed home the reminder to the Prime Minister that Peel had offered Albert an actual role – to chair a Royal Commission to decide on the decoration of the newly built Houses of Parliament, and the choice of artists for the murals. This task will be discussed in the next chapter.

Clearly, it was one for which Albert was well suited. As the autumn advanced, however, Albert had more on his mind than royal farms or the murals at Westminster. Victoria's second confinement approached. It was natural to hope that the next baby would be a boy. Albert, his mind on the future, quizzed Peel about a whole variety of matters. If the baby were male, how would he be styled? Would he be Prince of Wales? And an Elector of Hanover? How would he be described when mentioned in the liturgy? Peel patiently pointed out that the last Prince of Wales had been born in 1762, and that the sovereign had been present at the Council meeting in which his styles and titles were agreed. Peel added 'that there was no recognition... of any foreign Title to which they might lay claim, and that these titles would not be included in any prayers in the liturgy'.[34]

Albert was not especially interested in the Church of England, and never bothered to master much of a sense of that mysterious organization which at that date played so large a role in the life of the nation. His question about the liturgy was indicative of his desire to spread his net wide; and he freely used Peel as a tutor, to instruct him in the arcane by-ways of English lore. The Bedchamber Crisis had demonstrated to Albert not merely that Peel was a man to be reckoned with; it had also revealed the Byzantine complexion of the Royal Household, as well as the inefficiency with which it was controlled by Lehzen. Albert could begin to see in Peel a potential ally. He conceded that 'the ancient institutions and prescriptive usages in the Court, ought never to be touched by the Queen, but with the mature reflexion & caution'. In that sentence, although he wrote 'the Queen', he really meant himself. 'Something should be done to introduce into the present system (or rather confusion) improvements so peremptorily necessary'.[35] Quite clearly, he had reached for the dictionary here and where he should have written 'urgently', he had chosen the word 'peremptorily'.

The end game was to get rid of Lehzen, but Albert could scarcely do this when his wife was eight months pregnant, and in a depressed state, and more dependent than ever on her old governess. Meanwhile, there was yet another matter weighing on Albert. That was money. Victoria, as well as the Civil List, received an income from the Duchy of Lancaster. This, by the end of her long reign, would make her the richest woman in Britain, in her own right. The decision by the Exchequer to allow her to enjoy this income had been nicely balanced, and before the reign began, they were poised to take the Duchy income away from her. At this early stage, 1841, the income from the

Duchy was still agricultural rent: substantial and welcome money, but not prodigious, as it would become when Victoria found herself the owner of coal mines in South Wales, of the Mersey dockyards, of the newly created boom spa town of Harrogate, causing her fortune to grow to unrivalled, Croesian heights. She never fully recognized how different this made her from all her predecessors as monarchs in the previous century, who had been obliged to beg Parliament for money. Every Prime Minister and every Chancellor of the Exchequer throughout her long reign would receive letters complaining about the cost of entertaining foreign Heads of State, about the poverty of her children and their need for Civil List allowances, and about the cost of running her Household.

Albert primarily concerned himself at this stage, not with the accumulation of private wealth which was transforming the life of the monarch, and the monarchy, but rather with the narrower question, of how the Civil List, the money voted by Parliament to the sovereign and her family and Household, could be made to cover the necessary royal expenditure. He wrote to Peel, 'I should be greatly oblidged [*sic*] to you if you would take the trouble of explaining sufficiently to me the substance of the Civil List Act, so that I may be able to understand for myself the precise nature of the responsibility, which is contracted by the Great Offices of State to the Treasury.'[36]

Albert had not chosen his moment well. Nineteenth-century British governments, unlike their counterparts in the twenty-first century, did not like spending money. There was, by modern standards, a tiny tax revenue, and in that very month two considerable drains on the Exchequer had sprung. One was an expensive fire at the Tower of London on 31 October, which would cost £200,000 to repair. The other, much more embarrassing to the Government, was caused by a forger named Rapallo, who had succeeded in drawing down a colossal, as yet unspecified, fortune with counterfeit Treasury Bills. Possibly, he used the money to finance Louis Napoleon's Boulogne expedition in 1840. If so, there was even more embarrassment for the incompetent Comptroller of the Exchequer, Lord Monteagle, who had been very slow to notice the forgeries, and, when told of them, slow to act.[37]

No one in the Government, at this juncture, would look kindly on the royal pair holding out the begging bowl. What interested Albert, in the event of the Queen having a boy, was whether, as Duke of Cornwall, the child would be entitled to receive the incomes of the Duchy, at present held by the Treasury. Peel was not inclined to tell him.

The politicians had noticed how firmly, since the Queen's first pregnancy, Prince Albert had tightened his grip on the executive powers. When she was confined for the birth of the Princess Royal, for example, the Foreign Office stopped sending her the boxes of State Papers to sign. As the second confinement approached, they were instructed to send the boxes, regardless of the Queen's state of health, and to supply Albert with the necessary keys.

Albert was also impatient with the British convention, dating from Stuart times when it was feared a spurious 'heir' might be brought into the Queen's chamber in a warming-pan or similar container, of summoning various public figures to assist at the birth: the Archbishop of Canterbury, the Lord President of the Privy Council. Some looked askance at the 'crotchet of Prince Albert's',[38] as Greville called it. On 9 November, Albert scribbled a note to Peel – in a hurried hand, very different from his copy-book epistle about the *Nibelungenlied*: 'My dear Sir, the Queen has been taken ill and Dr L----- says it may be short. Believe me always, yours truly, Albert'.[39]

'Dr L', by the way, was Sir Charles Locock, first baronet, physician/accoucheur, who had delivered the Princess Royal. Albert waited until the Queen was in labour before summoning the Archbishop and the Lord President of the Privy Council, and by the time they arrived at Buckingham Palace on that Tuesday morning, at forty-eight minutes after ten, the son and heir had been born.[40]

At 2 p.m., the Privy Council assembled. Albert had taken the chair on the last occasion, a year earlier. Greville, Clerk in Ordinary to the Council, was determined this should not happen again and steered the Prince to the top of the table at the left. None of the Royal Dukes were invited. The Establishment was thereby making very clear that, in a constitutional monarchy, they remained in charge and power was not to be delivered to the Queen's husband who had no constitutional role. He had performed his function, the most basic in an hereditary monarchy; as far as the Constitution was concerned, however, he had no official role. Barring a calamitous illness, such as had carried off Queen Adelaide's babies in the previous reign, the succession was now secure. There was now a Prince of Wales. For the foreseeable future, the state had a titular head. The state itself and the monarchy were going to have to adapt themselves to a fast-changing world. The extent to which the state did so was largely thanks to the Prime Minister, Robert Peel. The speed with which Prince Albert recognized Peel's qualities as a statesman ensured that the monarchy would continue to have a role to play in the years to come. It would require a delicate balance, however. Albert would never completely see that the advisory and symbolic roles of a modern monarch were very different things from executive power. Had someone else been on the throne, or married to Victoria, however, the story might well have been very different.

We have said that it is in the next chapter that we shall begin to look at Prince Albert's patronage of the arts and, in particular, the role he played in decorating that highly symbolic place the new Palace of Westminster. We will, however, pause beside the statue of the Queen which adorns the Prince's Chamber there, the robing room of the Prince of Wales for ceremonies of state. It is one of the finest works by John Gibson. The figure of Queen Victoria is flanked by Justice and Clemency. The plinth on which she stands, however, is adorned with three reliefs, depicting Commerce, Science and Industry, or the Useful Arts. Here is a marble embodiment of what Albert wanted the monarchy to be.

When Peel took office, his Government, after the genial years in which Lord M. had played whist and charmed his sovereign with witty gossip, the country had faced a series of crises any one of which would have darkened a Prime Minister's or a Foreign Secretary's brow. There was a dangerous war in China. British troops had catastrophically invaded Afghanistan and those who had not been massacred would make an ignominious retreat from Kabul in 1842. Disputes over the north-east corner of Canada and the United States led to periodic bouts of violence; and there were squabbles with France about Syria.

Above all, however, the problem facing Peel when he took office after the indolent Lord M. was the state of the economy and the socio-political consequences arising from it. In the year ending April 1842, there was a Budget deficit of over £2 million, an accumulated deficit over £7.5 million. The population was increasing by 15 per cent per decade.[41] These were not simply figures in a spreadsheet. Human lives were affected. The continuation of high tariffs on imported corn meant that, in order to please the great landowners, many of them Whigs, the price of bread was kept artificially high.

There was a growing movement to repeal the Corn Laws, but if Peel were to succeed in doing so, he would split his Tory Party and alienate the Tory landowners and the diehards. The tightened economy manifested itself in town and country. Agricultural wages were low and labourers went hungry. This was also the decade when Lord Ashley (later 7th Earl of Shaftesbury) was campaigning for a maximum ten-hour day for workers in mills and factories, with respite and education offered to the child-labourers. The Chartist Movement was growing, its leaders attacking with equal vigour the free trade capitalists of Manchester and the old Tory defenders of Church and Queen. They wanted root and branch reform of the system, with universal suffrage, some even calling for women to have the vote, though many Chartists believed this was a step too far. Ireland was a tinder-box waiting to explode. The girl-Queen's popularity, high for a short period after her accession, had slumped, and the British public had reverted to its habitual distaste for the monarchy. It was not Peel's job to save it. Could Albert but harness the fortunes of the monarchy to Peel's high-risk, forward-looking Conservatism, however, much mutual benefit would ensue.

So, when Parliament was opened by the Queen on 3 February 1842, it was greatly to Peel's advantage to do it with as much pomp and pageantry as possible; and were his economic measures a success, the young Queen and her husband stood to play a role in the free trade success story, or at least to be caught up in the happiness such success would bring. Enormous economic potential existed in Britain at this date if it could only release itself from the trammels of tariffs, if it could only keep the poor busy and well fed, if it could only, eventually, bring not merely prosperity but political power to the growing middle classes.

The State Opening was more than customarily splendid, because it coincided with a state visit from the King of Prussia, who had come over to be a godfather at the christening of the Prince of Wales. The jolly old King ('common-looking' in Greville's eyes[42]) described himself as 'an ally of Old England… a true admirer of Her British Majesty, & a good Protestant'.[43]

Albert stage-managed the visit, which was a triumphant success, even if it was just as well that the politicians did not read King Friedrich Wilhelm's thank-you letter to the royal pair. He wrote that the Queen had 'completely turned my head'. He described her, accurately, as 'a Princess, born a Guelph, and belonging now by marriage once more to the Saxons… a German lady, notwithstanding her wearing three crowns and notwithstanding she writes like a native of Germany'.[44]

During her first two pregnancies, Victoria could not fail to see that Albert took an ever more active role in political affairs. When exhausted, and laid low by postnatal depression ('I have been suffering so from lowness that it made me quite miserable'[45]), she actually welcomed his readiness to help her. Help, however, was all she wanted. She did not wish to be replaced.

'Oh, my dearest Uncle, I am sure if you knew how happy, how blessed I feel, and how proud I feel in possessing such a perfect being as my husband,'[46] she could rhapsodize, less than a month after Bertie was born. She had, however, done her utmost to show to her ministers that she was capable of exercising her duties only days after giving birth. This defied the nineteenth-century convention that women should lie up for weeks after a confinement. Albert wrote to the Prime Minister on 30 November, clearly on Victoria's instruction, 'The Queen is so far recovered that you may recommence your communications to her. She only begs that in matters which require more explanation you would for the present still write to me or see me.' This from Buckingham Palace only a day before the Queen, still at Windsor, invited Peel down to the Castle to dine and stay the night – 'as the Queen has as yet not been out of her private Apartments of an eveng [*sic*], consequently not been in company, she will not ask any of the other ministers (except the Lord Chancellor if he should come) to stay to dinner'.[47]

Something approaching the beginnings of a friendship with Peel was to be seen here. Intense and introvert as she was, Victoria could only operate, professionally, with friends. Perhaps one of the differences between the royal pair was that, for her, all human relationships, however formal, aspired, ideally, to the condition of friendship; whereas for Albert, all relationships, however intimate, aspired to the nature of business. She was learning to abandon her prejudices and caprices – these were weapons in her political armoury – and Albert had begun to find in Peel one of the father substitutes which we can watch him crave throughout his life. Albert could open her eyes to Peel's virtues, even though the Prime Minister, reserved, unfrivolous, unamusing, was so little her 'type'. When she issued that invitation to Windsor, however, she was unaware of the machinations which had occupied Albert and Stockmar during the last stages of her pregnancy and during her confinement.

Two of Victoria's oldest confidantes and intimates stood between Albert and his mastery of her. They were Lord M. and Lehzen. Stockmar's activities behind the arras did not let up. From Albert's German secretary, Praetorius, Stockmar learned that the Queen was in 'daily correspondence' with Lord Melbourne. Stockmar made an excuse to see the Prime Minister about Belgian affairs. He slyly introduced into the conversation the hope that 'the Queen would seek and find her real happiness in her domestic relations only',[48] that is, she would leave the politics to Peel, but not solely to him, to Peel in conjunction with Albert, an Albert who was in daily contact with Stockmar. Peel's antennae were immediately alerted to the fact that Stockmar was warning him of something. Peel knew that in replying to Stockmar, he was writing words which would be read, eventually, by Melbourne. Peel was no fool, and he had probably heard the rumour that the Queen was still discussing with her old mentor the political situation at home and abroad. 'Becoming suddenly emphatic',[49] Peel told Stockmar that he was prepared to indulge the Queen, 'in all her wishes relating to matters of a private nature... But on this I must insist, and I do assure you, that that moment I was to learn that the Queen takes advice upon public matters in another place, I shall throw up'.[50]

He did not mean he would vomit; he meant he would resign the premiership. Victoria had, of course, corresponded with Melbourne, asking his advice about everything, before and after the birth of the Prince of Wales. She had received a refresher course on the history of political offices in Britain, and Lord M. had told her when the offices of Prime Minister and Home Secretary had come into being.[51]

He had mused with her about what was passing through the mind of Prince Metternich (that cunning Austrian statesman whose good looks, earlier in the century, had so captured the fancy of Prince Albert's mother) and on the suitability of the British Ambassador in Vienna, Sir Robert Gordon.[52] The old chums, Victoria and Lord M., would discuss the dispositions of the Kings of Hanover, Prussia and Holland.[53] In January 1842, he explained to her the theological convulsions at Oxford, over the so-called Tracts for the Times, High Churchery, which smacked of Rome and, for its most celebrated exponent, Newman, would lead to Roman conversion. They discussed the situation in Spain. They weighed up one politician against another – 'Palmerston dislikes Aberdeen and has a low opinion of him. He thinks him weak and timid, and likely to let down the character and influence of the country'.[54] These were the sort of exchanges which Lord M. had been having with his devoted sovereign since, as a teenager, she had acceded to the throne five years earlier. Only now, he was not her Prime Minister, Peel was, and Aberdeen, whom Melbourne so glibly vilified, was Peel's appointment as Secretary of State for War and Colonies. Sent from a variety of country houses, from his own at Brocket Hall, from Castle Howard, from Broadlands, as well as from his London house in South Street, Lord M.'s epistles to his little friend were shameless, sunny, Whiggish effrontery.

Stockmar confronted him, warning him that, if he continued to exercise his role as the Queen's political adviser, there would be a 'catastrophe'. Stockmar, with that sharp-edged slyness which so often bordered on blackmail, had the temerity to remind Lord Melbourne that he had once described the Queen to him as 'a most passionate, giddy, imprudent and dangerous woman'. He reminded Melbourne of a conversation they had had in September 1841 on the same subject, 'though you did not avow it in direct words, I could read from your countenance and manner that you assented in your head and heart to all I said'.[55]

Melbourne was not going openly to concede a point to the diminutive German busybody. For a while, the correspondence between Lord M. and the Queen continued. The point, however, had been made, and the pair of old friends had learned their lesson. They were firmly given the unpleasant intelligence that they were being watched. From now onwards, it was unthinkable that a British sovereign could openly side with one political party against another, or take advice on political matters other than from her elected ministers. (From now onwards, at any rate, as long as Albert lived. The Queen's friendship with Disraeli, and the days when John Brown was consulted on questions of foreign policy in the Balkans, lay in an as yet unimaginable future.) A principle had been established. Even when she seemed to stray from it in late middle age, Queen Victoria would never really deviate from it, namely that she was a modern constitutional monarch. In such a political world, the monarch was above politics. Stockmar had insisted upon this principle, and Albert was to be the instrument by which it was achieved; even though he always thought the monarch should retain more executive power than the British system allowed, he was a realist.

It was inevitable, and therefore relatively straightforward, that Albert (Stockmar pulling the puppet strings) should replace Lord M. as principal adviser to the sovereign. Victoria's other influential intimate, Lehzen, would be harder to dislodge. Louise Lehzen had been her constant companion since Victoria was five years old. She had been her protectress against Conroy, and a buffer between Victoria and her mother. She had been responsible for Victoria's education, and her lack of it. She was, whatever her defects, Victoria's closest female companion. Albert was fiercely jealous of Lehzen. It is easy to see how the matter appeared in his eyes. As an attentive father, he believed his wife's upbringing had been catastrophic, educationally and emotionally. As a born administrator, he saw Lehzen as an inefficient manager of the Royal Household and as a potentially dangerous supervisor of the nursery.

So many of their early marital rows were about Lehzen. In January 1842, Albert took Victoria for a few days away at Claremont in the hope of lifting her depression. She was still tired after the birth of Bertie, a child with whom she never truly bonded, and both parents had been made anxious by the sickliness of their firstborn, Pussy, now fourteen months old. When they returned from their little trip, both young parents rushed to the nursery. Pussy had been put on a diet, by Sir James Clark, of asses' milk and chicken broth. They found her worryingly pale and thin. Albert said that the child's sickliness was attributable to Lehzen, supposedly in charge of the nursery. There was the inevitable explosion. As Victoria screamed at him, he responded as he usually did to her fits of bad temper. He silently left the room, and sat down to write his wife a letter, coldly making clear that from now on their firstborn was entirely Victoria's responsibility. He washed his hands of the matter. If their daughter died, it would be the responsibility of Victoria and Lehzen.[56] He always knew, at just the right moment of a squabble's cycle, how to play on Victoria's vulnerability and to increase her sense of inadequacy and guilt. When these rows broke out, Victoria would go to Lehzen to pour out her heart. Albert went to Stockmar, who was not content to be a mere confidant. The little doctor would round the circle and write directly to the Queen.

Lehzen, Albert told Stockmar, 'had never been away from her and like a very good pupil [she] is accustomed to regard her governess as an oracle. Besides this, the unfortunate experiences they went through together at Kensington had bound them still closer and Lehzen in her madness has made Victoria believe that whatever good qualities she possesses are due to her.'[57]

Stockmar, in his own letter to the Queen, was calmly blunt. If Victoria could not control her temper, he would leave the Court. It was an effective threat. Had she been in charge at this stage – in charge of her own emotions, in charge within the marriage – she would have called Stockmar's bluff and sent him packing. He knew she was depressed, defeated, abject. She walked further into the snare. She told Stockmar that if Albert were displeased, he should explain his reasons directly. By writing this, she was in effect admitting that she must now choose between her husband and her trusted companion of eighteen years.

Albert had disliked Lehzen from the first. He was also genuinely and justifiably appalled by the inefficiency of the Royal Household, and by the sheer waste. At Windsor Castle, for example, a dozen large joints of meat were roasted daily, even when the Queen was in London, as she generally was: before she was widowed, she loved the capital, and it was obviously more practical, while Parliament sat, for the sovereign to be near her Prime Minister, rather than being separated by a three-hour carriage drive. Prince Albert discovered innumerable additional extravagances. Hundreds of candles were placed in candelabra and candlesticks daily, and then replaced, even if never lit. Since Queen Anne's day, the privilege of selling 'Palace ends' had been a perk for the servants. Victoria knew nothing of these things. Lehzen either knew nothing or felt herself powerless to change them.

At every turn, now that he had Peel on side, Albert told the Prime Minister of all the domestic economies which he was practising. In general, of course, the more rational the economies and reforms, the more they made the Prince enemies in the Household, and neither servants nor courtiers ever liked him. As for the nursery, without informing Lehzen, who was notionally in charge of it, he appointed Lady Lyttelton to take charge of the royal children. She was a widow with five children of her own, born Lady Sarah Spencer. Albert prepared to take over personally the work of the Queen's Private Secretary, and for some of the more trivial work, which had always been undertaken by Lehzen, he appointed Marianne Skerret, the Queen's dresser. Skerret, the niece of Queen Adelaide's sub-treasurer, had been appointed upon the Queen's accession, and would remain in royal service until 1862.[58] It was Skerret's task to copy Victoria's outgoing correspondence to tradespeople. Albert himself undertook the political correspondence.

Thus, Lehzen, by the slow erosion of her various roles, was rendered redundant. There remained only the question of when she would be disposed of, and how much she could expect by way of a pension.

Stockmar buzzed busily from flower to flower. 'He did not think it judicious to tell Her Majesty that he had previously showed Stockmar a memorandum left by Lord M. on the subject [i.e. the subject of Lehzen]. Would it embarrass Sir Robert to be told its contents?' And, no, the Queen, in whose character candour was such a marked feature, replied, 'I shall beg Sir Robert not to say to any one that he has seen it'.[59] In the event, they settled £800 per annum on the Baroness and gave her a carriage. (Peel's pensions to Wordsworth and to Tennyson from the Civil List were £300 and £200 per annum respectively. Both lived comfortably.[60]) On 25 July, Albert informed his wife that Lehzen had gone abroad for her health. Greville recorded, 'the Baroness Lehzen has left Windsor Castle and is gone… (as she says) to stay five or six months, but it is supposed never to return.'[61] He was right. She settled in Bückeberg and lived until the age of eighty-six, dying in 1870. She left the

Queen 'in high spirits'.[62] Both the women believed she was merely visiting Germany, but she never returned.

For the rest of her marriage, Victoria had no really intimate female friend and such a lack makes any woman vulnerable. Any woman who enters a relationship with a man needs a female ally, and Lehzen's banishment would leave the Queen vulnerable, exposed. When Lehzen had been banished, and Albert had triumphed, it was natural that, under his influence, she would begin to see the whole relationship with Lehzen differently. She came to believe that Lehzen had exploited differences with her mother, even that she had deliberately held back her royal pupil to strengthen her own power over her.

Some months before Lehzen's departure, there was a second assassination attempt on the Queen. The first had been in 1840 by a man called Oxford. Now, another failed attack occurred, in more or less the same spot in the Mall, as they were returning from the Chapel Royal on Sunday, 29 May. It was Oak Apple Day, the anniversary of the Restoration of Charles II – 'when the King shall enjoy his own again', as the song had it. They passed through cheering crowds. Albert noticed a man step forward with a pistol. He heard the trigger snap, but the gun misfired.

'I turned to Victoria,' he told his father, 'who was seated on my right, and asked her "Did you hear that?" She had been bowing to the people on the right and had observed nothing.'[63]

When they returned to the Palace, Albert questioned the footmen who had been at the back of the landau. They had not noticed anything either. He did not mention it to anyone except Colonel Arbuthnot, an equerry, asking him to inform the Inspector of Police, the Prime Minister and the Home Secretary.

The next day, the police brought a fourteen-year-old boy named Pearse to the Palace. In spite of a bad stammer, he was able to say he had seen a man point a pistol at the royal couple and then exclaim, 'Fool that I was, not to fire!' The Queen was very much upset. A doctor was sent for, and it was agreed that fresh air would do them good. The alternative would have been to skulk indoors while the police searched for potential suspects.

The royal party therefore set off for an outing to Hampstead Heath at four that afternoon. Albert did not pretend to any heroic sangfroid.

> You may imagine that our minds were not very easy. We looked behind every tree, and I cast my eyes round in search of the rascal's face. We, however, got safely through the Parks, and drove towards Hampstead. The weather was superb, and hosts of people on foot. On our way home, as we were approaching the Palace, between the Green Park and the Garden Wall, a shot was fired at us about five paces off. It was the fellow with the same pistol – a little, swarthy, ill-looking rascal. The shot must have passed under the carriage, for he lowered his hand. We felt as if a load had been taken off our hearts, and we thanked the Almighty for having preserved us a second time from so great a danger.[64]

The party trotted home: Victoria and Albert, the Duchess of Kent, Uncle Mensdorff-Pouilly (married to the Duchess's sister Sophie; in spite of his name he was a French Count) and the two equerries, colonels Wylde and Arbuthnot. The 'ill-looking rascal', a twenty-two-year-old joiner named John Francis, was arrested and within less than three weeks he had been tried and condemned for High Treason. It was a capital offence. When it was pronounced, Francis swooned and collapsed into the turnkey's arms. The Queen was anxious that the death penalty be commuted to transportation for life. So it was that Francis joined the boatloads of Magwitches bound for Australia.

Before Francis's trial, Parliament had hurried through a measure altering the legislation about attacks or threats to the person of the sovereign. Hitherto such miscreants could only be tried for treason. The new bill, introduced on 12 June, made attempted attacks punishable by transportation for seven months, or imprisonment with or without hard labour for as long as three years; or to be whipped 'as the Court shall direct, not exceeding thrice'.

On 3 July, as they drove to the Chapel of St James's, this time with King Leopold, a hunchbacked chemist's assistant, Bean by name, tried to shoot at their carriage. The pistol misfired. A boy named Dassett took the gun from Bean's hand. Bean would be condemned to eighteen months' imprisonment.

When Peel heard of Bean's action, he was in Cambridge, but hurried at once to London for an interview with the Prince to discuss what further measures could be taken to make the Royal Family safe. The new bill became law on 12 July. In a sentence clearly dictated, if not actually written, by Victoria herself, 'Sir Theodore Martin' completed the dramatic tale: 'During their interview, Her Majesty entered the room, when the Minister in public so cold and self-commanding, in reality so full of genuine feeling, out of his very manliness was unable to control his emotion, and burst into tears.'[65]

The unlikely emotional bond between Peel and the royal pair was essential. These were tough and confusing times, both for Peel and for the monarchy. Few, perhaps, wanted Victoria and Albert dead, but Albert was yet to win hearts. Peel, who was emerging as one of the great Prime Ministers, was prepared to address the dangerous state of the country – the Exchequer deep in debt, poverty and extreme political discontent widespread – by facing down the opposition of those who, since 1689, had run England, namely the landed classes. Albert had been taught by Stockmar to abhor Victoria's Whig partisanship. Politics, however, are not logical, and Peelite partisanship, at this juncture, really was a different matter, for Peel alone was in a position to break the deadlock of problems by breaking, in a sense destroying, his own party and his own career. The choice was real danger – of hunger, of revolution, even – and a non-violent way forward with a radically altered set of parliamentary alliances and oppositions.

Albert and Peel both risked much by their overt alliance. Were Peel to fail, the monarchy would have yoked itself to a political disaster.

Peel, in purely political terms, had enough difficulties to contend with, so it was in a sense foolhardy to champion a still unpopular foreign young prince. Both men, however, saw that the changed world needed a new politics, politics which addressed the appalling conditions of the working classes, and which liberated the vast potential of the manufacturing classes; a politics which wrested from the landed classes the inequitable benefits which they derived from tariffs on corn. Free trade, if the Government could bring it to pass, would change an indebted, rancorous, half-starved country into a, no, *the*, major economy, the major power in the world.

Peel signalled the direction in which he intended to lead the country by his first Budget in 1842. He reinstated income tax, which had hitherto only been used as a stopgap wartime expedient. (The Queen herself agreed to pay it.) It was charged at seven pence in the pound (when the pound had 240 pence!) for those whose income was over £150. It was a masterly Budget, for, not only did the new tax pay off the national debt quite quickly; it enabled Peel to provide the Exchequer with an alternative source of income to tariffs. He greatly reduced the tariffs on 750 of the 1,500 imported items on which tariffs were still levied, and, most controversially of all, he reduced tariffs on imported corn to the point where it was now possible for a poor family to buy a loaf of bread. The Duke of Buckingham resigned from the Cabinet. (It was this extravagant peer who, when told by his banker that it was unnecessary to employ a special pastry-cook at Woburn Abbey, had replied, 'Can't a fella eat a biscuit?') Eighty-five Conservative MPs rebelled against Peel's budget, but it was carried by Liberals, Radicals and Peelite Tories. After a year in which mobs had attacked mills and factories throughout northern England and Scotland, in which the poor were hungry, in which the Chartists were demanding revolutionary change, Peel had offered a vision of a future in which revolution was not necessary to achieve peace, stability and prosperity. Albert's closeness to Peel had not gone unnoticed in the press, or among the political classes. He had publicly hitched the monarchy to the Peelite wagon. The gamble looked like paying off.

NINE

PUBLIC ART, PUBLIC LIFE

EOS, ALBERT'S FAVOURITE, glossy black and white greyhound bitch, could have laid claim to be his closest intimate. Albert's drawings of the animal, preserved in the Print Room at Windsor Castle, are beautiful, but they are those of an amateur, whereas Victoria's studies of the same subject show her semi-professional flair. Lucian Freud would have admired the way she caught the flirtatious sideways look of the dog, and of the greyhound's marvellous ability to curl herself into a ball. But then, Victoria had been given drawing and painting lessons by one of the finest of all animal painters. Sir Edwin Landseer's portrait of Eos is one of the most glorious animal paintings of the nineteenth century. She stands looking upward. We have no doubt into whose eyes she is gazing with such abject love. Surely it is the owner of the tall black hat, as silky as herself, and the kid white gloves which are cast suggestively upon the stool at her feet. The year after this masterpiece was executed, Landseer depicted Eos's royal companions, dressed not in the tall silk hat or bonnet of the period, but in the supposed fourteenth-century costumes of Edward III and his Queen Philippa of Hainault. The Latin inscription on Edward's tomb in Westminster Abbey may be translated, 'Here is the glory of the English, the paragon of past kings, the model of future kings, a merciful king, the peace of the peoples...' He was one of the most successful of English monarchs, reigning over fifty years, after the deposition of his dissolute father.

The decision to depict themselves as Edward and Philippa, therefore, was an advertisement that Victoria and Albert intended to stay put until they were old and revered. If some regarded it as tactless, at a time of economic hardship and political strife, to hold an extravagant costume ball with over two thousand guests in the throne room in Buckingham Palace, its idea had been a worthy one. Victoria had genuinely believed that, by having her costume made from Spitalfields silk and encouraging the guests to do the same, she would revive the pride of one of London's most illustrious industries, now flagging from cheap foreign competition.

The Victorian desire to see themselves in the guise of other ages, and in particular in the setting of medieval architecture and design, is a phenomenon with which modern taste finds it hard to come to terms. Their paradoxical need to cloak modernity in fancy dress, to build railway stations in the manner of Gothic cathedrals, or to make brand-new schools look as if they had been built in the time of Shakespeare, is probably part of their jumble of uncertainties: bourgeois uncertainty about their social status making them build bogus medieval schools, shooting boxes and villas for themselves; doubts about religion feeding, rather than diminishing, their desire for fake-medieval edifices, their pointed lancet windows and gaudy, factory-made stained glass gently numbing their awareness of the extent to which they did not share any of the medieval beliefs. Victoria and Albert, anxious to demonstrate their credentials as modern constitutional monarchs who supported the introduction of income tax and the repeal of tariffs, held a party in which Albert dressed as the longest-serving English King of the Middle Ages, and Victoria as his consort. The costumes were designed by James Robinson Planché without a slavish medievalist authenticity. Planché who, as Sir Roy Strong has written, 'combined a somewhat unlikely career as a highly successful writer of burlesques and harlequinades with that of an antiquary and scholar of heraldry and costume',[1] was frequently consulted by Victorian genre painters about the details of their work, but they did not always achieve entirely academic accuracy. For example, Landseer depicted Albert wearing the jewelled Sword of Offering, which had been made, by the London metalworkers and jewellers Rundells, for the coronation of George IV in 1821. Victoria's dress, shaped by stays and multiple petticoats, plainly follows the fashions of the 1840s. They are no more convincingly medieval than some nouveau-riche family, housed in a modern crenelated 'castle' adding a French preposition – à or de – to their ordinary English name in the hope that future generations will suppose them to have come over with the Conqueror.

The costumes and the jewels are something in which Albert took deep interest. At the christenings of his children, he had designed the decorations of the font. In the same year that Landseer painted Albert's much-loved Eos, Winterhalter painted the Queen wearing a delicate sapphire tiara, designed in that year by Albert to match the sapphire engagement ring which he had given her in 1840. It is one of his finest designs.

The coronet, in which some of the largest sapphires in the Queen's collection had been cleverly arranged in a row, matches the sapphire and diamond brooch which he had designed as a present for her on the eve of their wedding. The relationship between Victoria and Albert was often expressed in the gift and design of jewels. Albert introduced his wife to Germanic forms of commemorative jewellery, incorporating images, and even the teeth, of their children, as well as stag's teeth, stones and pebbles from their favourite places such as Claremont and later Balmoral. The exquisite coronet, however, is perhaps the loveliest thing he ever wrought for her, and represents in the most delicate way a poem in stones about the gentleness of their royal aspiration. They were to be as true to one another and as long-reigning as King Edward III and his Queen. But here was no Iron Crown or emblem of absolutism; rather, a romantic, understated and beautiful reflection of her sovereignty. The coronet was passed down to Edward VII and to George V, who gave it as a wedding present to his daughter Princess Mary, the Princess Royal, when she married Lord Lascelles, later Earl of Harewood, in 1922. It has recently been sold, and, very appropriately, thanks to generous donors, it is now housed in the Victoria and Albert Museum.

Peel was an aesthete and an art collector. Goethe had been full of admiration when, in 1829, Peel had paused from his controversial pressing-forward of the cause of Catholic Emancipation to buy one of Claude Lorraine's finest paintings for the staggering sum of £4,000.[2] At the same time as sharing Albert's connoisseurship, however, the statesman could see that, by encouraging the young man's interest in art, he was also distracting him from an incautiously overt involvement in politics. It was during the premiership of Peel, and the early years of Albert's marriage to Victoria, that the position of the modern, constitutional monarchy was established. Albert, guided by Stockmar, continued to hope for an increased political influence. Peel, and all subsequent Prime Ministers whatever their politics, made sure that this did not happen. Naturally, as the Queen had more and more children, there were periods when Albert would take charge of the red boxes. His interest in the English political scene became better-informed. But in the 1840s, we see the monarchy strengthening the position of the political class, more than the other way about; and in so far as the position of the monarchy was honing its political identity, it was being guided – above all by Peel – into civic and cultural concerns which acknowledged the nullity of its political power.

Albert, an obsessively busy man, as well as full of political ambition, found it difficult to turn down any appointment – whether it was as the chairman of a committee, or as the administrator of royal estates and farms, or as a consultant to charities. The more of such tasks Peel could find to occupy the workaholic Prince, the less chance there would be of the monarchy ever causing the problems which had beset the previous three reigns, either in the form of the monarchy's political bias or in its scandalous way of life, both of which encouraged the march of radicalism. Albert's high moral rectitude made it inconceivable that he would cause scandal. His equally powerful political ambition could, with the careful and subtle manipulation of Peel, be directed in channels which would enforce and strengthen the existing political order. The symbiosis, personal between Peel and Albert, political between the Crown and the political class in general, rather than exclusively an alliance between Victoria and the Whigs, was going to turn into a system that worked, and that endured.

Given Albert's marked interest in the arts, and his recent, albeit condensed, version of the Grand Tour just prior to his marriage, it was natural that Peel should have wished to involve him in the politicized, but not directly political, question of Public Art. And an immediate matter to hand was the question of how to decorate the newly built Palace of Westminster.

Three years before Victoria became Queen, fire swept through the old Houses of Parliament. It was more than a simple fire. It was an emblem of the old world, and the old way of doing things, being reduced to ash. The architect Sir John Soane had described the old Parliament buildings as 'an extensive assemblage of combustible materials'.[3] The Court of the Exchequer, abolished in 1826, had kept records of its accounts with tallies and foils, notched elmwood sticks. One half would be kept by the Exchequer, the other presented as a receipt to those paying money to the Court. These sticks looked like, and became, firewood. In his justly celebrated speech about the fire, Dickens spoke about the obvious ease with which the tallies could have been given away to the poor of Westminster to warm their homes. Instead, they were amassed at Westminster until the pile of them grew so huge that orders were given to burn the lot in the big furnace beneath the House of Lords. The fire got out of control, and by the following morning, Lords, Commons and much of the old building which had been known by the title of the Palace of Westminster was in ruins.

It was the chance, in that age of Reform and Change, to build something entirely new. Architects were invited to submit plans 'in the Gothic and Elizabethan style'. Ninety-seven entries were submitted and the winner was Charles Barry, who enlisted the help of Augustus Welby Pugin for the Gothic embellishments of the building. Construction began, again symbolically enough, in the very year that Victoria became Queen, 1837, and a decade later, the House of Commons was complete, though it was not until 1858 that the Clock Tower was finished.

In other words, for most of Albert's life in England, the Parliament buildings were 'work in progress'. With obvious symbolism, the new buildings were erected during the critical change of temper in political life. Britain gently changed from being a purely aristocratic society to one in which the cleverer or richer middle classes shared, together with Radicals and the multifarious Irish members (who belonged, severally, to all these categories – aristocratic, middle class, clever, rich and radical!), in the political process. At about the time that Albert and Peel's acquaintanceship ripened into a very friendly working partnership – when the Prince lent the Prime Minister the *Nibelungenlied* – Parliament was discussing the possibility of decorating the corridors of its new chambers with mural paintings and it was agreed to set up a Royal Commission to commission and supervise the works. Albert wrote to Peel congratulating him on the warmth with which this idea was received in the House. 'I only give you my crude views,' wrote Albert, who was learning the English art of saying all but the opposite of what he meant, when trying to get something for himself, 'and have no wish whatever to press them against the experience of others.'[4] Peel took this for what it was, a job application, and saw that Albert would be ideally suited to chair the Commission, with the painter Charles Eastlake as Secretary.

Albert urged Peel to consult an artist, rather than merely leaving the matter in the hands of public men. Peel wrote back on 4 October 1841 to say that 'M. Cornelius is arrived in London'.[5] Albert was anxious about this, since Victoria was about to go into labour to produce her first son, Bertie. As it happened, however, Bertie was not born for a month, and there was a chance to consult Peter Cornelius, who was believed to be the greatest living *frescanto*.

In the previous two decades, Cornelius, who had learned the art of fresco in Rome, had enjoyed the patronage of Ludwig I of Bavaria, and transformed Munich, his capital. His paintings of the Life of Christ, of German History and Mythology, now adorned the Hofgarten, the Neue Residenz, the Glyptothek, the Pinakothek, the Ludwigskirche, the Allerheiligen-Hofkirche, the Bonifazkirche and the Isartor. Munich was now famed as the centre of Cornelius's so-called Nazarene school of painting. The city was the 'unrivalled queen of modern art'.[6] Ludwig could be seen as a role model for Albert – though only as a royal patron of the arts, not as a political template. Ludwig (grandfather of his more famous mad namesake) became increasingly reactionary during the 1830s, an ultramontane in religion, an absolutist in his concept of monarchy: he would be forced to abdicate in 1848.

Cornelius was sniffy about the abilities of English artists, and doubtful whether the climate of London was suitable for fresco painting. Albert reported to Peel that he 'can understand that he [Cornelius] may be disgusted with' the English school's 'mistaken tendency: to obtain mere effect & to bribe the senses & acquire applause by rich colouring & the representation of voluptuous forms overlooking at the same time its deficiency in real poetical imagination & invention & the importance of Correct Drawing'. Cornelius, for his part, returned to the Prince the thought, 'The power of a London smoke, he can hardly have had the means of appreciating sufficiently.'[7]

Albert proposed a competition. To calm the rumours that a German Prince was threatening to use German artists to decorate the good English Parliament, the competition would be open to British artists only; and they would have to demonstrate their drawing skills, because Albert wanted a cartoon only, 'so as to necessitate the artist to exert his powers of conception & imagination & force him to the greatest correctness in drawing'.[8] These weird German sentences which Albert, presumably with the aid of a dictionary, had rendered into a sort-of English would characterize all his correspondence with Peel and other statesmen in his first few years in England. We watch him develop, however, learning the structure of English sentences and becoming, if never exactly idiomatic, at least more plausibly English.

The competition artists were invited to choose subjects 'taken from British history or from Spenser, Shakespeare or Milton'. They submitted a bumper crop of entries and produced, according to Baron Christian Karl von Bunsen, the Prussian Minister to the Court of St James's, 'a wave of national optimism'.[9] There was an exhibition in Westminster Hall in the summer of 1843 which was well attended; this led to a second exhibition, the following summer, this time illustrative of the artists' skill in fresco. Following the popularity of this, the Commissioners decided to commission six of the artists to furnish new fresco subjects, this time illustrative of the function of the House of Lords in relation to the sovereign. The earnestness, the seriousness of it all, still has the power, 180 years later, to be moving.

Equally, no dispassionate consideration of the story can ignore its absurdity. For a start, Barry, the architect, was always opposed to 'decorating' his corridors with murals. He wanted the corridors to be sculpture galleries and had designed the 220 niches to fit them.[10] Cornelius, an indifferent-to-bad painter, had come to a nation which could produce Turner and Constable, two of the finest painters in Europe, and questioned the English ability to draw. Albert saw, among the English – as Cornelius had blazed in the Fatherland – 'if an encouraging opportunity were afforded them, if some were sent to study at Munic [*sic*], Florence & Rome, I have not the slightest doubt, they would produce works fully equal to the present school of the Germans'.[11]

Inevitably, there was criticism of Prince Albert's choices. As well as Barry's opposition to murals,[12] Henry Cole, who would later be such a friend and colleague of the Prince, doubted the wisdom of having no artists on the panel. Even if they had done so, Cole wrote in the *Westminster Review*, 'it would be a stretch of the imagination far too visionary to suppose that, in case of any differences of opinion, the views of the artists would be suffered to outweigh those of a prince, a prime minister and a duke'.[13]

By doggedly looking at British art through his German spectacles, and by seeking in the land of Etty, Constable and Turner for a new Cornelius, Albert naturally concluded that the most promising artist of his generation was William Dyce, whose *Baptism of King Ethelbert* was completed in the summer of 1846, and set the tone of the other Westminster murals. The only painter with a claim to greatness who won a prize in the Westminster fresco competitions was G. F. Watts for his *Caractacus Led in Triumph through the Streets of Rome*. His later painting, not a fresco, but oil on canvas, *Alfred Inciting the Saxons to Prevent the Landing of the Danes*, is easily the best of the murals in Westminster. Today the other pictures – Edward Armitage's *Caesar's Invasion of Britain*, or Charles West Cope's *The First Trial by Jury* or his *Charles I Raising His Standard at Nottingham, 1642* – seem like charming period pieces more than great works of art.

It is time to mention Ludwig Grüner, who would become to Prince Albert the patron of the arts something of what Stockmar had been to Prince Albert the politician. It is sometimes supposed that Albert met his artistic mentor in Rome during the *Kunstreise*, but there is no evidence for this. In all Albert's letters to Ernst from Rome – and they are copious – he never mentioned Grüner once. It seems much likelier that Albert first encountered Grüner when he had become President of the Royal Commission of the Fine Arts in 1841.[14] Two months earlier, the *Athenæum* had reported, 'There has arrived among us a German artist, M. Grüner, whose enthusiasm, though exclusive, all must respect and regard with admiration; he graves from two painters alone, Raphael and Overbeck… his engravings after these favourite models are supereminent… Raphael himself appears to have stood over M. Grüner as he did over Marc Antonio.'[15]

Grüner was born in Dresden in 1801, so he was eighteen years Albert's senior. His ambition had been to become a painter, but adolescent illness forced him to abandon this hope, and he was apprenticed as an engraver in his teens. He went to Italy to perfect his craft, and was commissioned to make eight engravings after Raphael to illustrate *Rafael von Urbino und sein Vater Giovanni Santi*, published in Leipzig in 1839 by J. D. Passavant. In Rome, Grüner had joined the group of German archaeologists and artists whom Albert encountered during his visit there, and who were patronised by the Counsellor to the Prussian Legation (now Minister to the Court of St James's), Baron von Bunsen. The English Catholic Bishop, Nicholas Wiseman (who was planning the establishment of a Roman Catholic hierarchy of bishops in England), commissioned Grüner to engrave the illustrations to his *Four Lectures on the Offices and Ceremonies of Holy Week*. Wiseman had introduced Overbeck and Dyce. Given Albert's religious views – Liberal Protestant with pronounced distaste for Catholicism – there is something deeply paradoxical in his championing art from this quarter. (Dyce was a High Church Anglican. James Hope-Scott, of similar persuasion, would eventually, under Wiseman's influence, become a full Roman Catholic. Hope-Scott, a church-mad lawyer, Oxford friend of Gladstone's who had married the granddaughter of Sir Walter Scott, Charlotte Lockhart (hence his addition of the surname Scott to his own of Hope), and founder of the High Church Scottish public school Glenalmond, was a key figure in the so-called Oxford Movement, which tried to emphasize the Catholic nature of the Anglican Church.)

Grüner met Hope-Scott in 1840. Hope-Scott, still an Anglican, wanted to acquire cheap religious prints for the Society for Promoting Christian Knowledge. It was through Hope-Scott's encouragement that Grüner came to London, where he was welcomed by Charles Eastlake at the Royal Institute of British Architects and eventually met Prince Albert. The Prince's first commission to Grüner was to ask him to produce an enlarged edition of his *Sammlung von Denkmälern in Rom im 15 und 16 Jahrhunderts*. It came out as *Fresco Decorations and Stuccoes in the Churches and Palaces in Italy during the 15th and 16th Centuries*, and in May 1843 Grüner presented a finished copy of the book to the Queen. In his preface, Grüner wrote, 'At a moment when… the erection of the Houses of Parliament… gives additional interest to every kind of architectural embellishment, it cannot be doubted that the access afforded to compositions of such skill and beauty will be gratefully acknowledged even by those Painters whose efforts are directed to the higher branches of the profession.'[16] (By 'even', he clearly meant 'especially' or 'precisely', German '*eben*'.)

Ludwig I of Bavaria, to a smaller extent the Wettins of Coburg, in Ehrenburg and Rosenau, Frederick the Great at Sanssouci... they had all seen the value of the monarchy-as-theatre; theatre which provided a stage set. Grüner would be the set designer for Albert's particular way of being royal. His memorials today, which we shall discuss in due course, were, above all things, the seaside palazzo of Osborne, and the last resting-place of Victoria and Albert, the mausoleum at Frogmore. His first excursus on English soil, sadly, does not survive: this was the Garden Pavilion which he designed for Buckingham Palace, which would provide an opportunity for fresco painters to perfect their technique. The chosen theme for the paintings would be scenes from Milton's *Comus*. It is not clear who chose this theme. If it was Albert himself, it would reflect his preoccupation, as the Royal Family grew, with sexual purity. Milton had written *Comus* for the family of the Earl of Bridgewater, which had been rocked by sexual scandal. The masque, in which the children of the family took part in Ludlow Castle in 1634, was designed to advertise to the world that the dissolute ways of the parents were not to be repeated in the second and third generations. In very much the same spirit, Victoria and Albert, with their growing family, were repudiating the dissipations and dissolutions of George IV, and of Duke Ernst I of Coburg.

Grüner conceived the Garden Pavilion as the equivalent of Ludwig I's Hofgarten in Munich.[17] Victoria recalled how delighted Albert was by the whole scheme, and how frequently he went down to watch as the pavilion was constructed, and then to observe the fresco artists at work. It was a glad day, 7 July 1844, when Landseer brought his fresco to the pavilion. On 30 August, Grüner arranged with the Italian painter Aglio that he should paint one of the side rooms in Pompeian style in encaustic, for the sum of £100.[18] When it was complete, he commissioned a guide entitled *The Decorations of the Garden Pavilion in the Grounds of Buckingham Palace*, with a preface by Anna Jameson, author of *Memoirs of the Early Italian Painters* (1845).

It was a contribution to the cultural revivification, as he saw it, of his adopted country. In July 1845, Grüner was appointed 'artistic adviser' to the Prince. It was a public display of the Prince's wish to gather around himself a working group, 'a day to day secretariat, all men of expertise and standing in their own worlds'.[19] And it is in such a setting that we see Albert at his best – happy to be among clever and gifted people, and, unlike his dealings with the Household or with his children, not at all pompous. Eastlake remarked that while discussing fine arts issues with Prince Albert, 'two or three times I quite forgot who he was – he talked so naturally and argued so fairly'.[20]

Equally, the commissioning of an imitation in the middle of London of a German Hofgarten by a German designer was a sign of Albert's yearning for the *Heimat*.

Visits to and from Germany were essential to Albert's happiness. He was still extremely young, and his homesickness was intense. When his daughter Vicky left England to live in Germany, Albert deeply sympathized with her agonizing *Heimweh*. By way of consoling advice, he described to her his discovery that it was possible to maintain a superficial level of happiness, while nursing the everlasting sense of displacement and sorrow which is the lot of the exile.

> The identity of one's personality is, so to say, interrupted; and there is a kind of Dualism, in which the earlier 'I', with all its impressions, memories, experiences, feelings etc [*sic*] which were simultaneously bound up with the years of one's youth, with their special locality, with the sense of place, and with personal relationships of those times, is coexistent with the new 'I'. It is as if the self has lost its covering, and yet the new 'I' cannot be separated from the old, which as a matter of actual fact always remains there, unchanged.[21]

How well the father understood that Vicky would always, from the moment she arrived in Berlin, to her death, be '*die Engländerin*'; because Albert never ceased to be the German.

Every family event brought it home to him, especially the births of his children. In April 1843, the third child was born, Alice, with the old King of Hanover and Albert's father Ernst being godfathers. It is interesting that, although he named one of his daughters after his mother Luise, there were no Ernests or Ernestines. This perhaps needs no explanation. It would certainly be wrong, however, to suggest that Albert's feelings for his father, tied up as they were with all the 'impressions, memories, experiences, feelings etc.' of childhood, were not deep. His father did not come to London for the christening of his god-daughter Alice, and that summer, which might have given Albert the chance to go home for a while, was taken up with a visit to France, to visit old King Louis-Philippe.

'Sir Theodore Martin' emphasized that the visit was a purely personal one. It was motivated simply by 'a desire on both sides to cement by personal intercourse a friendship, that had grown up through years of correspondence, and had been strengthened by the ties of intermarriage'.[22] It is perfectly true that the Coburgs and the Orléanists were intermarried. Uncle Leopold had married the daughter of Louis-Philippe, Louise-Marie. The Duc de Nemours, Louis-Philippe's son, was also married to a Coburg cousin. Marriage for these people was a political act, so, despite 'Sir Theodore Martin's' assurances, historians have seen the visit to Louis-Philippe's summer retreat, Château d'Eu, as one charged with hopes and significances: an attempt to forge something like an entente.

True, it was a great personal success. The newly built royal yacht, the *Victoria and Albert*, made the short voyage from the Solent to the French coast. The royal pair were saluted by kisses and warm hospitality. Albert rose at seven each morning for sea bathing while Victoria basked in the beauty of 'my favourite air-castle of so many years'.[23] Albert told Stockmar that 'the old King was in the third heaven of rapture'[24] when in Victoria's company.

The diplomats and politicians, however, saw it as a chance to pace round one another like slightly suspicious dogs, wondering whether or not they were friends or foes. Aberdeen, who had been at school with Palmerston and always disliked him, fully understood the desire by Guizot – his opposite number as Foreign Minister – to undermine Palmerston's foreign policy decisions. It now seemed in the interests of France and Britain to come together over the future of Turkey and the 'Eastern question'. This was the beginning of the fatal alliance between the two European powers over the 'sick man of Europe', which would lead in the following decade to the Crimean War. At the same time, the British Foreign Office was acutely aware of Louis-Philippe's ambitions nearer home. The King wanted to unite the crowns of Spain and France. The Spanish Infanta, left fatherless since 1833, was now thirteen and the question arose: whom should she marry? Albert and Victoria favoured their Catholic Coburg cousin Leopold, but Guizot made it clear to Aberdeen that the French could not accept this. Louis-Philippe assured Victoria that he had no ambitions in Spain, while everyone knew that this was the opposite of the truth; and eventually Maria Luisa married one of his sons, the Duc de Montpensier, three years later, an event which annoyed the British Establishment and put paid to a Coburg King of Spain.

In 1843, in the bright weather, and over the jolly picnics (Louis-Philippe served them beer and cheese to make them feel at home), such calamities were not imagined. They returned to England in a glow, and this happy feeling persisted during the closing months of the year. Albert boldly elected to visit Birmingham, a Chartist stronghold, despite the head-shakings of the politicians. The gamble paid off. As he drove through the streets, beside the Radical Mayor, the crowds cheered wildly. At the opposite end of the social and political spectrum, Albert and Victoria stayed at Chatsworth (Albert called it Chatsford in his diary)[25] and went on to a hunting party at the Duke of Rutland's seat, Belvoir. Anson, Albert's Secretary, a passionate fox-hunter, was pleasantly surprised that Albert bravely kept his seat in the saddle. 'One can scarcely credit the absurdity of people,' wrote the Queen to Uncle Leopold, 'but Albert's riding so boldly has made such a sensation, it has been written all over the country, and they make much more of it than if he had done some great act!'[26]

While all these successes, social and political, were enjoyed on the surface of life, we may conclude from his later confession to his daughter Vicky that he was inwardly homesick, and longing for the opportunity to revisit the Coburg of his childhood. Shortly after Christmas, an excuse to do so arose, but it was not one which he either expected or wanted. His father died on 29 January 1844.

The all-knowing physician Stockmar had foreseen the death. Although he had attempted to warn the Prince, for Albert, the news came as a total shock. Ernst, after all, was not quite sixty years old, Albert just twenty-four.

The human heart is unpredictable. The reader so far in our story has not found much to love about Duke Ernst of Saxe-Coburg. Although he took intelligent care that his two sons should be educated, he was not close to them, and it would be strange had not his second marriage, to his much younger niece, widened the gulf between Ernst and the boys.

Nevertheless, for Albert, the demise of a parent whom he had not seen on any regular basis for years was painful in the extreme. 'Not to have seen him, not to have been present, to close his eyes, to help to comfort those he leaves behind, and to be comforted by them, is very hard. Poor mama here [the Duchess of Kent, Ernst's sister], Victoria and I sit and weep'.[27]

Albert had the commonly experienced sense, in early bereavement, that the death was simply impossible to take in. 'He cannot accustom himself to the sad truth,' he told his diary, writing of himself, as he always did in this document, in the third person.[28] Victoria wrote to Stockmar, 'My darling stands so alone, and his grief is so great and touching… He says (forgive my bad writing but my tears blind me), I am now all to him. Oh, if I can be, I shall be only too happy, but I am so disturbed and affected myself, I fear I can be but of little use.'[29]

Albert was given leave of absence by the Queen and the Prime Minister, and allowed to return to Germany to visit his brother, and – for there was no chance of reaching Coburg in time for the funeral – to see his father's grave. He decided to go during the ten-day Easter parliamentary recess, an indication that, although he was prostrate with grief, the Government of Great Britain and Ireland could not be expected to continue in his absence.

No sooner had Albert departed than he was followed by a hurricane of letters from his besotted wife, pouring out her love in her characteristic, slightly ungrammatical German, peppered with English words. 'Angel beloved, beloved husband, the dreadful hour is past. The good Louise and I have wept together, but not for long. I am very peaceful, and cheerful but now I am bewildered that I must endure these things without you!' ['*Engelsgeliebter, angebeteter Mann, Die schreckliche Stunde ist vorüber! Die gute Louise und ich haben zusammen geweint, aber nicht recht lange; ich bin sehr ruhig und sogar* cheerful, *jetzt aber sehr* bewildered *dass ich die Sachen bestellen muss ohne dir* [*sic*]']... And, next morning, 'Every time I hear a man's footstep or a door go, *es giebt mir einen Dolchstich im Herzen, ich glaube, ich hoffe, du kommst,– und ach! Die Wahrheit, die schreckliche Wahrheit* opens upon me. *Ich war recht low nach* luncheon.' ['Every time... it gives me a stab in the heart, I believe, I hope, that you are coming,– and oh! The truth, the terrible truth opens upon me. I was really low after luncheon.'] The rooms of Windsor Castle felt so empty without him! But above all, the emptiness was felt in bed, where she yearned for him. It made her so sad that her bed '*jungfraüliche ist*' – is virginal. Yes, she missed the angelic voice, the angelic presence; but above all, she missed the sex. 'Your little wife, your slave-girl loves you so indescribably much!' ['*Dein Weibchen, deine Schlawin, liebt dich – ach! Ganz unbeschreiblich!*'][30]

While Victoria pined in the empty English marriage-bed, Albert was revisiting, as is common when one loses a parent, the vanished regions of his own childhood. The death of his father and the ten-day break in the home-country ('*in der Heimath*') allowed Albert to take stock. His long letters, both to the Queen and to Stockmar, have left us a record of his very distinctive state of mind. On the one hand, he had never felt more German, never felt more keenly the hostility of the British towards him. The recent triumphs, on the streets of Birmingham and the hunting-fields of the 'Dukeries', were forgotten. As the carriage swayed and rumbled over the icy potholes of bad European roads, the Prince, always prone to persecution mania, could think only of hostile remarks in the press. All his good works seemed, in this vulnerable moment, to be received with ingratitude and vilification. 'Every imaginable calumny is heaped upon us, especially upon me, and although a pure nature, conscious of its nobleness of purpose, is and ought to be lifted above such attacks, still it is painful to be misrepresented by people of whom one has believed better things.'[31]

On the other hand, this feeling of alienation from Britain, and his wounded sense of British ingratitude, only fired him the more to yet more work, yet more busyness.

He felt that his emotional life (*Gefühlsleben*) and his political existence were now inextricably linked. 'My youth with all the recollections linked with it, has been buried with him around whom they centred. From that world I am forcibly torn away, and my whole thoughts diverted to my life here and my own separate family. For these I will live wholly from this time forth, and be to it the father whose loss I mourn for myself.'[32] He was mourning, therefore, not his dissolute, extravagant father, but an earlier self, who now had to be buried or forgotten, as he forced the new self to follow its ambitions and fulfil its duties. He saw himself as a machine. The visit to Germany was therefore an act of renunciation, and of self-preparation. 'I will, therefore, at once close accounts there, and set about putting the machine into a state in which it may go on working for the future.'[33]

His wife left behind, Albert's spirits appeared to revive. Together with Seymour, British Minister in Brussels, he called on King Leopold, and the two Coburgs, with Seymour and a couple of others (General d'Hane and Count Gersdorff), headed for Germany by train. 'Prince Albert was in good humour and talked all the time,' Seymour recorded. 'Among other things, he told me that the

Ministers hardly ever see the Queen, that all their conversns are carried on in writing.'[34]

Victoria, back in England, who had never been apart from Albert for a single day or night since their wedding, was desolate at his absence. His emotional life was engaged, by contrast, entirely in the German childhood he had lost. His dedication to her, and his children, and to Britain, were acts of will. The weather was cold as he and his entourage of equerries and servants crossed the Channel to Ostend. They took the railroad ('wonderfully beautiful') from Liège to Aachen. Thence to Cologne where they put up in the Imperial Hotel, and on to Gotha the next day, a journey of two hours. 'Mama' – that is, his cousin Marie, Ernst's second wife – 'has grown much stouter, and at the same time looks older'. He evidently did not realize that mourning clothes allowed women to dispense with corsets. 'She wears the black point and the long veil of a German widow... Oh! How many varied emotions overwhelm me! Remembrance, sorrow, joy, all these together produce a peculiar sadness. Tomorrow I shall make the trying journey to the Palace. Could you have witnessed the happiness my return gave my family,' he told Victoria, 'you would have been amply repaid for the sacrifice of our separation... Now I am rather tired after my night journey'.[35]

He went on to Coburg, by way of Meiningen, having bought souvenirs for his wife – German views painted on porcelain – and dolls for the children. When he reached Coburg, he stayed with his brother Ernst and his wife Alexandrine at Kalenberg, the least attractive of their castles, but with stupendous views, about three miles from the centre of town. On Sunday morning they attended the service in the Staatkirche, the great old church, with its reredos of their Wettin ancestor, the patron of Luther. 'The beautiful devotional singing of the congregation, as well as the admirable sermon of General-Superintendent Genzler, moved me to tears.'[36] (Genzler had preached at Albert and Ernst's confirmation.) All too soon, the sojourn in the *Heimat* was over.

On 24 May, when the Queen celebrated her twenty-fifth birthday at Claremont, Albert gave her two pictures. One was by Eastlake, a detail of the angels he had painted in his fresco, which Victoria had much admired. The angels are offering a medallion emblazoned with the words '*Heil und Segen*', 'Hail and Blessing'. The other was a miniature portrait of Albert himself by Thorburn. When she was dictating the biography of the Prince Consort to Sir Theodore Martin, Victoria declared that this picture was 'a work fit to take rank with the masterpieces of the great Venetian School'.[37] Albert, at the Queen's request, had been represented in medieval armour. She always maintained that this portrait was the best likeness of her Angel and that 'during the fatal illness, and on the last morning of his life, he was wonderfully like this picture'.[38] The 'machine' was clad in fancy dress. The whole busy programme of his life was to be as much of a performance as the *bal costumé* in which he had dressed as Edward III.

It was a time for stocktaking, for visions and revisions. 1844 was the year when he lost not only his father, but those two companions of his youth and early manhood, his favourite dog, and his hair. 'The Prince's dear old dog Eos dies suddenly,' noted the diary on 31 July. And, a week later, on 7 August, the day after the birth of Prince Alfred, Albert decided to wage one last desperate war on his swiftly developing baldness. In the hope of thickening his hair, he had his head shaved and took to wearing a wig.[39] It had no effect. From now onward, the Angel with whom Victoria had fallen in love when he was still little more than a teenager had become prematurely aged. At twenty-five, he was already bald and paunchy.

Although it involves us stepping out of chronological sequence, we cannot end a chapter that has been chiefly concerned with Albert and the visual arts without reference to one of his greatest achievements, namely his collection of illustrations of the works of Raphael. Even if he had fathered no children, taken no part in the political evolution of constitutional monarchy, done nothing to benefit the housing of the labouring classes and failed to interest himself in the

Great Exhibition, his art historical work on Raphael would alone make him worth remembering. It demonstrates that on top of all his multifarious concerns, he was a serious art historian on the German pattern of thoroughness, and, indeed, that he was something of a pioneer in the use of the most modern, photographic, techniques, in the cataloguing of old paintings and drawings.

For Albert, one of the most exciting consequences of becoming the occupant of Windsor Castle was the presence there of the stupendous collection of drawings and paintings assembled by the Kings and Queens of England. His imagination was caught by the huge collection of prints, and he early conceived the notion of using prints to create a catalogue raisonné of the works of Raphael. J. D. Passavant's monograph *Rafael von Urbino und sein Vater Giovanni Santi* had made a great impression on the Prince when he had begun his art historical studies, and, always being a man who took large views, he decided to concentrate, once installed at Windsor, on a survey of one of the greatest of all European artists, Raphael. His idea of assembling a complete collection of prints of all the artist's works began towards the end of 1852. Using Passavant as his template, Albert consulted experts as diverse as G. F. Waagen, J. C. Robinson, Sir Charles Eastlake and Frédéric Reiset. The decade since Albert's marriage, and his close appreciation of so many works of Raphael near at hand in Windsor, had coincided with the growth of photography. It is typical of Albert, of his passion for new technology combined with his profoundly informed sense of history, that he should have put the two together and realized that the most faithful method of recording Raphael's oeuvre should be not through engravings but through photographs.

In all the ensuing years of Albert's life, and throughout the pages of this biography, we must imagine him – while he discussed army reform, toured the industrial heartlands of the British Midlands and the north, entertained foreign Heads of State and kept up an all-but-full-time feud with Lord Palmerston – thinking about his Raphael project. His eye became ever sharper. When he saw the painting of *St Michael Triumphing over Satan*, a work in the collection of the Reverend Frederick H. Sutton, exhibited as by Perugino, at the British Institution, he realized it was a Raphael – a view in which he was followed by learned opinion.

Not every owner welcomed the Prince's insistence that their Raphael should be photographed. Frédéric Reiset, curator at the Louvre, objected when he found that photographic reproductions of Raphaels in the Louvre were on sale in London. Photography in those early days was only possible in natural light, and this often meant taking the pictures outdoors. (The photographer sent to the Museo Borbonico had to wait for weeks in Naples before he was allowed to set to work.) Lake Price, the photographer tasked with reproducing the painting of Cardinal Cesare Borgia in Rome, announced himself defeated by poor light and by 'the yellow varnish of three centuries'.

It was no doubt sometimes enraging for the museum curators, and the owners of great houses, to be invaded by a photographer, and even to have their great treasures reproduced and sold – which was only done with their permission. Some owners insisted on keeping the negatives, only allowing Albert himself to have prints. But as the collection grew, so did the craze for photographic reproduction of great art. It was typical of the Prince, and of his mentor Grüner. Art was not the sole possession of those lucky enough, the Kings and Queens of England included, to own priceless works. Albert's photographic reproductions of Raphael would lead to such enlightened developments as the Arundel Society, which promoted the wide distribution of reproductions of great art. The work of building up the Windsor Raphael Collection, initiated by Prince Albert, was continued by Ernst Becker and later by Carl Ruland. Even today, it is a collection that is of academic use to art historians throughout the world. Jennifer Montagu wrote in 1986:

> We are accustomed to monographs illustrating all the known works of an artist in good photographic reproductions, and most scholars expect to have access to photograph collections

of works of art which, if they are fortunate, may also include engravings, both old and more recent. In the mid nineteenth-century there were no illustrated monographs, no photograph collections, and no print rooms in the world that could claim so full a coverage of any one artist.

Yet with all our modern resources of widespread and relatively cheap photography, and the activities of the publishing houses, which flooded the market with books on Raphael during his centenary year of 1983, there is no assemblage of his works anywhere that can rival the collection brought together by the efforts of the Prince Consort.[40]

Above left: Albert's father, Duke Ernst I of Saxe-Coburg-Gotha.

Above right: Albert's mother, Princess Luise of Saxe-Gotha-Altenberg. The undignified collapse of his parents' marriage scarred the whole of Albert's childhood.

Left: After the parents' divorce, Albert's closest companion was his brother Ernst (painted by Paul Emil Jacobs).

Schloss Ehrenburg, the official residence of the Dukes of Saxe-Coburg until 1918.

Schloss Rosenau, where Albert was born.

Left: Prince Albert as a boy.

Below: Albert's watercolours as a boy often depicted military subjects.

Both Victoria and Albert were keen draughtsmen. She was much the more accomplished of the two: this shows her picture of him…

… and his of her, in 1840, the year of their marriage.

Sir George Hayter's painting of the marriage of Prince Albert to Queen Victoria, 10 February 1840, in the Chapel Royal.

Sir Robert Peel, the Prime Minister who completed Prince Albert's political education.

Lord John Russell, the Liberal Prime Minister with whom Albert felt less empathy.

Lord Palmerston, Albert's bête noire.

Lord Aberdeen, a real friend to the royal pair. It was through his offices that they acquired Balmoral.

One of Albert's many sketches of Eos, his faithful greyhound, who accompanied him from Germany.

Five sketches by Victoria of the children.

Victoria and Albert's bedroom at Windsor, 1847. Their parents and grandparents look down from the walls.

Victoria's exuberant, colourful evocation of her children at Osborne.

Scotland was a release from court life. Here, Victoria's drawing of men fishing in Loch Muick shows Albert fishing, assisted by his gillie John Macdonald.

Victoria's painting of Balmoral, a view from the Approach (Abergeldie side), 4 October 1852. These were the happiest times of their lives together.

TEN

NEPTUNE RESIGNING HIS EMPIRE TO BRITANNIA

THE PREVIOUS SUMMER, in August 1843, Victoria and Albert had visited France together and stayed with Louis-Philippe at Château d'Eu.

In addition to the old King being 'in the third heaven of rapture', Albert told Stockmar that 'the whole family received us with a heartiness, I might say an affection, which was quite touching'.[1] Moreover, they had Louis-Philippe's reassurance about the Spanish marriages, of Queen Isabella and the Infanta,[2] assurances which the royal pair were naïve enough to believe.

One of Louis-Philippe's sons, the Prince de Joinville, accompanied the royal pair on their journey back to England, and they took him to spend two nights at the Pavilion in Brighton. 'He was very much struck with the strangeness of the building,' the Queen observed.[3]

She herself had never liked it – 'a strange, odd, Chinese looking thing, both inside and outside; most rooms low, and I only see a little morsel of the sea from one of my sitting-room windows'.[4] She found Brighton as a place 'cold'[5] and 'dull'.[6] Moreover, the Pavilion being in the middle of the town meant that the Royal Family felt exposed to snoopers. It was easy to look over the garden walls, and when the royal pair went for walks, for example to the Chain Pier, they 'were mobbed by all the shopboys in the town, who ran and looked under my bonnet, treating us just as they do the Band, when it goes to the Parade! We walked home as fast as we could.'[7]

Clearly, as their family was growing, the young pair felt the need of a retreat, a place of their own. Claremont was still officially leased to King Leopold. Buckingham Palace, uncomfortable and smelly, and Windsor were official residences more than they were 'home'. Albert, by nature a countryman, felt a Rosenau-shaped cavity in his heart.

Little by little, Victoria and Albert took leave of Brighton and its Pavilion. By 1846, they started to strip it of its contents. Vanloads of the gilded furniture, rich carpets and crystal chandeliers which her uncle George IV had installed in the place were removed to Buckingham Palace, Windsor Castle and Kensington Palace. In 1847, they held a sale of the lesser items on the garden lawn: furniture, objets d'art, garden furniture and even garden implements went under the hammer. By the summer of 1849, the Pavilion was empty, and permission was given for visitors to look around; 27,563 people did so. Cubitt, the builder-speculator who had started to build Belgravia and had restored much of Buckingham Palace, offered to buy the site for £100,000. The vandalistic Board of Woods and Forests, which owned the fixtures of the Pavilion, tore them out, removing the copper bell-wires from the walls and selling them for threepence a pound; gouging out skirting boards and attacking the floorboards with pickaxes; and reclaiming the Chinese wallpapers, mirrors and other items of value. In their clumsy desire to rip out things of value, they destroyed plinths, statues, pedestals and friezes. The Pavilion looked ready for demolition. It was rescued by the Brighton Town Council, which bought it with a loan from the Bank of England of £60,000. Once this had happened, the Queen repented of the destruction which had gone on, and donated chandeliers and paintings which had never been used at Brighton. The Duke of Wellington gave a complete batterie de cuisine from Apsley House. Carpets were replaced with help from the Regency

Society. It was now a tourist attraction, but it had been saved. The Queen would never live there again.

This was because Peel had found a solution to their desire for a seaside retreat. Back in 1843, Victoria had begun to remember the childhood holidays which she had spent on the Isle of Wight. It is strange that these memories should have pleased her since they were invariably spent with the hated Conroy. Nevertheless, she recalled Norris Castle, the 'charming toy fort'[8] overlooking the Solent in East Cowes. It is a lovely little castellated fantasy, built by James Wyatt in 1799 for Lord Henry Seymour. When Victoria became Queen, she was offered it for sale at £15,000, declined the offer and had regretted doing so ever since, remembering the bliss, as a child and teenager, of sailing in the little schooner which her mother had given her. In 1843, the chance arose to buy Osborne House, with an estate of 1,000 acres. The owner was Lady Isabella Blachford, daughter of the 3rd Duke of Grafton. Peel made thorough searches, including investigations into public rights of way through the park and grounds. Clearly it was essential for the royal pair to be private. They also needed a decent water supply and a sea view. Osborne appeared to match all these demands. Lady Isabella drove a hard bargain. When Peel let it be known that the Queen would accept the asking price of £28,000, Lady Isabella raised it to £30,000. Albert and Victoria offered £26,000, without furniture or crops. Lady Isabella accepted, and then changed her mind – egged on, Peel supposed, by a greedy auctioneer. ('Lady Blachford [*sic*] behaves extremely ill,' Albert told his diary.[9]) She had, however, signed the contract, which, if necessary, Albert was prepared to pursue through the Court of Chancery, embarrassing as this would have been. The royal pair accepted Peel's advice to stand firm, and on 3 April they took possession.

Victoria wrote to Lord Melbourne that it 'is impossible to imagine a prettier spot – valleys and woods which would be beautiful anywhere; but all this near the sea (the woods grow into the sea) is quite perfection: we have a charming beach quite to ourselves without fear of being followed and mobbed, which Lord Melbourne will easily understand is delightful. And last, not least, we have Portsmouth and Spithead so close to hand, that we shall be able to watch what is going on, which will please the Navy and be hereafter very useful for our boys.'[10]

She was not exaggerating. It was, and remains, one of the most beautiful places in Britain, and it is, simply, a perfect royal residence. Had it not been for the extraordinarily strong antipathy felt by Edward VII for his mother, it would have remained so, but he moved out of the place the minute she died there, propped on the wonky arm of her grandson the Emperor of Germany, in January 1901. Until then, it was her favourite home, and it was, of course, supremely the creation of her Angel.

For that first summer, of 1845, they simply enjoyed their new-found paradise. A journal entry for a bright Sunday in June gives the flavour of how happy Victoria was.

> Another beautiful morning. — After our breakfast we walked down to the sea, where we found (as daily) the Children sitting under the tent, & we picked up shells with them. It was dreadfully hot when we came home at 11. — Little Alice lunches with us every day, as well as Vicky & Bertie. — We attended afternoon Service at Whippingham Church, where Mama & I used to go so often 12 & 14 years ago. On returning from Church, Albert walked out with Ld Aberdeen & Col: Bowles, & I drove with Ly Portman, Col: Grey riding, to meet with them at King's Key. It was a delicious evening, & the sea so blue & viewed so lovely. We walked down to the beach & remained some time there. Ld Aberdeen was charmed. Ly Portman & I, drove home again, & I went to see the Children having their baths. — Albert showed me a complete cuttle fish, which produces sepia, & is a very curious animal; he had found it on the beach. — I heard Vicky say her Prayers. — Dinner the same as yesterday, excepting Anson. I wore a wreath of cornflower, & have been wearing natural flowers every evening. — had some conversation with Ld Aberdeen.[11]

The next day, they laid the foundation stone for the new house. For Albert's restless and creative spirit could not be happy with Lady Isabella Blachford's original house. Charming as they might have found it as a simple family holiday house, it was not entirely suitable as a royal residence. When it became clear how much time Victoria and Albert intended to spend there, holding meetings of the Privy Council and entertaining members of the Cabinet and visiting foreign Heads of State, it was seen that the house was too small.

Albert had originally thought 'the house as it is requires no alteration, only the addition of a few rooms to make it a very suitable and comfortable residence for the Queen and the children and part of the suite. This addition can easily be made and if done in a plain, unassuming style conformable to the rest of the house ought not to cost a great deal'.[12]

When Cubitt was brought down to survey the house, however, he recommended that they pull it down and do a complete rebuild. Albert was 'bursting with plans... full of alterations he wants to make, and so happy'.[13] Clearly, as was shown by their laying a foundation stone in June 1845, by that summer, they were poised to make the Osborne we see today. Cubitt was the builder, but the design was in Albert's hands. The house bears the most striking resemblance to Trentham, the Staffordshire seat of Queen Victoria's great friends the Duke and Duchess of Sutherland. This cannot be accidental, even though I have found no record of Albert visiting Trentham. As for its interiors, Grüner was by now the Prince's artistic adviser, and Hermione Hobhouse states that 'it seems likely' that the German advised Albert about the interiors of Osborne, especially the colours.[14] Since so few letters by Grüner survive today in the Royal Archive, it is difficult to be sure how much active involvement he had with Albert's aesthetic choices. What is certainly clear is the lineage of taste which led to the birth of Osborne.

When the Palace of Westminster was destroyed by fire and a new Parliament building was required in the 1830s, the debate hinged on whether it should follow Greek or Gothic Revival. As it happened,

Charles Barry, one of the most imaginative and inventive of early nineteenth-century architects, came up with a sort of Tudor-Gothic which was both unmistakably nineteenth century, but somehow summoned up the reassuring age of Shakespeare and Queen Elizabeth I. When designing in a classical mode, Barry, similarly, took his inspiration less from the Doric or Ionic severities on display in, for example, Wilkins's The Grange, the Baring palace which appropriately enough looks like a clearing bank surprisingly located in the verdant Hampshire countryside. Wilkins's original house was remodelled by Charles Cockerell, who went on to create the Grecian splendours of both the Fitzwilliam Museum in Cambridge and the Ashmolean Museum in Oxford. While Cockerell was establishing his career in England, Barry, seven years younger, was in Italy, mixing with many of the people who had been Grüner's associates in Rome and Florence. Barry brought back to England an essentially loose, romantic conception of Italian architecture, in which the canvases of Claude Lorraine, set in imaginary landscapes and bathed in dreamy dawn or twilight, played as strong a role as the Palladian rule-books of form. His most glorious exercise in the Italianate manner was the building of the Travellers' Club in Pall Mall, undertaken when he was thirty-five years old, which immediately demonstrated an alternative to 'the extremes of Greek and Gothic'.[15] Here was a Roman palazzo set down in the heart of London, cool, elegant, spacious, and yet somehow not just a pastiche, but instantaneously at home. Barry was to repeat this trick in his work for the 2nd Duke of Sutherland, most notably at Trentham Hall in Staffordshire (1834–40) and at Cliveden in Buckinghamshire (1850–51).

In both of these cases – alas, only Cliveden survives – Barry was able to domesticate the Italian vision. A similar translation of visions experienced in Italy can be seen in the Potsdam of Friedrich Wilhelm IV, where the Pfingstberg (1849) and the Orangery (1851), or the Albrechtsberg in Dresden, show a comparable set of inspirations. These German examples, however, point up how very distinctive Osborne is by comparison. These German buildings are relentlessly symmetrical. What Albert learned from Barry was the dizzying joy of asymmetry. Like the paintings of Claude Lorraine which inspired it, Osborne injects the calm of classical order with the sweetness of romantic disorder. And the house is perfectly placed, so that from its terrace and drawing room, or its upstairs apartments, the view of the Solent, painted in watercolour so lovingly and so often by Victoria and Albert, is perfectly framed.

Visitors today must demolish in their minds' eye the Durbar Wing, added by Queen Victoria in 1890 at the height of her passion for her new friend Abdul Karim, and her enthusiasm for the British Empire, and designed in part by John Lockwood Kipling – father of the writer. It is so much at variance with Prince Albert's original vision that, amusing as it is, one must ultimately see it as an excrescence. To see Albert's Osborne, imagine only what is called the Pavilion Wing of the house, which was completed in 1846. This contains the state rooms, and the private apartments above them, and the flag tower 107 feet high, with its loggia and deep Italianate eaves. The house was brick-built and rendered. Greville, who visited in September 1845 when work had begun, was told by Anson that it would cost the Queen £200,000 'but it is her own money and not the nation's'.[16] It reflected, after only nine years, the enormous benefit which Victoria was enjoying of being allowed to retain the incomes from the Duchy of Lancaster. Had that not been allowed at the beginning of the reign, there would have been no Osborne and no Balmoral. She had the Midas touch. In addition to the wealth generated by the Duchy of Lancaster, Victoria inherited about £560,000 in 1852 from an eccentric miser named John Camden Neild. It was a prodigious fortune and confirmed her private wealth as, eventually, by far the richest person in her kingdom.

Every detail of Osborne was Albert's inspiration: the layout of the gardens; the construction of the Household Wing (finished in 1851); the planning of the modern drains, which both provided the house with up-to-date plumbing and irrigated the gardens; the garden statues (concrete supplied by Austin and Seeley, or zinc purchased for the Prince from Geiss of Berlin or Miroy Frères of Paris, and coated with bronze, cast in moulds); the bathing machine on its scroll brackets; even the cheerful coloured gravel in the paths to the Swiss Cottage, purchased for the children in 1853.[17]

And inside the house, with the help of Grüner, he made an eclectic and distinctive statement of his domesticated European aesthetic. Rightly has it been said, 'they began their married life with everything that surrounded them belonging to her, at Osborne she came to be surrounded by everything that was chosen and arranged by him'.[18]

The sculpture corridor, completed in 1852, is set at regular intervals with niches for classical sculpture or modern pastiches. On the floor are bright Minton tiles. At the far end, a glass door opens out with views of the sea. In one of Albert's watercolours of this scene we see three of the children and a dachshund making their way to the garden or the beach. The whole is bathed with clear sea light.

The walls of the drawing room and the domestic quarters are adorned with paintings by Winterhalter. In the state rooms may be seen exquisite copies of Winterhalter portraits of Victoria and Albert, he in dress uniform, she wearing the Garter sash – the gifts of Louis-Philippe in 1846. Perhaps, though, the truly remarkable iconographies of Osborne will strike visitors to other parts of the house.

At the top of the nursery landing, in a tall niche, was a statue of Albert by the Prussian sculptor Emil Wolff, which Albert had given to Victoria as a birthday present in 1841. (Wolff had been brought to England by Grüner.[19]) When the statue had been delivered to Windsor in 1844, Victoria called it 'very beautiful'[20] but admitted, 'we know not yet where to place it'. Two years later she recorded that the statue had been moved to Osborne, explaining in her journal, 'Albert thinking the Greek armour, with bare legs and feet, looked too undressed to place in a room'.[21] It depicted Albert in the skimpy garb of an Homeric warrior, legs and thighs naked. It became the custom on his birthday, if he were absent, for his children to come and lay posies of flowers at his bare feet. This striking and notably erotic marble – one was almost going to write homoerotic – gazes across the stairwell to a fresco commissioned for the house. It is Dyce's *Neptune*

Resigning the Empire of the Seas to Britannia, which fast became Victoria's favourite painting. We could spend a profitable hour on this landing, staring from one image to the next, and contemplating all the strange things which they told us, both about the couple who had come to inhabit the new-built Osborne House, and the political reality which they heralded.

The naked sea-god's three horses just cover the lower half of his body, but the mane of the third grey on the right, a sprawl of black tresses, at first sight looks like Neptune's abundant pubic hair. From the turbulent waters, where Triton blows his horn and smooth-limbed nereids strain in ecstasy towards the calm emblem of Britain, the Roman deity surrenders his crown of power. Britannia herself wears a golden short-sleeved gown, behind whose exaggerated folds a lion nuzzles, rubbing his nose affectionately at the back of her left knee, while gazing threateningly across the water, presumably at France. Behind the feminine image of nationhood who is both Britannia, clutching the trident, and Victoria herself, stand three human figures, a bearded man with thick crinkly hair, a female of forbidding aspect, perhaps representative of Victorian propriety, and, perhaps most striking of all, a naked young man, the left hand on his hip, the right lolling round the neck of the older beardy. The pronounced buttocks of the young man, the rippling muscles of his biceps and shoulders, above all the angle at which he leans backwards, convey a message which no reader of nineteenth-century British school stories would find surprising but which was quite possibly, on the artist's part, unintentional. The painting seems to be saying, which would undoubtedly turn out to be an historical reality, that for the second half of the century, the world was to be taken over by Victorian chaps, some of whom were, some of whom were not, aware of the sexual character which nurture or nature had bestowed.

Before we leave the subject of Osborne and its iconography, it is worth making the journey to the Prince's bathroom, to see the fresco which Albert commissioned from Joseph Anton von Gegenbaur. It depicts Hercules and Omphale.

In the middle of the fresco we see Hercules, seated, naked but for a loincloth; he holds a distaff, the symbol of feminine toil, in his left hand and holds the naked Omphale by the waist with his right; she stands holding the reel of thread which spins from the distaff; two winged putti hold his club. He is not an emasculated god, but he is one who has been subjugated to feminine domination.

As punishment for murdering his friend Iphitus, Hercules was, by the command of the Delphic Oracle, entrusted as a slave to Omphale, Queen of the Kingdom of Lydia, in Asia Minor. There are many late Hellenistic and Roman textual and visual-art references to Hercules' obligation to perform women's tasks, and even wear women's clothes. In this painting Hercules holds the spinning distaff and Omphale is depicted wearing the skin of the Nemean lion. That is, their roles are reversed. The Queen is depicted wearing one of the trophies from Hercules' famous Labours, while her man sits and spins at home. In Renaissance and particularly Baroque painting this often-depicted reversal was interpreted as an illustration of the theme of woman's domination of man.

So, although Osborne demonstrates, as has been said, the first space shared by Albert and Victoria in which everything had been chosen and arranged by him, it also demonstrates, pretty unsubtly, the tense dynamic of the marriage. Surely in the majority of nineteenth-century marriages, it would have been the wife who chose the carpets and curtains and paint colours. Albert's taste predominated, however, at Osborne, and it is both a tribute to his vision, and evidence of the fact that, while he might be free to decree how to decorate the corridors of a seaside house, as far as the Constitution of Great Britain was concerned, he had no official role.

The early years of the marriage of Albert and Victoria were characterized by intense sexual passion between the pair, and at the same time, the Prince's political frustration, greatly exacerbated by Stockmar's unrealistic expectations about the remit of Albert's political influence. As more and more children came to be born, Victoria would vacillate between exhaustion, nervous collapses and postnatal depressions, and a steely determination to continue exercising her political powers, such as they were. Albert could take over the administrative functions of constitutional monarchy while she was in labour, and she would sometimes express the wish to abandon all her political roles to him. Even if this wish of hers had been consistent, however, it would still not have left him with anything like the power he had been taught to imagine would one day be his. He was always doomed to feel politically frustrated. Within the marriage, both as father and husband, he could, however, exercise control. Yet even here he could take nothing for granted. That, surely, is the message of this bizarre bathroom mural.

Victoria was a very strong personality, and a past mistress at emotional manipulation. The mural more than hints at the notion that she rejoiced in the 'feminine' side of Albert's character. Even if we dismiss the idea that there was anything *conscious* in Albert of the homoerotic, he certainly seems, to the outward eye, to be gay in all but sexual preference – the obsession with interior decoration, music, clothes, the desire for order and control, all seem like stereotypical 'gay' characteristics. Far from being disturbed by these qualities in Albert, Victoria rejoiced in them. 'He would have made an excellent nursery-maid. I, not,' she commented.[22] She became jealous and petulant if he spent too much time away from her company. She was frequently cross with him. But she knew that by the standards of most royal marriages, she was lucky. Though her feelings for him were passionate and uncontrolled, and his for her more dutiful – he conceived it his 'duty' to be a loving husband, but showed very few signs of being in love – she did not allow herself to notice this. She knew she had no rivals for her sexual or emotional claims on Albert. Unlike his brother or father, he had no mistresses; indeed, as she noted after his death, appeared to have no interest in any women, or in any female being apart from his alliance with Victoria, and his yearning for an all-but-forgotten mother. The closest he had to friendships of any kind outside the marriage were with males whose interest he shared – with Peel, politics; with Grüner, carpets, tiles, tableware, paintings. These were not friendships in the term normally used. We do not find Albert going on holidays with university friends, or having a circle of drinking-pals or men with whom he played cards – because he did not do any of these things. True he loved shooting and he took part in hunting, but he did not have hunting or shooting cronies. The schoolma'amish interest Albert took, and would increasingly take, in every detail of the children's upbringing, from the timetables of their lessons, to their physical exercise, their gardening, painting and mental skills, would have seemed strange to many royal or aristocratic fathers of the time. They were the sort of interest usually taken by a pushy mother. He had, like Hercules, embraced the distaff side, with gusto.

To this extent, however, the dynamic of the relationship developed the intensity which is absent in many heterosexual marriages, where the 'boy jobs' and 'girl jobs' are clearly delineated in the household, and where men go off to their clubs or regiments or girlfriends, while the women occupy themselves quite separately. In the case of Victoria and Albert, they were both domestic, both vying for political influence, both obsessed (though Albert far more than Victoria) with the children. To this extent, if the 'gay' idea is pursued, the marriage was less a relationship between a woman and a man who did not realize he was gay; it was more like an intense relationship between two women. And this is what von Gegenbaur's fresco so disconcertingly lays bare. When their domestic life was bathed in sunlight and joy, it had the serenity of their happy contemporaries the Ladies of Llangollen. When it began to go sour, it cascaded into the psychodrama of *The Killing of Sister George*.

While Albert and Victoria were purchasing and redesigning Osborne, the country was undergoing one of the deepest political crises in its history, and the career of their friend Sir Robert Peel reached its most desperate, and at the same time its most heroic, apogee. This was the beginning of the great Irish Hunger, and the devastation wrought upon the divided Tories by the question of the Corn Laws, which protected the interests of landowners by imposing tariffs on imported corn – thereby, in straitened times, keeping the price of a loaf of bread indecently high. The politics of 1845–6 was therefore both as esoteric as it could get – hanging on the technicalities of economics, and the curious internal complexities of the two major parties in the Commons and the Lords – and also as basic and as simple as it was possible to be. How could the population eat? And – a fundamental question – how could the Government make it easier, or more difficult, for the population to eat? Since revolutions have nearly all begun with hunger, these were desperate questions, desperate times.

The first rumble of the crisis, however, could be said to have nothing to do with either Westminster politics or food. It was the question, in the spring of 1845, of whether the Government should increase a grant to the Roman Catholic seminary of Maynooth, near Dublin. Peel had come round to the idea, back in 1829, that Catholics should be allowed to study at universities or Inns of Court, join the professions, become Members of Parliament or take their seats, if peers, in the House of Lords. Until 1829, all this had been denied them. Many diehard Tories resented Catholic Emancipation, as it was called. And the devout High Churchmen, especially in Oxford, as we have seen, objected to the measure on religious grounds. This was because they believed the true Catholic Church was represented in Great Britain by the Anglicans. W. E. Gladstone (Eton and Christ Church, Oxford), took this High Church view and had therefore always opposed giving public money to support the chief training college for priests in Ireland, Maynooth. At this point in history, Maynooth was not merely the main seminary for priests: it was the *only* place of higher education available to Catholics.

Since Ireland was a mainly Catholic country, the bias against Catholics by the British Establishment and by the Protestant Ascendancy in Ireland particularly, was a source of deep resentment. The increase of the Maynooth grant was therefore an issue which aroused feelings, on both sides, which went far beyond the consideration of how Catholic priests could best be trained. Peel had a struggle persuading both Houses of Parliament to accept the increase, but he succeeded.

The Queen told her uncle Leopold, 'Our Maynooth Bill is through the Second Reading. I think, if you read Sir Robert's admirable speeches, you will see how good his plan is... The Protestants behave shockingly, and display a narrow-mindedness and want of sense on the subject of religion which is quite a disgrace to the nation.'[23]

This crisis passed. Peel knew that the real crisis, both economic and political, which faced the country was that of the Corn Laws. In all other branches of trade, the lifting of tariffs and the promotion of free trade had increased prosperity, created jobs and made the poorer classes more comfortable. It was absurd, as he could see by now, to make agriculture a special case, 'whatever country squires may think'.[24] The country squires, and with them the great landowners who made up the most powerful members of the aristocracy, thought very firmly. This was *not* simply a question of economic logic. Had it been, the existence of tariffs on imported wheat and oats, and the maintenance of an artificially high price for bread, would have been obviously indefensible. This was really a question about the future of Britain. Free trade had helped to enrich the emergent middle classes. The wealth from trade and manufacture had created what was in effect a new aristocracy, the upper middle class, an entirely new social and economic entity in whom power increasingly resided. This was the class which, educated at public schools and the universities and Inns of Court, and sustained by huge amounts of invested money, would dominate British life until the Second World War, providing it with its professionals, its higher civil servants, its bankers and financiers, its (in many cases) intellectuals and politicians. Hitherto in history, wealth had depended largely, or entirely, on land. From now on, the rentier classes could be as rich and as powerful as the minor aristocracy, and infinitely richer and infinitely more powerful than the squires, by living off investments.

The squires and the landed aristocracy who had governed England since the Reformation were therefore fighting for their life. Historically speaking, many of these – though by no means all – had been those who fought for the King against Parliament in the Civil Wars of the seventeenth century. In the nineteenth century, it might have been expected that they would be the natural allies of the monarchy. Partly thanks to Victoria's rooted hatred of Conroy's party – the Tories – partly thanks to the friendship forged by the 'moderate' Tory Peel and the royal pair, this would not be the case. The Corn Laws crisis and the way it was handled by Peel would have a lasting effect, not only on trade and agriculture, but also on the economy.

Peel and his allies, a distinct minority, in that Tory Cabinet could see, by 1845, that the Corn Laws would have to be scrapped. They hoped that it would be possible to fudge the issue, perhaps by scrapping them gradually, or by reducing the tariffs on imported corn, or by extending the freedom at present allowed to imported American maize to oats and wheat. But this envisaged gradualism was not in the end possible.

It was a cold, bleak summer, and in August, Peel wrote to Prince Albert to say that he had seen cornfields between Watford and Coventry flattened by the rain as he passed them by railroad.[25] In that same month, August 1845, two scientists, Dr Montagne in Paris and the Reverend M. J. Berkeley in Northamptonshire, were identifying the cause of a deadly potato blight which had swept across Central and Western Europe. A minute fungus of the genus botrytis, previously unidentified by science, *Phytophthora infestans* – got into the cell-growths of potato leaves and killed their growth. It would be fifty years before anyone came near finding a cure for this blight.

The Irish poor depended upon their potato crop in order to eat. In October, the Lord Lieutenant confirmed that the potato crop that year had failed completely. The price of cereals, and with it panic, had begun a sharp rise. Peel authorized a scientific report on the stock of remaining healthy potato tubers and the result was pessimistic.

There was now an overwhelming argument to suspend the tariffs on imported corn, if only temporarily.

Peel held an emergency Cabinet Meeting at the end of October, having spent three days with Albert and Victoria at Windsor explaining what he intended to do. He was by now physically exhausted, suffering acutely from gout and sleep-deprivation. Although he valued the friendship with the royal pair, and in particular the confidence of the Queen, which had been won only with a struggle, their need to be closely involved with all the day-to-day decisions of his crisis-ridden life greatly added to his workload. He complained of 'the constant communication with the Queen and the Prince'. Because Albert was so intensely interested in politics, but also so emotionally in need of Peel's approval, and so anxious to assert his power alongside or instead of his wife, the Prince was oblivious of the extent to which he was adding to Peel's burdens at this all-but-impossible time. Peel further confided – to Colonel Arbuthnot, the Queen's equerry – that the work, the involvement with all the minutiae of governance, the requirement 'to write with his own hand to every person of note who chooses to write to him; to be prepared for every debate, including the most trumpery concerns; to do all these indispensable things, and also sit in the House of Commons eight hours a day for 118 days. It is impossible for me not to feel that the duties are incompatible and above all human strength – at least above mine.'[26]

It was in this state of frailty that Peel attempted to deal with the first wave of horrific crisis in Ireland and to confront its implications for the Corn Laws. The Cabinet were overwhelmingly against Peel. As a stopgap measure to prevent famine, Peel authorized Barings bank to purchase £100,000 worth of maize and meal in the USA. The ships carrying this relief reached Cork at the end of January 1846. Peel had bought enough food for a million people to last four weeks. Professor Robert Kane of the Queen's College, Cork, author of a recent work on *The Industrial Resources of Ireland*, was brought in to advise the Lord Lieutenant. He calculated that the famine would last not days but months, and that the numbers affected were not one million, but four million. A calamity of unparalleled magnitude was about to unfold.

Since Peel was unable to persuade his Cabinet that the Corn Laws should be suspended, or repealed altogether, which was what he now so urgently wanted, he felt that the only course was to resign the premiership, and to advise the Queen to choose the Liberal leader, Lord John Russell, as her Prime Minister. This happened in December 1845.

Both the Queen and Prince Albert took an emotional farewell of Peel. The Cabinet went down to Osborne en masse in December for a last Privy Council meeting, and it was then assumed that Russell would take on the responsibility of government. Instead, he dithered. First, because he would have led a minority government, he tried to extract from Peel assurances that the Peelites would vote with the Liberals to ensure the repeal of the Corn Laws. Although this would almost certainly have happened in practice, it would have been impolitic and unconstitutional of Peel to give such assurance formally. Then the Liberals themselves began to squabble about the composition of the Cabinet, Lord Grey saying he could not serve if Palmerston were returned to the Foreign Office.

In this, the country's worst crisis since the Napoleonic Wars, the political classes showed more interest in themselves than either in serving the Queen or doing what was best for the people. Peel had felt elation at giving up the burden of office which was making him ill, but his freedom did not last long. At the Queen's pleading, he returned, feeling, as he said, 'like a man restored to life after his funeral service had been preached'.[27]

Peel made it clear to Albert and Victoria that he did not intend to go into another General Election with the Corn Laws in place. He would do what it took, even if it meant the sacrifice of his future career and the future eligibility of his party, to get the necessary legislation through Parliament.

His big day in the Commons was 27 January 1846. He was on his feet for three and a half hours, for every one of which Prince

Albert was sitting in the public gallery, listening to his hero. The royal presence caused huge offence among the protectionist Tories, most of whom had always disliked Albert anyway. They had voted to reduce his grant even before he arrived. They distrusted his interference in British domestic affairs. And they now deplored his open partisanship in the matter of free trade and the Corn Laws.

Peel's speech was one of the most decisive moments of British history. It spelt the end of the old order. Dramatic as all its implications were, it was not a great piece of oratory. For the most part, it was a mass of detail. It began with a long list of all the items on which he intended to reduce duties – tallow, timber, cloth, carriages, candles, soap, straw hats, sugar, tobacco. Then he proposed lifting duties on imported meat. Finally, he reached corn. Only at the end did his speech become emotional, as he outlined the reasons for his conversion to lifting the tariffs, and his concern for reducing the cost of living for the poor.

The fat was now in the fire. Tory members of the Royal Household resigned, and many MPs resigned the Tory whip. Lord George Bentinck and Benjamin Disraeli began their campaign to eviscerate Peel when his measures were debated throughout February. All the attacks on the Prime Minister, many of which were highly personal, and some of them coming not from protectionists but from lifelong abolitionists such as Cobden (who bore a grudge against Peel for his earlier hostility to abolition and almost resented his conversion), only had the effect of increasing Peel's popularity in the country at large. The public mood wanted the Corn Laws abolished. Most of the Liberals voted for Peel – 227 of them, with only 11 voting against – while 28 Tories were absent or pairing.[28] Of the rest, 231 voted against Peel and only 112 voted with him. The Tories were now a party which would be out of office for a generation, the Peelites, such as Gladstone, eventually joining the Liberals. The Corn Laws had been repealed. In the Commons. By June, after seemingly interminable difficulties, the measure was passed through the Lords.

The Tory diehards were not slow, however, to exact their revenge on Peel. On the very evening that the Repeal of the Corn Laws was effected by the Lords, Lord George Bentinck and the diehards voted with the Irish nationalists and the Radicals to defeat Peel's Irish Coercion Bill. Wellington called it 'a blackguard combination'. Similar Coercion measures, to suspend Habeas Corpus where there was a danger of public disorder, had all been supported by the diehards. Peel was only introducing this bill to keep the peace while the tragic situation in Ireland unfolded. Now his enemies used it as an instrument to get rid of him. He was not sorry to go. He had done his great work. In July, Lord John Russell kissed hands as the new Prime Minister. The Queen considered that his 'Government is weak, and I think Lord J. does not possess the talent of keeping his people together'.[29]

ELEVEN

MALTHUSIAN CALAMITY

ALBERT SAW THAT the profound political crisis occasioned by the repeal of the Corn Laws was a turning point. The Tory diehards and the landowners dreaded the direction which political life would now take. Seen from this perspective, the country was no longer 'at ease with itself'. Factionalism could surely destroy it. The interests of English Radicals, Irish nationalists, the landed aristocracy, northern mill-owners, displaced Scottish peasants, the aspirant urbanized lower middle classes – to name only some of the categories who were only partially represented by the franchise – were so profoundly at odds.

When the Corn Laws had finally been repealed, and the anti-Peelite Tories, led by Disraeli and Lord George Bentinck, were baying for Peel's blood, Albert told Stockmar, 'the party conflict is carried on with a bitterness, a fury, & a want of common sense which never, perhaps, had a parallel in history'.[1] Yet, paradoxically, Albert believed, this could work to the advantage of the monarchy. Even the radical paper *The Examiner* saw 'the advantage for the country of the existence of a third power so free from partisanship'.[2]

Albert subscribed to this view even though his Peelite partisanship had been made obvious throughout the mid-1840s. Hitherto in British political life, certainly in living memory, the sovereigns had been forces of reaction. William IV had opposed

Catholic Emancipation and the Great Reform Act. George III, when sane, had always been a reactionary. Now a new phenomenon was observable. A monarchy which was young, and, in the political climate of the time, progressive: not on the side of Radicals, but on the side of the burgeoning, property-owning bourgeoisie. After Peel's fall, Albert and Victoria could retreat from openly siding with any party; the Crown could now be a 'third power… free from partisanship'. This was because even the editor of *The Examiner* could see that, technically at least, the monarchy really was above party; and in the case of the Conservatives, after the repeal of the corn tariffs – the measure was finally passed through the Commons on 25 June 1846, after *five months* of debates – it was open to question whether there still was a party to be above. Four days later Disraeli and Bentinck forced a vote on the Irish Coercion Bill in which they voted with the Liberals and the Irish members, and defeated the Government. Peel resigned as Prime Minister, to be replaced by Lord John Russell and the Liberals. Palmerston, Prince Albert's bête noire, was once more Foreign Secretary.

In his resignation speech in Parliament, Peel praised one of the chief campaigners for the repeal of the Corn Laws, the Manchester liberal Richard Cobden, whose name lives on to this day in so many English towns – Cobden Crescents, Squares and Streets being a reminder of how much the people owed to the man who brought them cheap bread. Nor was it the urban poor alone who revered Cobden. The statue of him at the bottom of Camden High Street in London was paid for by public subscription, one of whose donors was Louis Napoleon, the future Emperor Napoleon III. Yet two of Peel's closest supporters and admirers, Aberdeen and W. E. Gladstone, deplored Peel's praise of a man who had argued for the repeal of the Corn Laws 'on the principle of holding up the landlords of England to the people as plunderers and knaves'.[3] Since Aberdeen owned thousands of acres in the north of Scotland, this was not surprising. As for Gladstone, though the family money derived from Liverpool trade, his Scottish father had acquired an estate at Fasque, Kincardineshire, and Gladstone himself had married into the landed family of the Glynnes, and he lived on his wife's estates at Hawarden in North Wales. Gladstone also inherited his father's estate at Seaforth, near Liverpool.[4]

We have already mentioned the enormous personal wealth which Victoria would accumulate from the Duchy of Lancaster, and this would be enhanced by the acquisitions of the estates at Osborne and Balmoral. The latter, as we shall see, came into royal possession through the agency of Aberdeen himself.

Aberdeen was a loyal friend to the Crown, but there were moments where the friendship was strained. As Foreign Secretary, he accompanied Victoria and Albert on their visit to Germany in August 1845. It provided him with the chance, for example, to discuss at first hand with the old King of Prussia the relations between Prussia and Austria, and the position of France.

For Albert, the central question of his times was the future of Germany. Since 1815, when the smaller states, and some of the larger, of German-speaking lands had been united under a loose federation, the question had been – would there ever be a united Germany? With the help of the Customs Union, the *Zollverein*, would it be possible to move forward to a political union? And if so what shape would this united Germany have? Would it be a state under the shelter of an Austrian Emperor? Or would the powerful Kingdom of Prussia be the dominant partner? At this stage, a decade before Bismarck came on the scene, Albert saw Prussia as an essentially benign force. Already, he and Stockmar were dreaming that, by a dynastic marriage, Britain/the Coburgs could marry into the Prussian Royal House. Albert saw the old King, Friedrich Wilhelm IV, an admirer of Protestant England and her Constitution, as Germany's brightest hope.[5] He thought that the King would lead Prussia into the direction of being a constitutional monarchy on the English pattern, with a representative Parliament. In April 1847, however, when the King made an off-the-cuff speech to this effect, Albert, though admiring of the King's candour, felt that it was offering too many hostages to fortune. He read the newspaper transcripts of the speech with 'alarm'. It was, he said, lapsing into English, or his version of it, as he wrote to Stockmar in German, 'slab-dash'.[6]

Intensely interested as the royal pair were in such questions, the Foreign Office, even when their friend Aberdeen was in office, did not wish to commit Britain to any particular template of what a future Germany would look like. The more Albert interfered, or tried to, in matters that were strictly and constitutionally the business of the Foreign Secretary, the more Aberdeen withheld papers from the royal scrutiny, or sent them to the Palace in a red box when they had already become a fait accompli. To this extent, although Albert and Victoria found Aberdeen as congenial as Palmerston was offhand, the Foreign Office treatment of royal busybodydom remained the same under both Secretaries of State. Back in 1845, the Queen complained 'that for some time past, the drafts are sent to her from the Foreign Office when they are already gone, so that, if the Queen wished to make any observation or ask any explanations respecting them, it would be too late'. She claimed that, in times past, all such papers had been shown to her well in advance of dispatch. Aberdeen immediately replied:

> The practice has usually been to submit to Your Majesty the drafts of despatches at the same time that they are sent from this office. Should Your Majesty then be pleased to make any remark or objection, it would be immediately attended to by Lord Aberdeen, who would forthwith either make any necessary alterations, by additional instructions, or he would humbly represent to Your Majesty the reasons which induce him to think that the interest of Your Majesty's service required an adherence to what had already been done.[7]

This was polite, but it made clear that neither the sovereign nor her husband were to be allowed to dictate British foreign policy. This would (nearly) always be the case for the rest of Albert's life, though there is one tragic instance of it, when the Hamlet-like Aberdeen, in an agony about the possibility of war with Russia in 1854, misread the Queen's mood and would, perhaps, have pulled back from the brink of war had he understood her position.

Apart from differences in foreign policy, there existed personal gulfs between Albert and the mild Aberdeen. One of Landseer's more celebrated canvases of this decade was *The Otter Speared*, a picture exhibited at the Royal Academy in 1844. It depicted Aberdeen's famous otter hounds. Those who visited him at his estate, Haddo, in Aberdeenshire, enjoyed the sight of the kilted statesman, thigh-deep in water, following his dogs to chase the beautiful water-mammals. He considered himself lucky if he killed two per season. The chief sport was in the exercise, the excitement of the dogs, and the cold and wet which is clearly the delight of the northern nobility. Hearing of these hounds, Albert asked if he could hunt with Aberdeen's otter hounds while staying at Blair, the seat of the Duke of Atholl. The otter hounds were brought down, but Aberdeen was revolted to discover that the Prince did not take them to the local river, but rather simply set them on a live otter which had been 'bagged' and released for the purpose.[8]

When Aberdeen accompanied the royal party on one of their trips to Germany in 1845, the Prince's cruelty to animals was something which became public, and caused a journalistic furore. It was, in any case, a trip which did not go well. The Queen was in a bad mood, and believed that neither she nor Albert were given enough deference in the various German courts which they visited. When they reached Coburg, Albert's brother arranged a deer shoot. Once again, the differences between Scottish and German ideas of good sport were thrown into relief.

The London press had followed the Queen and Prince Albert to his ancestral land, and reported in detail on the *battue*. Whereas, on a Scottish deer-stalking expedition, the stalkers clamber up mountain-sides and often walk for miles without seeing an animal, the Germans arranged things differently. The deer were driven down by keepers into an enclosed area and simply massacred. The comic magazine *Punch* was soon carrying cartoons of the Prince with his gun, surrounded by dead birds and animals. Greville, ready, as always, with a disobliging observation, was unconvinced by Victoria's expressions of disgust at the spectacle. 'The truth is, her sensibilities are not acute, and though she is not at all ill-natured, perhaps the reverse, she is hard-hearted, selfish and self-willed.'[9] Moreover, the deer-slaughter was not the only instance of what to the eyes of English urban newspaper-readers looked like barbarity. Victoria and Albert found themselves placed, by his hospitable brother, on an elevated stand in an enclosure where thirty wild boar were released to be speared or shot before the eyes of an appreciative audience. Victoria's journal did not express much horror at the experience, merely stating that the audience had been 'charmingly arranged... we walked away & then got into our carriage & drove home', comments which make Greville's disobliging remarks about his sovereign seem fair.[10]

The deaths of boars, otters and deer were distressing to the English readers of *Punch* and *The Times*, but a calamity of infinitely greater magnitude was now looming. It was the two years of Irish famine, the tragedy which made it inevitable that tariffs on corn would eventually be lifted, but which also set the seal on Anglo-Irish relations for the rest of history. It is probably true to say that, even to this day, and even knowing all that is known about the disaster, most English people do not really grasp its significance. It is not only in Sir Theodore Martin's Albertiad, but in most biographies, and most general histories of the period, that the famine is seen as a tragic episode. The potato crop failed, first in 1845, then again in 1846. In Scotland, where very many poor people depended upon the potato, there was widespread hunger and some cases of actual starvation. In Ireland it was all much worse. Out of a population of eight million, a million died and four million were displaced.

Behind it, however, lurked an attitude which can still be read in non-Irish narratives, however well intentioned. The famine was the result of the potato blight. There followed Lord John Russell's famous remark, when Prime Minister, that 'a famine of the thirteenth had fallen upon a people of the nineteenth century'.[11]

This, however, was not the case. It was a peculiarly, disgustingly, nineteenth-century famine. The landlords, and the British Government, did not cause the potato blight. They reacted, however, with unpardonable lateness and inadequacy to the scale of the crisis. And the Government did nothing to force the landlords who, throughout the two years of famine, were exporting their grain at huge profit to themselves. They did not see, until it was pointed out to them by journalists, that having vast state banquets at Dublin Castle during the famine was tactless. They thought that the 'thirteenth-century' famine could be solved by sending 'aid', far too little far too late, and deporting half the Irish population to America. Thereafter, when the good old spud started to grow again, the status quo could be resumed. The English could go on mocking and belittling Irish people, ignoring their political aspirations, and using them, when they came to the mainland, as navvies, maids, prostitutes and cannon-fodder. The Irish middle classes in some senses – the Phineas Finns – actually colluded unconsciously in this process, because they were either Protestants or supporters of watered-down 'Home Rule' notions of how Ireland could be governed. The famine was a nineteenth-century famine, not a medieval one, because the lacklustre response by the Government was determined by their mind-set. They believed, for one thing, in the (now totally discredited) economic ideas of Thomas Malthus. They thought that there was only a limited amount of food on the planet and that, when it had been eaten up, starvation or war was a simple inevitability. In fact, as the population in Victorian Britain soared, so did the food supply – by agricultural ingenuity, and by import.

The Irish starved because the Malthusians thought it was nature's way – God's way, even. For, little as most Victorian intellectuals believed in God, the God to whom they paid lip service was a Protestant God. And the feckless Irish, as well as being dirty and fecund, owed their allegiance to the Pope. The gut anti-Catholic prejudice displayed by public figures in Victorian England at every turn can hardly be described as 'thirteenth-century'; but it played a deadly part in the British response to the famine. Many clergymen and politicians expressed the view that the famine was the Judgement of Heaven against the Roman Catholics.

Before, during and after the famines, therefore, Ireland when viewed by the Irish, and Ireland when viewed by the British, was not the same place. Albert, who had never been to Ireland until 1849, and who shared a fierce anti-Catholic prejudice with the majority of British people, saw that other island as a 'problem'. Seeing something as a problem means seeing it as something which can be solved. Albert saw himself as a problem-solver. All his letters to Stockmar about the situation in Ireland see this 'problem' as one comparable to 'problems' in India – problems which could be solved if only the native population would settle down and be obedient to their English lords and masters. Any evidence that the natives did not accept this version of events was to be seen as a simple demonstration that they were unreasonable, and did not know what was good for them. We cannot blame Albert individually for having these views, when we remember that they were shared by every single member of the governing class in the nineteenth century.

TWELVE

CAMBRIDGE

PRINCE ALBERT AND the Queen visited Cambridge together on 25 October 1843, thoroughly enjoying themselves. They attended the evening service at King's College, where, after dinner, they held a Levee. In the Senate House, the Prince was made an Honorary Doctor of Civil Law. They saw St John's, Christ's, Corpus Christi, the Round Church.[1] Oxford had shocked them: the Tory undergraduates hooting and booing the Liberal politicians who were brought to the Encaenia, the glowering of the Queen's Tory bête noire, the Chancellor, the Duke of Wellington, the rather alarming grandeur of Nuneham Courtenay and the household of the Archbishop of York. Cambridge was a complete contrast. 'I seldom remember more enthusiasm than was shown at Cambridge, and in particular by the undergraduates,' Victoria told her journal. They stayed at Trinity College, and slept in a tiny bed, which must have been cosy for the little Queen but uncomfortable for the Prince. (Their son Alfred was born a little over nine months later.) Very characteristically, Albert sent word that he would like to visit the Woodwardian (geological) Museum with the Professor of Geology, Adam Sedgwick. This was the bluff Yorkshire (Sedbergh and Trinity) clergyman-geologist who, thirteen years earlier, had taken Charles Darwin on his first serious geological expedition (in Wales), and whose hostile reaction sixteen years later to *On the Origin of Species* Darwin would find so wounding.

Sedgwick was thrown into some panic by this request, finding in the museum as many 'mops, slop-pails, chimney-pots, ladders, broken benches, rejected broken cabinets, two long ladders and a rusty scythe', but having cleared the place up, he was able to show them the picture of the Megatherium, the skeleton of a Plesiosaurus and a giant stag, and other treasures. Although the Queen seemed completely ignorant of geology, Sedgwick was glad to find in Albert a ready and well-informed listener. When they were shown the Ichthyosaurus, a recent acquisition which Sedgwick had only just unpacked, Victoria asked what it was, and when Sedgwick explained, she asked where it came from. He whimsically replied that it had come 'from the monsters of the lower world to greet Her Majesty on her arrival at the University'.[2]

Sedgwick was a serious man of science. When appointed to the Chair of Geology, he had no knowledge of the subject whatsoever, but he had felt morally bound to get it up, and as he did so, his interest had quickened, and after a number of geological expeditions, he had built up an impressive collection of fossils and palaeontological specimens, many of them collected in Germany.

This was the period in which science was beginning to come to terms with the great age of the planet, to ask how the rocks and the mountains were formed, to study the evidence of fossils, and to investigate, through such investigations, the origin of species. It was, in fact, the period when modern science, proper, may be said to have begun. Like so many men of science of that generation, Sedgwick thought that God had made each individual species exactly as they were visible at that moment in history, and that development or mutation of forms somehow suggested an impiety. Nevertheless, Sedgwick was a serious man of science, and he had absorbed much of the new wisdom. Toweringly the most influential English geologist of the time was Charles Lyell, and Sedgwick had read Lyell's works. He was proud to be able to show to the Queen and to Prince Albert what he considered to be 'the finest collection of German fossils to be seen in England'.[3]

In spite of what people may say, there was, and is, a world, a universe, of difference, between Oxford and Cambridge. Oxford was always more conservative in temperament than Cambridge, also more worldly. Of the Victorian Prime Ministers, Lords Melbourne, Aberdeen and Palmerston were Cambridge men, but with the exception of Lord John Russell, who went to Edinburgh, and Disraeli, who did not attend a university, it was a story, not merely of Oxford men, but of Christ Church men ruling the country: Peel, Derby, Gladstone, Salisbury and Rosebery were all at Christ Church, Oxford.

The ethos of the two universities was mysteriously but palpably different. After the collective nervous breakdown caused by the parties which had raged for and against Newman,[4] Oxford never had another guru. Cambridge has always had gurus, whose devotees, like those of the Desert Fathers clustering at the base of their towers, have cherished some version of Truth not vouchsafed to the majority of mankind. In the twentieth century, one saw this at work among the followers of F. R. Leavis and Ludwig Wittgenstein, but it was a palpable phenomenon long before that in Cambridge, as the lives of Oscar Browning, George Moore and others show.[5] Moreover, the esoteric, ethereal society known as the Apostles has no equivalent in the history of Oxford.

Prince Albert, though educated abroad, was far closer to the spirit of Cambridge than he ever was to that of Oxford. His intense seriousness, his wish to embrace the Modern, his lack of formality, his distrust of worldliness, all made him seem like an honorary Cambridge man, long before the idea was ever proposed that he should become the Chancellor of the university.

On 12 February 1847, the Duke of Northumberland, Chancellor of the university, died, and a group of Fellows of St John's College immediately proposed one of their own, the Earl of Powis, a former Tory MP for Ludlow, to succeed him. Powis was a staggeringly undistinguished person. Over the garden wall from St John's, in Trinity College, the Master, who was hoping for the modernization of Cambridge, viewed the Earl of Powis with some dismay.

Whewell (pronounced Hugh-ull) was, like Sedgwick, a man of relatively humble origins, the son of a master carpenter in Lancaster. He was prodigiously brilliant, a philosopher, linguist, clergyman, mathematician. Whewell invented the word 'scientist' to describe the emergence of this new type of nineteenth-century intellectual. He was a Kantian, who had mastered the works of the sage of Königsberg. His studies in German literature and philosophy led him to Berlin, where he had an appointment to meet the great polymath, Alexander Humboldt. After a conversation with Whewell, Humboldt complained that he was waiting for the arrival of a young Englishman – Whewell's German was so perfect that Humboldt had not realized he had come over specially from England to meet him. This was the man, who, having been Professor of Mineralogy at Cambridge, became the Master of Trinity, in succession to Christopher Wordsworth (son of the Poet Laureate).

Whewell hit upon the idea that Prince Albert would be the ideal candidate to become the Chancellor. Albert replied, having taken the advice of Stockmar, that he would do so if this were the unanimous wish of the university. The Vice Chancellor, Dr Henry Philpott, Master of St Catherine's, assured Albert that it was the unanimous wish of the Heads of Colleges that he should succeed the Duke of Northumberland.

Philpott and Whewell were not lying, but they had gambled that, when St John's heard of a royal candidate being proposed, they would withdraw the ridiculous Earl of Powis from the contest. This was not the case. In fact, the attempt by Trinity College, the great rival of St John's for 300 years, to impose a known liberal and, to boot, a foreigner on the university excited the Earl of Powis's spirit of controversy.

Albert was now in the embarrassing position of being a candidate in a contested election. He consulted Peel, who advised him to stay in the contest, but his previous Private Secretary, Anson, and his new one, Colonel Phipps, concealed from the Prince the vigour with which the election was contested. When Whewell sent to the Prince an election circular written on his behalf, Phipps did not show it to

Albert.[6] In one of Whewell's pamphlets there was the fair but devastating thrust – 'In Lord Powis, we should only have a Chancellor of St John's.'

In the first day of polling, Lord Ernest Bruce reported to General Bruce, 'It is a neck and neck race': 582 votes had been cast for Albert, 572 for Powis. Many of these voters, those who held the degree of Master of Arts, had descended on the university from crowded London trains. The next day, Albert was edging ahead – 875 to 789. By the time the polls closed on the third day of voting – on Saturday evening – Albert had won the vote by 953 to 837. It was something of a shock to Albert, and he asked Peel whether, given the relative closeness of the fight, he should in fact accept the chancellorship. Peel was in no doubt. 'The acceptance of the office without reluctance or delay has about it a character of firmness and decision, of supporting friends instead of giving a triumph to opponents.'[7] The Queen was also adamant. 'All the cleverest men were amongst those on my beloved Albert's side.'[8]

On 25 March, a party of Cambridge notables, chiefly Heads of Colleges, came to Buckingham Palace and Albert was inaugurated as their Chancellor. In July, Albert and Victoria travelled to Cambridge for the ceremonies to welcome him. There was a special ode, composed by the Poet Laureate, Wordsworth, who had done his homework. He imagined the star of Brunswick shining brightly on 'Gotha's ducal roof' when the Prince was born, and reminded the assembly that Albert sprang from 'that wise ancestor… Who threw the Saxon shield o'er Luther's life'.

Resound, resound the strain
That hails him for our own!
Again, again, and yet again,
For the Church, the State, the Throne![9]

When the evening was over, Albert and Victoria walked along the Backs, as they call the gardens by the River Cam at the back of the colleges. He was still wearing his Chancellor's cap, but he had a mackintosh over his formal clothes. She was in full evening dress, but with a cap over her diadem, hiding some stupendous diamonds. They were very happy, their only regret being the absence of music after the dinner. They slept that night in another tiny bed at Trinity College, adequate for the small sovereign, but Albert's feet must have poked out into the night air. Their sixth child, Princess Louise, was born nearly nine months later.

If there were any at Cambridge, however, who supposed that Albert would see his place there as purely ceremonial, they were mistaken. The story of his chancellorship repays attention, because, in microcosm, it mirrors his approach to all the tasks he undertook in England, not least the chief one, of being the Queen's husband and preserving and transforming the monarchy.

Two things – firstly, he regarded it as a job to be accomplished, and, whatever he undertook, he tended to do thoroughly, but rapidly, before moving on to something else. The bulk of Prince Albert's activities as Chancellor happened in his first three years in post.[10] This was characteristic of the way he went to work. He would take on a new project; analyse what needed to be done; do it; move on. He could move on with confidence because, as in this case, he had given royal approval to a group of highly efficient men who had an end in view and used Albert as a way of helping them.

Secondly, he clearly, at the start, mistook the job description, but it was a creative mistake, greatly to Cambridge's advantage. Being Chancellor does involve executive roles, but these are more limited than he realized. A lazier person, or one less well informed about the condition of universities in Germany, would surely not have done so much to encourage progressive reform of the syllabus, and with that, the modernization not only of Cambridge, but of what came to be called 'The Idea of a University'.

In the 1840s, Durham and London were fledgling universities, and, as far as England was concerned, nearly all university life was Oxford and Cambridge. Both the older universities were dominated by the Church of England, and it was only lately (since the Catholic Emancipation Act of 1829) that non-Anglicans were admitted. The Chancellor of Oxford was the Duke of Wellington, a dyed-in-the-wool Tory who would not have dreamed of imposing modern educational ideas on the dons. The only major disruption to the academic quiet of Oxford in the Duke's time as Chancellor had been a theological one. His support, in 1829, for Catholic Emancipation and, subsequently, for a grant in Ireland for the Roman Catholic seminary at Maynooth (the only place of higher education open to Catholics) had scandalized the High Church faction of Oxford. This was because they believed that the Established Church was the true Catholic Church in this land; its bishops true bishops; its sacraments true sacraments. To subsidize the Roman Catholic seminarians of Ireland, and to allow Roman Catholics to come to Oxford, without assenting to the Thirty-nine Articles of the Church of England, was, for them, a diminution in the public awareness of what the Church was. Instead of being Christ's Body in their midst, the Established Church seemed to be little more than a religious branch of the Civil Service. Wellington's – to their eyes – casual attitude to these esoteric questions called forth a denunciation of Catholic Emancipation as a 'National Apostasy', and the development of what was called the Oxford Movement: a movement to make more explicit the Catholic nature of the Church of England. The paradox of this movement, which really got going in 1833, had been that, when they actually studied the matter more carefully, some of the Oxford dons decided that the Church of England was not so Catholic as they had supposed. The most charismatically attractive of these figures, John Henry Newman, had decided, in fact, that the Church of England was not, after all, Catholic in the least, and had made his submission to the Church of Rome.

That was in 1845. As far as the syllabus or the administration of the university was concerned, however, there were very few dons who interested themselves in reform, though two of the more distinguished liberals – Jowett and Stanley of Balliol – set off to Germany, shortly after Newman's conversion, precisely to see how things were done over there.

In Cambridge in the 1840s, the colleges were by today's standards very small. Most of those who came up to study there were intending either to become lawyers or parsons. They could choose between two courses. One was a simple course, at the end of which there was no Honours Degree. It involved a rudimentary study of Mathematics – the elements of geometry, algebra, mechanics and hydrostatics; in Latin they were merely required to read Horace's Epistles, Book 2; in Greek, Herodotus Book 9 and some bits of the New Testament; a very little 'Moral Philosophy', more properly, Theology (a book by a don at Christ's named Paley). The other, more ambitious, possibility was that the young man (for the students were, of course, all male) could read for an Honours Degree in Mathematics and Classical Languages and Literature. About forty undergraduates a year read for an Honours Degree. Only graduates studied Theology, which consisted of doing the Book of Deuteronomy in Hebrew. That was the extent of academic study on offer to the English young, when Prince Albert took on the role as Chancellor of Cambridge.[11]

It was not true that no other subjects were studied, but these were not studied at an undergraduate level. There were twenty professorial chairs, which had been endowed by benefactors over past years. These included professors of science – the Regius (that is, appointed by the Crown, in reality by the Prime Minister) Professor of Medicine, of Chemistry, of Anatomy, of Botany. There was the Downing Professor of Medical Jurisprudence. The Lowndean Professor of Astronomy and Geometry lectured for one term on the construction and use of astronomical instruments.

Albert quickly came to see that the majority of Cambridge undergraduates did not behave as he had done at Bonn, getting up at five in the morning so that his brain was fresh to wrestle with the philosophical conundrums of Fichte. He consulted Chevalier Bunsen, the Prussian Minister to the Court of St James's (a figure regarded with horror by the High Church dons of Oxford, because he had championed a joint Lutheran-Anglican bishopric in Jerusalem). Bunsen pointed out to the Prince that one of the drawbacks of the Cambridge system was that the actual tutors were young men who had only recently graduated, and who were greatly 'inferior to the teachers in the great schools like Rugby, Eton and Harrow... Is it reasonable to oblige almost all the young men coming from schools like Rugby, to give themselves up, *the first two years*, to habits of idleness, with all their consequences, because what is offered to them is below what they know: because the really studious young men are kept back by the crowd of those who have the same tutor, & besides, knowing little or nothing, are decided to learn nothing?'[12]

When Albert consulted Peel about the matter, the Prime Minister wondered how realistic it was to suppose that anyone actually should attend the lectures given by the Cambridge professors of science. Since the science lectures were voluntary, it was reasonable to ask, 'to what extent the attendance really takes place... Is there the same encouragement to the Lectures in these departments of science, the same prospect of academical advantages, which is held out in the more regular and accustomed studies of the University'? Peel asked.[13]

Ruefully, Albert looked at the syllabus for his own two semesters at Bonn. He wrote to Lord John Russell, in November 1847:

> History is entirely excluded, there is not a lecture on history of any kind given out at the moment, and this has been the case for some time, political Economy, Constitutional Law, Law of Nations, Metaphysics, Psychology, Comparative Philology, Modern Languages, Oriental Languages, Old Languages with the exception of Greek and Latin, Geography, Chemistry, Astronomy, Natural History (with the exception of Geology), History of Art, Esthetics and Countrepoint [? Illegible] are quite excluded.[14]

He conceded that there were professorships in some of these subjects but they were not actually *taught*.

Philpott, the Vice Chancellor, came to Windsor on several occasions to discuss syllabus reform over a number of days. Albert 'acknowledged the defects [of the Cambridge system] but said that the Heads of Colleges were such a nervous and essentially Conservative body that it required the greatest caution in proposing any improvement not to rouse an insurmountable opposition'.[15] Lord John Russell was also expressing 'grave doubts', in letters from Woburn Abbey, about the feasibility of the reforms, urging Prince Albert, in January 1848, not to go ahead until he had consulted the Archbishop of York and Mr Macaulay.[16]

They did, however, make changes. A Syndicate of Professors was set up, and by April 1848, the Moral Sciences Tripos and the Natural Sciences Tripos were offered as alternatives to the everlasting diet of Classics and Mathematics. As early as February 1848, the Prince could tell his diary, 'the newly proposed studies in Cambridge have gone through, not without opposition'.[17]

Albert had entered deeply not only into the spirit of reform, but into its practicalities. He calculated the rents which were coming in, both to the universities, and where information was available to him (as Visitor to individual houses) to the colleges. He proposed that all Foundation Fellowships, that is endowed for a particular subject or college, should be open to all, and he proposed making some of the moneys accrued by the university's investments available to fund between 600 and 700 free places for poor students.[18] When you remember that only forty undergraduates read for the Honours schools, this was a revolutionary proposal which would transform Cambridge, and indeed transform the idea of British universities.

Albert who was young and clever was in tune with the spirit of the times. Despite the Prime Minister (Russell)'s caution, an impressive petition was presented to him demanding a reform of the universities. It was signed by, among others, J. S. Henshaw, Charles Lyell, Erasmus and Charles Darwin, Hensleigh Wedgwood (their brother-in-law), Francis Newman (John Henry's brother), Thackeray, Arthur Hugh

Clough, James Clark, the Physician to the Queen – well over a hundred names.

As the list showed (Clough, for example, was a Fellow of Oriel College, Oxford when the petition was signed), the impetus was not limited to Cambridge. In addition to Jowett and Stanley's fact-finding mission to German universities, Samuel Wilberforce, son of the abolitionist, and Bishop of Oxford, cheered on reform from his palace at Cuddesdon. He wrote a letter to Prince Albert sneaking on the diehards who were trying to prevent change at Oxford – 'the Dean of Christ Church, and the Heads of Corpus, Jesus, Lincoln, Balliol, Merton and up to a certain time, St Alban's Hall'.[19]

On 1 November 1848, the *Morning Chronicle* wrote that 'The University has, by a single stroke, taken the ground from under the feet of its adversaries. They accused it of insensibility to the progress of events, insensibility to the progress of knowledge... each of these reproaches will in future be an impossibility.'

Albert cannot take the credit for transforming university life single-handedly. His contribution, however, was palpable. Not only did royal approval help the reformers face the diehards. His policy, in defiance of Russell's caution, to act fast and boldly had played a decisive role.

THIRTEEN

THE YEAR OF REVOLUTIONS

THERE IS A famous letter written by Prince Albert, when the huge Chartist demonstration in London, in April 1848, fizzled out: partly because of appalling rain, partly because the Government of Lord John Russell had prudently called in the old Duke of Wellington to barricade all the great public buildings – the Bank of England, the British Museum – against the mob, to enlist 80,000 special policemen, and make it clear to the Chartists that they would, if the Duke deemed it necessary, be shot. Albert, who, with the Queen and their children, had been bundled off to the Isle of Wight for safekeeping, wrote to his stepmother, on 11 April 1848, 'We had our revolution yesterday and it ended in smoke.'[1]

It is sometimes quoted as proof of the fact that, unlike the French, the Germans, the Sicilians, the Italians, the Austrians, the Portuguese and the Spanish who all underwent convulsions during the so-called Year of Revolutions, the British somehow, by some smug inner genius, escaped the troubles which overcame the countries that had not possessed the advantages of having Victoria as their Queen or the Duke of Wellington as their national hero.

As a matter of fact, Albert was obliged to admit on 5 June, only six weeks after writing about 'our revolution' ending in smoke, 'We have Chartist riots every night, which lead to numbers of broken heads. The organization of these people is incredible: they have secret signals and correspond from town to town by means of carrier pigeons. In London they are between 10,000 and 20,000 strong, which is not much out of a population of two millions, but if they could through their organization throw themselves in a body upon any one point, they might be successful in a coup de main.'[2]

He was also obliged to admit that 'Ireland still looks dangerous',[3] and in many ways this was an understatement. A truer, as well as wittier, assessment of the situation was made by a modern historian, a cynical Irish Protestant, when he noted, 'So Britain's 1848 revolution happened in Ireland, with an incoherent conspiracy followed by a rising inescapably connected with a "cabbage patch" in Tipperary, led by Smith O'Brien.'[4]

1848 compelled Victoria and Albert to confront the reality of things, to stare into the face of their Orléanist relations and connections as, bedraggled and penniless, they staggered off the Channel packet, King Louis-Philippe having abdicated, and his family having skedaddled with what jewels and belongings they could bundle together in a matter of minutes. There was no magic formula which would stop the British mob getting rid of their monarchy if they were hungry enough or the politicians were sufficiently incompetent.

Never able to be still, even in peaceful times, Albert frenziedly spent the opening months of the Year of Revolutions writing proposals for how the peoples of Europe might better be governed. To Stockmar, in March, he had written, 'Today I send you a plan for the new Germany, as I picture it to myself. It is the duty of every German to contribute his quota, then something may come out of the discussion.'[5] He added that 'I have sent my plan to Vienna, Berlin, Dresden and Munich'. Victoria touchingly believed that the reason Albert's Grand Plan for Germany had not been acted upon at once was that it had arrived 'too late'.[6] On 18 March, three weeks before the Chartist agitations began to seem truly threatening, the Queen gave birth to their sixth child, a daughter – named Louise after Albert's mother.

At this moment of political crisis in Britain and Europe, Albert reverted to the Gotha childhood. His brother Ernst and his wife

Alexandrine were staying with Albert and Victoria that spring when the news came that their Gotha 'grandmother' – that is, Luise's stepmother – had died. Ernst rushed back to Germany for the funeral. Albert stayed to see Victoria through her accouchement, his mind straying through scenes of his own childhood. The naming of their children reflected the *égoïsme à deux* of the pair and the list of all the nine, when it was complete, is notable, by the time of the ninth birth, for its omission of any Ernest. The first two babies were named after their parents – Vicky and Bertie. Then came Alice, 'an old English name', as Albert helpfully explained to Uncle Leopold. Then Alfred, Helena (Lenchen) and now Louise. (Arthur, Leopold and Beatrice would follow in the 1850s.) Brother Ernst noted with pain that his, and his father's name, was not used. Nor were the names of any of the 'grandmothers', either the Coburg grandmother, Auguste/Augusta, nor the name Charlotte, which had been one of the Gotha grandmother's names, and was also the name of the sainted first wife of Uncle Leopold. When their grandchildren came to be born, Victoria would always insist that every child, in its list of names, contained a Victoria/Victor or an Albert. In naming the children, with impersonal, non-family names like Alice and Alfred, or repetitions of their own names, the choice of 'Louise' for their sixth child, in the Year of Revolutions, is all the more striking. Until they came to name their eighth child Leopold, they eschewed all their male German relations. It was as though, in a rewrite of history, Albert's vanished mother, who was not actually present for most of his childhood, had replaced the father Ernst who had been there.

While Victoria recovered from the birth of Louise, Albert took the older children to 'a performance at Drury Lane'; but lest anyone should consider he had spent the day frivolously, his awestruck wife noted, 'Albert wrote down excellent proposals for a Constitution'. The Government of Lord John Russell was not in the least inclined to accept the Prince's political advice, any more than Metternich in Austria or the Kings of Prussia would do so. Albert still could not stop himself working. 'I have never been so sorely pressed as now; Events, business, feelings, thoughts almost bow me to the earth.'[7]

In May, Victoria and Albert became apprised of the Constitution of Dahlumber 'which would place all Sovereigns [in Germany] under Prussia, in a completely medialised position, which is really too painful and would be almost impossible to maintain. The King of Prussia to be Hereditary Emperor of Germany and all the other Princes under him, not being allowed to appoint any Officer in the Army or any Diplomat – all appointments to be made by the Emperor. This can never do.'[8] Hindsight allows us to know that this was, from the 1850s onwards (and the rise of Otto von Bismarck), what became inescapable; it was what, after 1870, would become the political reality. Victoria and Albert hated it. They worried for what it would mean for small German duchies and principalities such as Gotha and Coburg. There was nothing they could do to resist it. All too seldom, that summer of revolutions, were there moments such as the pleasant one on 12 May when she could write, 'I took a delightful drive with Albert in the "environs" of London and then we dawdled about in the garden.'[9] The preoccupation with what was happening on the Continent, and in particular what was happening in Germany, never left them. Even when they were at the opera, watching Jenny Lind perform in *La Sonnambula*, 'we could not thoroughly enjoy it, our thoughts being elsewhere, and so taken up with the awful and sad state of affairs and the entire dislocation of everything'.[10]

Albert could not accept the fact, probably could not even see that it was a fact, that he had no power and no influence in Germany and no real power in Britain either. He spent their breakfast on 3 May reading aloud the new German Constitution to the Queen. Victoria's journal described him spending the better part of two days drafting yet another 'counter proposal to the German Constitution'.[11]

The exhausting work of drafting political plans and constitutions for a state in which he had absolutely no role was at least a useful 'displacement' therapy. It helped to distract him – not her – from the plight of their Orléans relations. She bluntly described Princess

Victoire, their Coburg cousin, now Duchess of Nemours, touching her for money and admitting that they could not afford to live. Albert and Victoria drove out to Claremont ('the country in such beauty'[12]) to see the old King, Louis-Philippe. He told them that when he had arrived in England he did not know whether he owned *anything*.[13]

Her contemporary politicians and journalists could wonder whether there would be a revolution in Britain, just as historians could decide why there wasn't one. Victoria, understandably enough, could not but contemplate the personal fates of the dethroned, the undignified and hurried exit from a palace, the dangerous journey into exile; Lear on the heath, suddenly exposed to 'feel what wretches feel'. And there were plenty of 'wretches' in Victorian Britain, not only in Ireland.

It was with them in mind that Albert agreed to become the President of the Society for Improving the Condition of the Labouring Classes, which had been the brainchild of the evangelical Tory philanthropist Lord Ashley. When Ashley had introduced his Mines and Collieries Bill in Parliament, which increased the age at which children could be employed in the pits, Albert had written to him, 'I know you do not wish for praise, and I therefore withhold it, but God's best blessing will rest with you and support you in your arduous but glorious task.'[14] The Prince was vilified by the pit-owners, the press and the stupider politicians for 'interfering' in political life, and for showing 'bias'.

His bias was instinctual and praiseworthy. He could see the vile condition in which many of his fellow citizens were compelled to live, and the Society, although its name might make modern readers smile, was, in those pre-socialist days, the most vigorous instrument for its purpose: which was both to heighten public awareness of such evils as child labour and truly horrific housing conditions for the urban poor: and to do something to correct this. Ashley's Ten Hours Bill reduced the numbers of hours when it was legal for mill- and pit-owners to force labour. Albert's schemes for building sanitary and affordable housing for the poor were of a comparably practical nature and inspired others to do the same.

One says that the Society was pre-socialist. It was also, it perhaps goes without saying, specifically anti-socialist. The aim of Ashley and Albert was to allow working people the dignity of helping themselves, rather than becoming 'welfare-dependent', to use an anachronistic piece of modern jargon. In his first speech to the Society, at a public meeting on 18 May 1848, Albert spelt out the disaster which would ensue if antagonism were fomented between labour and capital, employer and employed.

> Depend upon it, the interests of classes too often contrasted are identical, and it is only ignorance which prevents their uniting for each other's advantage. To dispel that ignorance, to show how man can help man, notwithstanding the complicated state of civilised society, ought to be the aim of every philanthropic person... Let them be careful, however, to avoid any dictatorial interference with labour and employment, which frightens away capital.[15]

He went on to enunciate what could be seen, by our generation, as the quintessentially Victorian free-market creed:

> God has created man imperfect, and left him with many wants, as it were to stimulate each to individual exertion, and to make all feel that it is only by united exertions and combined action that these imperfections can be supplied, and these wants satisfied. This presupposes self-reliance and confidence in each other. To show the way how these individual exertions can be directed with the greatest benefit, and to foster that confidence upon which the readiness to assist each other depends, this Society deems its most sacred duty.[16]

Later in the century, John Ruskin, in *Unto This Last*, would question whether philanthropy alone could correct the unfairness of the market; the idea of benign socialism as the only remedy against the unfairness of things, and the concept of public ownership of resources or even of whole industries, would eventually take hold. But William Morris's socialism and the Independent Labour Party lay in the future, and would not, in any event, ever have recommended themselves to Prince Albert. In making so public an alliance with Peel's campaign to abolish the Corn Laws and with Lord Ashley's belief in the possibility of bettering the system, without destroying it, Albert had linked the monarchy itself to the early-to-mid-Victorian ideal of free-market capitalism. It was a bold stroke, and, as it happened, a fortunate one. He was harnessing the monarchy to the Victorian success story.

Surveying the situation in Europe, the Prince could see that 'in the long run, it is the troops who must be the instruments for restoring peace and order in Germany… [as in Italy]. In Paris there has been a bloody conflict, and I should like to learn something of the impression which this has made in Germany. As monkey-like imitation of the French is unhappily a leading characteristic of the Teutomania-inflated heroes of the new epoch, military restraint should be introduced, and the workers' union scattered to the winds'.[17] He made no secret of believing that the Chartists should be treated with similar severity at home, if they got out of hand. The point of supporting Ashley's society was to neutralize working-class discontent and to remove its causes.

The first model houses created under Albert's scheme for finding decent accommodation for the poor were designed by Henry Roberts, a pioneer of model housing, and they were to be located near the site of the big 1848 Chartist meeting, in Kennington Park. (As an example of model housing, it was re-erected outside the Great Exhibition in Hyde Park in 1851, and was later used to house the park-keepers.) The block contained four dwellings, two on each floor. Each had a living room, three bedrooms, a lavatory, a scullery fitted with a sink, a coal bin, a plate rack, a dust shaft, a meat safe and an airing cupboard in the living room which was heated from the fire. Hollow bricks were used, believed to be dry, long-lasting, fireproof and cheap. The interior facing of the walls was so smooth as to make plastering unnecessary. The total cost of the building was £458 14s. 7d. The design was later used in Cowley Gardens, Stepney, and Fenelon Place, Kensington.

The wide variety of Albert's talents was now on display: his fascination with design and modern technology; his intense intellectual seriousness, and his belief in the university as a vehicle for social change; his sense of change in the air; his awareness of Britain's evolving class system; his social conscience. All these factors reshaped the relationship between the British and their monarchy. He was making a role for the monarchy which would be reproduced through subsequent generations. It is difficult to think of any other individual who could have fulfilled so many roles at this date in British history, with so much energy and aplomb. These roles – his success at Cambridge, as a patron of social housing for the poor, as a follower of and encourager of modern technology, as well as a designer and an overhauler of the royal estates – were roles which he had created for himself. And he showed, in his friendly collaborations with Peel, with Grüner, with Whewell, and with so many others, his perfect suitedness for them.

Where he was doomed to perpetual frustration was in the role which had been prepared for him since birth by Baron Stockmar and by the Coburgs. They had never fully grasped – because they did not understand Britain – the very strict limitations which had been placed on the monarchy by the ruling power – in reality since 1689, and ever more so since the madness of King George. Because Britain has an unwritten constitution, and because the political climate in Britain had undergone such changes in the 1840s, this lack of understanding was very comprehensible. In the tussle for influence between Queen and consort, Albert could sometimes feel himself victorious – though Peel, in his exhaustion, had complained of the unnecessary effort involved with answering *both* Victoria and Albert's letters. (Many a government minister at this time would have echoed the sentiment.) In real terms, while the views of Victoria and Albert might be taken into account (sometimes) by the politicians, they exercised nothing which Stockmar would have recognized as real power.

In the area of foreign policy, the impotence was all the more marked. This was partly because Albert felt so strongly about the future of Germany, and had, in reality, almost no influence upon any of the major players – not upon the Emperor of Austria, not on the King of Bavaria, nor on the King of Prussia. He deprived himself of sleep writing out drafts of what should happen in Berlin or Schleswig-Holstein, but no one seriously was going to follow his counsels.

There was another factor at work, however, when Albert looked abroad, and that is summed up in two words: Lord Palmerston.

Palmerston's career as a meddler in international affairs might have been chequered with colourful failures, but it was a good deal more successful than Prince Albert's, whose success as an international meddler never really began. Palmerston was quarter of a century older than Albert. Where Albert was punctilious, Palmerston was slapdash. Where Albert was a Puritan, Palmerston was a roué, an unashamed Regency rake, deep into the Victorian age. Slapdash as he could be, Palmerston nonetheless mastered his brief when this was necessary. He was a good linguist. He had known, or lived through the careers of, most of the major statesmen in Europe. His prejudices – distrust of France and Russia, hatred of the autocracy of Metternich and the Austrian Emperor, sympathy with republicans, so long as they were foreign republicans – did not match Albert's own.

A hatred of Palmerston, and a longing for him to quit the political scene, dominated all Albert's adult life. Though he scored a few hits, most notably in 1851, when he helped to force Palmerston's resignation over a characteristic bit of diplomatic indiscretion, Albert was doomed to failure. Palmerston, who had been a lord of the admiralty in the Napoleonic Wars and whose career went back to an era which antedated Albert's life by a decade, would live to be a Prime Minister.

Palmerston had second-guessed what would happen in Austria – which he called 'an old woman', 'a European China' – during the Year of Revolutions, and somewhat crowingly made clear that he imagined Metternich and the forces of reaction would be routed. Thereafter he was hated by the Austrians, whom he described, in his bluff way, as 'really the greatest brutes that ever called themselves by the undeserved name of civilised men'.[18]

Typical of Palmerston's diplomacy was his reaction when Kossuth and the other leaders of the Hungarian revolt against Austria took refuge in Turkey with over three thousand men, who included the Polish 'rebels' against Russia. The Emperors of Russia and of Austria demanded from the Sultan of Turkey that the rebels be extradited. The Sultan refused.

Palmerston had already, in the Year of Revolutions, egged on the Italian, Polish and Hungarian nationalists against their Imperial oppressors, and encouraged the creation of an independent northern Italian kingdom. He instructed the British Ambassador in Vienna, Lord Ponsonby, to record the British indignant 'disgust' at the severity with which the rebellions were put down. When the Russians issued an ultimatum to the Sultan that there would be military reprisals if he failed to surrender the refugees, Palmerston and the British Minister at the Porte, Sir Stratford Canning, urged the Sultan to stand firm. Palmerston's favoured diplomatic device, the gunboat, was put into action. French and British squadrons sailed up and down the Dardanelles while Palmerston urged the Austrians and the Russians 'not to press the Sultan to do that which in a regard for his honour and the common dictates of humanity forbid him to do'.[19] When Baron Brunnow, for the Austrians, complained that they felt threatened by the British warships, Palmerston replied that sending the boats had been 'like holding a bottle of salts to the nose of a lady who had been frightened'.[20] It was a successful gamble. Within two weeks, Russia and Austria had withdrawn their demands for the extradition of the rebels.

Such antics – theatrical, showy, patriotic, dangerous and fundamentally on the side of the underdog – hugely endeared Palmerston to large sections of the British populace, while causing disquiet to all the more 'sensible' political classes in Westminster, especially to Palmerston's agonized old Harrow school contemporary, Lord Aberdeen, a statesman held in high regard by Albert and Victoria.

By antagonizing Austria, Palmerston had also made even less likely the prospect of Albert's dream coming true – a united Germany in which the Austrians would balance the power-mania of the Prussians. Albert himself always followed Stockmar's lead when German politics were in question, and, given the turbulence of the times, it was not surprising that Stockmar found himself performing a volte-face. Before 1848, the Stockmar line, followed by Albert, had been that all the German small states, together with Austria, Prussia and Bavaria, should somehow work towards a union. After 1848, Stockmar realized it was unrealistic to hope that Austria would ever be part of such a scheme and he pursued an alliance between the Three Kings – Prussia, Saxony and Hanover – and to oppose the sovereignty of the 'lesser kings'. Needless to say, in following this line, Albert found himself directly opposed by his brother Ernst in Coburg-Gotha and his brother-in-law Prince Charles of Leiningen. He and Stockmar continued to devise what they considered appropriate constitutions for the new Germany, but there was no reason why anyone in power in Berlin should heed them.

Albert believed he had a special influence since, in March 1848, the Prussian Crown Prince, Wilhelm (brother of Friedrich Wilhelm IV, future King of Prussia and first Emperor of Germany), came to London to escape the revolutionaries in Berlin. Palmerston was wary of the friendship between the two Princes and felt it was diplomatically unwise to ask Wilhelm to be a godfather to Princess Louise, who was christened on 13 May. (The Royal Family defied the Foreign Office over this, and Wilhelm was one of the godfathers.) Palmerston's reasoning was that Wilhelm at this point fell between three stools, being out of favour with his brother the King, because of his sympathy with liberal Stockmarish ideas for German unity, and with the radicals for being royal, and with the small states, such as Coburg and Leiningen, for being too autocratic. The truth is that there was never going to be a solution to the German question which satisfied everyone in Germany, let alone in Europe. Palmerston's task, as he saw it, was to protect British interests and he could not foresee the eventual success of the Prussians, through 'blood and iron', in uniting Germany.

Apart from the deep divergence between Palmerston and Albert over matters of foreign policy, however, there was something more instinctual, a visceral dislike. It was all the more painful that Pilgerstein, as Albert (a little heavily?) nicknamed him – Pilger being a Palmer or Pilgrim in German, Stein a stone – was closely related to those whom Victoria and Albert revered. Ashley, the future Lord Shaftesbury, whose philanthropic views so impressed Albert, was Palmerston's step-son-in-law.[21] That is, Ashley had married the daughter of Lord M.'s sister. His wife was the child of Emily Lamb, later Lady Cowper, later Lady Palmerston. So it was, when Lord Melbourne came to die at the end of 1848, the royal pair had to hear the news from their hated Palmerston. He wrote from Brocket Hall, as Lord M. was dying, 'Viscount Palmerston is here engaged in the melancholy occupation of watching the gradual extinction of the lamp of life of one who was not more distinguished by his brilliant talents, his warm affections, and his first-rate understanding, than by those sentiments of attachment to your Majesty which rendered him the most devoted subject who ever had the honour to serve a Sovereign.'[22]

It was a letter which announced the extinction, not only of Lord M.'s lamp of life, but also one of Victoria's two pre-Albertian allies and friends. Lehzen had been gone for six years from the Queen's life. Lord M., in reality, had been absent for just as long. His death could only serve as a reminder that, in Victoria's inner world, Albert was now totally in the ascendant. She had no figure who could be her confidante and friend. Only in her heartbroken widowhood would she ever regain such figures, who were nearly always in the position of trusted servants as well as confidential friends – Brown, Disraeli, Abdul Karim, to a lesser extent Augusta Bruce (later Stanley) and Randall Davidson. Key and shared features, present in all these figures, which had also been qualities shared by Lehzen and Lord M., were that they accepted Victoria's superior and royal status; they had spontaneously affectionate feelings for her;

and they had a sense of humour. Albert, all but devoid of humour, resented her royal superiority, and cut her down to size in all other respects. Though he might have been sexually passionate about her, at least in the first decade or so of the marriage, he was not spontaneously affectionate. In any event, affection, with him, always – whether directed to his wife or his children – went hand in hand with control. We are entering a phase, therefore, in which Victoria, while doting on her Angel, and looking up to him in every area, also deeply resented him. Each new pregnancy and birth occasioned deeper depressions and resentments. Her ups and downs, which had always been a feature of her character, which Lehzen had learned to manage with such aplomb, were always Cresta Runs of which the emotionally cautious Albert, fearful of 'scenes', could never attempt. The marriage was entering choppy waters.

Neither of them had anything which even approached therapy or help. Sir James Clark, the Queen's chief physician or 'physician-in-ordinary', was completely unsympathetic to what he regarded as 'hysterical' symptoms. As often happens in a domestic situation where all is not entirely well within a marriage, they restlessly looked to a change of stage scenery for the royal drama.

Prince Albert's diary for 1848 reads like a sped-up chronicle of the Year of Revolutions. It relates almost entirely to public events. '21 April Queen and Prince take the Sacrament at Whippingham. WARS DECLARED BETWEEN Germany and Denmark. Dreadful scenes in Poland.'[23] In May, we read, 'Elections in France moderate bloody risings in Rouen etc, risings in Germany.'[24] On 24 June, 'Dreadful rising in Paris between the ouvriers and national guard.'[25] 25 July, 'The Queen and Prince oppose strenuously Ld. Palmerston's wish to mediate with the French Republic in Italy.'[26] By 9 August, 'The Austrians beat the Sardinians and drive them across the Adda.'[27] By September, 'Dreadful revolution in Frankfurt',[28] and, on the 24th of that month, 'Louis Napoleon chosen in Paris.' On 9 October, 'The Prince and Queen have much trouble with Lord Palmerston', and on 11 October, 'New dreadful revolution in Vienna.'[29]

But in that month of October, in which the princely diary was so much concerned with the convulsions shaking the Continent, there is one entry of an entirely different character, an entry which is local and specific. 'October 21. The Prince prosecutes Mr Judge respecting the Publication of the Royal etchings. Meague is also found guilty.'[30]

It was a strange little story, highly revealing of a number of aspects of the Prince's character. The first thing of which it reminds us is his, and the Queen's, creativity. For some time, Victoria and Albert, who were keen painters and draughtsmen, and who had enjoyed taking likenesses of their family and of courtiers and friends, had begun to learn the process of etching. They made prints of their efforts, and from October 1840 to November 1847 they had sent the copper-plates to a Windsor printer, J. B. Brown, whose journeyman, Middleton (no relation, as far as research throws light on the matter, to the later Duchess of Cambridge), made the impressions.

Middleton, before returning the prints to the Castle, had been in the habit of running off copies of his own on card and paper. When a collection had been assembled, and it became known in the trade what he had been doing, he was approached by one Jasper Tomsett Judge, who arranged for these prints to be published. The publisher selected was William Strange, Paternoster Row, chiefly known, as emerged in the legal proceedings against him, for publications of 'a very Radical if not Red Republican nature'.[31]

Strange advertised the publication as 'A Collection of Etchings by Our Queen and her Royal Consort under the title THE ROYAL VICTORIA AND PRINCE ALBERT GALLERY'.[32]

Both the Queen and the Prince expressed themselves 'singularly astonished and sorely displeased' by the proposed publication, and it was, indeed, the most insolent act of piracy, since Strange had not consulted the artists nor sought their permission to publish their work, still less, of course, offered to pay for it. Judge, who was the go-between, was visited by the Town Clerk of Windsor on 15 August 1848 with a writ, demanding that he pay to Prince Albert 'or the bearer of these presents, £181 1s. 8d'. These were the legal costs which had hitherto accrued and some damages. The firm of White, Broughton and White in fact ran up costs of £970 11s. 9d. for the Prince and £869 4s. 11d. for the Queen, which included £1 14s. for the cost of a strongbox in which to contain all the necessary paperwork. Judge, being unable to find the cash, was incarcerated in Reading Gaol. The Queen declined to bring prosecutions, but the Prince proceeded and the case was heard in the Vice Chancellor's Courts on 6 November, with Albert's attorney, W. M. James, requesting that the defendants be enjoined from exhibiting the works or selling or otherwise distributing the catalogues, and that all impressions be gathered up and returned to their rightful owners, Victoria and Albert.

The Chancery suit was not fully wound up until the following summer, 1 June 1850, when the injunction, forbidding further publication of the works, was made permanent. The case, which was a highly satisfactory one to the learned members of the legal profession who were fortunate enough to be engaged upon it, raised a number of nice points. The Queen chose not to be a party to the action. Yet many of the etchings were reproductions of work by the Queen and the collection might be said to have been all her property. Albert had legal standing, as the Queen's husband, to go to law on her behalf to reclaim… her property, or his? This was one of the questions which occupied wise heads, but not before Strange had wisely conceded his mistake and got out with the minimum of costs and Judge had, less wisely advised, attempted to plead innocence.

The case, however, did little to enhance Albert's public reputation, even though he was plainly in the right from a purely legal point of view. 'I have been compelled to pawn the very blankets off our beds,' complained Judge,[33] whose wife wrote personally to the Queen begging for her to get the man released from prison (which Victoria did).[34] Albert had originally guessed that the etchings had been stolen from Windsor Castle by an unfaithful servant, and this was one of his reasons for pursuing the matter through the courts. Only as the case dragged on and the legal costs soared did it become plain that it was Middleton who had taken the copied etchings from Brown and sold them to Judge for £5. Judge claimed, rather improbably, that he had always intended to seek permission for publishing them, but he had already sold them to Strange without any such permission being sought.[35]

As it happens, it was an invasion of privacy, and it was an act of theft and piracy. Albert had been within his rights to sue and justice had been done. But, as well as seeming, in the eyes of many, to be a privileged German Prince being prepared to send a poverty-stricken British printer to prison for what was, after all, harmless, it was a case which shimmered with the ambiguity of Victoria and Albert's position. True, these particular etchings had been done privately for private consumption. On a broader scale, however, Albert would play the dangerous game of wanting images of the Royal Family, including some of these very etchings, to be made public. Pictures of himself and his children enjoying Christmas at home, when published in the *Illustrated London News* with his authority, would be in order. Photographs of the Royal Family were among the first photographs ever hung on the walls of a public exhibition. Neither he nor Victoria were above using intimate visual material to enhance their public and political position. Like many public figures since, especially in the Royal Family, they had wanted the benefits of public exposure but not its disadvantages. The 'victory' in the courts had not done their reputation any favours.

The diary was not a journal, it did not lay bare the Prince's soul, it merely chronicled and catalogued engagements and events. One entry – 'The Prince prosecutes Mr Judge' – is as neutral as another. But it was what he wrote on 13 July 1849 which would lift his reputation from the doldrums and ensure his place of honour in history: 'The Prince sees Mr Labouchere and discusses with him the place for a Great Exhibition.'[36]

FOURTEEN

BALMORAL

On 8 October 1847, Sir Robert Gordon – Lord Aberdeen's brother – was eating his breakfast at Balmoral, the modest Deeside castle which he had leased from the Earl of Fife for the previous seventeen years. He had been a conscientious tenant. He had cleared Glen Gelder to create a deer forest and engaged an Aberdeen architect, John Smith, to enlarge the late medieval tower house in the Jacobean style.

It was fish for breakfast, and Sir Robert, who was a healthy man of fifty-six years old, ate with a little too much appetite. A bone stuck in his throat. He choked to death.[1]

Lord Aberdeen, Foreign Secretary in Peel's Cabinet, had been out of office for over a year. He had inherited an enormous estate, Haddo, from his grandfather, when he was only sixteen. Looking back in 1858, he reckoned he had planted over a million trees there.[2] He had no need of the 20,000 acres of Balmoral; indeed, he became immediately anxious to find another tenant to take over the lease.

The royal pair had first visited Scotland in 1842. It had been a great success. They had returned two years later to stay at Blair Castle, then known as Atholl House. In 1847, they had gone again, this time as the guests of the Marquess of Abercorn, at his estate of Ardverikie on Loch Laggan. A fellow guest had been Sir Edwin Landseer, who, between and during the incessant showers, gave the Queen watercolour lessons. She was an accomplished, splashy, colourful painter, way above average amateur skill. Recognizing their enthusiasm for long walks in the rain, Highland Games, deer-stalking, bagpipe music and, in Victoria's case, the cold, Aberdeen recommended to them the joys of Deeside. Aberdeen is one of the hardest Victorian statesmen to read – as oblique as his Harrow contemporary Palmerston, whom he so much disliked, is transparent. Diffident, where Palmerston was impulsive; unwilling to intervene or take risks in foreign policy; personally gentle.

Yet few had been closer to Peel, or more able to see the toll which was taken on the Prime Minister's time and energy by his dealings with Prince Albert. As Peel observed, all the important correspondence of a Prime Minister at the time had to be written out by himself. It could not be left to secretaries. When the Prince insisted upon topping up the correspondence which Peel had already had with the Queen with clarifications and emendations, he hugely added to the workload. The Peel Papers are weighed down by the sheer volume of the Prince's outpourings – on his life as Chancellor of Cambridge, on the Schleswig-Holstein Question, on the future of German unification, on France, Spain, Italy, Portugal, on the condition of the working classes, as well as on the day-to-day running of the Government. Aberdeen, like Peel, was fond of the Prince, and of the Queen. This would not have stopped him from perceiving the advantages of 'peace, perfect peace, with lov'd ones far away'.

While Victoria and Albert were on the Isle of Wight, the railway service from London made it all too easy for them to summon government ministers, or on occasion whole Cabinets and Privy Councils, to their seaside retreat. Aberdeenshire was – who knew it better than Aberdeen? – a very long way from London. Would it not be worth the trouble caused to future statesmen to know that the occasions on which they could often be summoned so far north must, by the sheer geographical facts of the case, be fewer than those which could justify their being expected to journey to Southampton and East Cowes?

Aberdeen was a kind man, and he rejoiced in the genuine love which the Queen and the Prince had for the scenes, sports and pastimes of Scotland. Nevertheless, P. G. Wodehouse's ingenious

Jeeves himself would have been proud of himself if he had persuaded the workaholic Prince to purchase a holiday home almost as far from Westminster as it is possible to be in the main island of Britain.

Lord Aberdeen sent the Scottish landscape painter James Giles to London with a selection of tempting watercolours depicting Balmoral and its charming environs. The Queen and her husband were assured that Deeside was dry and sunny,[3] a meteorological judgement so at variance with experience that it is hard not to sense in it the urgency of Aberdeen's motives.

There had been talk, in the Year of Revolutions, of a royal visit to Ireland. As events unfolded, however, it was considered too great a risk. Lord Clarendon, Lord Lieutenant, came over to England and described the state of things to the Prince, and Albert dutifully copied it out as a memorandum: 'The rebellion has been put down, but the spirit among the people is still the same, and any agitator will have them all at his command. Arms are concealed, and murders and outrages of every kind happen daily – even highway robbery, a crime hitherto quite unknown in Ireland.'[4] It was obviously more tempting, in their zest for exploring Celtic lands, for the Queen and her husband to revisit Scotland. It was agreed that they should rent Balmoral, and, if the visit were a success, that they should take on the lease.

Ever since the failure of the 1745 Jacobite Rebellion, the lives and livelihoods of the Highlanders had been under threat. Walter Chalmers Smith's poem 'Glenaradale' tells the story of the emigrations of Scottish Highlanders to America. The glen has become the playground of the Victorian nouveaux riches.

> But the big-horn'd stag and his hinds, we know,
> In the high corries,
> And the salmon that swirls in the pool below
> Where the stream rushes
> Are more than the hearts of men, and so
> We leave thy green valley
> Glenaradale.[5]

'Scotland is overrun during the shooting season, not by English squires, for they have not the means, but by wealthy stockbrokers, by the heads of large establishments in London, by the owners of funded property'. This judgement, by Charles Milnes Gaskell, was expressed in 1882. He was looking back at a half century in which the new rich commercial classes had, in effect, destroyed the life of the Highlands. Victoria and Albert were part of this trend. Blood sport on the quasi-industrial scale it was practised by the millionaires (by 1872, one tenth of Inverness-shire had been converted to deer forest) had never been part of Scottish life until Victoria and Albert helped to popularize it.[6]

On 5 September, with the three elder children, they went aboard the royal yacht in intense heat and began the voyage north, Bertie and Affie dressed in sailor suits.[7] The weather became markedly more challenging as the ship went north, and by the time she docked at the city of Aberdeen, at 7 a.m. on 7 September, they had all been seasick. Victoria was torn between blaming the movement of the ocean and the behaviour of the Foreign Secretary for her bilious condition, expressing the view that she would be ill until Palmerston resigned. Although they were supposed to be on holiday, Albert and Victoria could not banish from their minds the many respects in which Palmerston's foreign policy enraged them. In particular they were infuriated by Pilgerstein's open encouragement of the French, who were siding with the Italian nationalists in their struggle to throw the Austrians out of northern Italy. Britain had offered to intervene, Austria declined the offer, and Palmerston omitted to tell the Queen – which was a further cause for rage. 'It is another question whether it is good policy for Austria to try to retain Lombardy, but that is for her and not for us to decide. Many people might think we would be happier without Ireland or Canada.'[8]

The Queen wrote to the Prime Minister about the matter on a daily basis – before she left England, on board the royal yacht, from Aberdeen, and from Balmoral itself. Such was the joy of being once more on Scottish soil, however, that they did compensate somewhat for the miseries of the political situation. On 8 September, they had their first glimpse of 'what is called Deeside', finding Balmoral to be 'a pretty little Castle in the old Scotch style'.[9]

They soon settled into a happy holiday routine. In the mornings, Victoria took upon the teaching of the children herself[10] while Albert went in pursuit of the deer. The first day they watched him fail to shoot a roe. A whole week passed of very cold wet weather, during which Albert tried every day, and failed, to shoot a deer. It was on 15 September that he shot a stag 'at last'.

Occasionally, the weather lifted. Then, 'we went for a delightful walk of about five miles. I felt as if I could have walked another five. We got up higher even than Craig Guran, and the views were glorious. It was wonderful, not seeing a human being, nor hearing a sound, excepting that of the wind or the call of blackcock or grouse.'[11]

On the 15th, the day the Prince shot his first stag, the Prime Minister arrived. Before he went down to dinner, Lord John Russell wrote a line to the Queen informing her that there had been a rising in Ireland in Waterford and Wexford. For the rest of her life, this routine would continue: for a short part of each autumnal visit to Deeside, she would be visited by her Prime Minister, and, except in the case of the real intimates, Disraeli and Rosebery, she would chiefly communicate with them – at any rate when matters of state were in question – on paper.

If they had hoped to share with Russell their love of the distant views, they were disappointed, for Lochnagar remained swathed in wet fog for the whole of his visit. He spoke to them 'at some length and very judiciously, about the measures to be tried for unfortunate Ireland'. There were other visitors. Charles Lyell came to dine on 19 September, and, noted Victoria, 'I knighted him afterwards with Albert's claymore'.[12] It was an appropriate instrument, for Lyell was a typical Prince Albert choice.

By knighting him, they bestowed a blessing not only on Lyell, but on his great enterprise. Lyell's *Geology*, almost more than any other book (save the anonymous *Vestiges of Natural Creation*, 1844, which popularized Lyell, as well as the evolutionary theories of Lamarck and the cosmology of Herschel), had laid to rest the view that the Bible was a reliable guide to science. Counting backwards through the different generations recorded in the Scriptures, Archbishop Ussher, in the seventeenth century, had concluded that the world was created in 4004 BC, followed shortly thereafter (9 October of that year, according to the learned Archbishop) by the Fall of Man. After Ussher's endeavours, it became customary for the printers to put down on the page the actual dates of the events described in the Book of Genesis. It therefore became customary, in the eighteenth century, for literalists to suppose that these dates were matters of faith. You have only to read Tennyson's response to Lyell's ideas, in his agonized verses in *In Memoriam*, to realize the impact which the great geologist had. Not only did Lyell reveal that the earth was infinitely older, millions of years older, than Ussher could ever have guessed. He also shared with readers the fact that the fossils told a story. Millions of years older than human beings, creatures had walked this earth which were now extinct. They had been, as Tennyson said, 'cast as rubbish to the void'. Where did this leave the idea of a loving Creator who cared so much for his world that not a sparrow fell to the ground without his noticing it?

Knighting Lyell was a sign that Victoria and Albert sided with the educated and enlightened minority. Some of this minority would continue, in a modified form, to accept the Bible as the word of God. Others, perhaps the larger share of Victorian intellectuals, would take leave of Christianity altogether. Albert was a believing liberal Protestant, a Lutheran who had absorbed much of the philosophy of Fichte, and via Fichte, Hegel – in particular, the Hegelian sense that in any generation the truths imparted from the last are reinterpreted. Truth is fluid.

All too soon, the weeks at Balmoral flew past. On one of the evenings the guests all waited for Albert to come to dinner, but the meal started late because of his 'struggles to dress in his kilt'.[13]

(What are valets for?) They had hoped, on their last day, to climb the brae and take in the glorious views, but 'the mist hung thickly on Lochnagar'[14] and Albert had caught a bad chill. When they left, the staff lined up with home-made banners, one of which read, 'More beloved than ever, haste ye back'.[15] It was an injunction which they heeded.

Having suffered so badly from seasickness on the way north, they thought to sail down only as far as the Firth of Forth and to go the rest of the way home by train. The weather was so foul, however, that they went by road to Perth and caught the train to Crewe – the children being 'very riotous and fidgety in the carriage'. The night of 30 September was spent in the Station Hotel, Crewe – 'very good, & quite sufficient for our needs'.[16] The train from Crewe got them to Euston at ten on Sunday morning and they came out through Philip Hardwicke's magnificent Doric arch into the beauty of London sunshine. They travelled to Buckingham Palace without an escort, so as not to be noticed. 'People are so very particular about not travelling on a Sunday in England,' Victoria told her journal. 'In my opinion it is overdone.'[17] They spent almost no time at all in London. The very next day, they sped to the Isle of Wight, where the three youngest children, Victoria's mother, Victoria's sister Feodore and her brood of four were all awaiting her. A real Coburg reunion.

As the Year of Revolutions drew to its close, Albert's reaction was summarized in a letter to his stepmother: 'You in Germany are entering upon a new era, stained with blood. Poor country! How many men have perished since March 1848! How many millions have suffered! And I should like to find one single person who was better off than he was before!'[18] To this could be added some words he wrote to his old tutor Florschütz in the autumn of 1849: 'In Germany, things still look dark & threatening. You will have to rekindle your light from England's torch; for she still has sound ideas of Freedom and Justice. Freedom is a negative idea & means to be "free from the caprice of men" whoever indulges in it – whether a tyrant of the people. If only men would remember this!'[19]

In that same letter, written in response to greetings from Florschütz on Albert's thirtieth birthday, the Prince had replied, 'I may count myself fortunate in having been able to be of use to the world early in life, at least I have tried to be so. And if my position demands that I should do the best I can behind the scenes. Yet I can already see some blossoms sprung from the seed I have sown, & have even gathered some fruit.'[20]

Over the next two years the busyness, far from letting up, intensified, and there would be a danger, from now onwards, of our story becoming a simple catalogue of memoranda drafted, foundation stones laid, civic banquets in provincial towns pushed dyspeptically around dinner plates, and meetings chaired. In addition, he kept abreast of world news as if he were the Head of State. In a long memorandum written in 1852, he utterly rejected the notion that sovereigns could or should cultivate indifference to political events. 'Is the Sovereign not the natural Guardian of the honour of his Country? Is he not necessarily a politician?'[21]

Nearly all the actual politicians of nineteenth-century England, and especially those who were Albert's friends such as Peel and Aberdeen, would have answered this question in the negative. The next eleven years, the last of the Prince's life, therefore become a description, not only of his own political ambitions (alongside his many other activities and achievements, as musician, philanthropist, patron of the arts, family man) but also of the politicians' aptitude in keeping those ambitions under control, diverting them, making use of them, but never allowing them to stray into the sphere of direct party or domestic politics. As far as foreign politics were concerned, things were a little different, especially when Albert's children became old enough to become pawns in his schemes for European geopolitics by means of dynastic marriage.

If it is true that the British 1848 Revolution happened in Ireland (see Foster), then the royal visit to that island, which took place immediately after Parliament was prorogued (1 August 1849), was one which provoked extreme anxiety, both among the statesmen and at Court.

Only in May, there had been yet another assassination attempt on the Queen, this time by an unemployed Irish labourer called William Hamilton. Some thought it was scarcely worthy of the name of an assassination attempt, since his gun had not been loaded, and the press on both sides of St George's Channel wanted to play down his ethnicity. The *Daily News* wrote, 'The accident, or the fact, of the man Hamilton's being an Irishman may be made the theme of animadversion, and conclusions may be drawn from it of the international hate or savage vindictiveness of the Celt. But the Irish elements which have contributed to his crime, will probably be found more those of poverty and vanity, than any thing more peculiarly malignant or Celtic.'[22]

Laying aside the offensive identification of 'Celtishness' and malignity, many Irish Victorians would have agreed with this. The Irish depicted by Trollope in *The Kellys and the O'Kellys* or in *Phineas Finn, the Irish Member*, were prosperous, intelligent people, as middle class as their English friends and no more likely to commit political outrages. Nevertheless, Lady Lyttelton, left behind on the Isle of Wight with the three younger children (the four eldest went with their parents), watched anxiously from the windows of Osborne House as the Royal Squadron steamed out of sight towards Cork. It arrived sooner than expected, in the fading light of a summer evening. As it did so, it was greeted from the shore by bonfires and showers of rockets. The feared hostility did not materialize, but the royal pair misread the situation. They thought that one week of hallooing and cheering by carefully selected Irish crowds meant that Irish problems had gone away. The next few decades would make clear that this was not the case. 'Oh Queen, dear!' screamed an old lady in the crowd, as the royal party trotted from Sandymount Station to the Viceregal Lodge in Phoenix Park in Dublin – Albert, Victoria, Vicky, Affie, Alice – 'make one of them Prince Patrick, and all Ireland would die for you.'[23]

'The reception of the Queen,' wrote her doctor, Sir James Clark, 'has been most enthusiastic from the moment of her first setting foot on Ireland to her quitting it... The effect produced by the visit was most salutary.'[24]

As far as Albert was concerned, the immediate consequence of the visit was that Lord Clarendon, the Lord Lieutenant, asked the Prince to be involved with the establishment of a new Irish university, to cater for the Presbyterians and Roman Catholics who were excluded from membership of the exclusively Anglican Trinity College Dublin. The Belfast Academical Institution, founded in 1810, had morphed into Queen's College, Belfast in 1845. Together with Queen's College, Cork and Queen's College, Galway, it formed part of the Queen's University of Ireland, which was formally opened in that year of the royal visit, 1849. Albert would have been an obvious choice as Chancellor, not least because he had demonstrated such a keen interest in university reform and administration at Cambridge. He regarded Clarendon's suggestion, however, as impracticable, not wishing to take on the chancellorship if he would be so seldom on the spot. On top of his many other commitments, frequent visits to the fledgling three colleges would have been impossible. Clarendon himself became Chancellor. Whereas Oxford, Cambridge, Durham and the Scottish universities remained autocephalous and independent of government, the Irish universities, until the twentieth century, were funded and administered at government level; it was an attempt by Westminster to undo the 150 years in which the law had excluded Catholics from higher education. In England, this affected the relatively small 'recusant' Catholic population who, since Catholic Emancipation (1829), could now attend universities and Inns of Court. In Ireland, it had excluded the majority, forcing Catholics either to forego a university career, or to apostasize. After 1829, Catholic MPs pointed out the many anomalies in the law which still existed, such as the illegality of holding Catholic funerals in public.[25]

One of the things which impressed Albert, during his short visit to Ireland, was the intelligence, and moderation, of the Catholic clergy. He had especially liked Archbishop Murray, the eighty-two-year-old Archbishop of Dublin, a venerable figure with white hair grown long over his shoulders, who had spoken up for the National Model

Schools, which Albert deemed excellent institutions. He saw Murray as a moderate who held up educational values 'against the bigoted opposition of others of his creed'.[26]

Shimmering through the whole story, however, like the pattern of the magenta watered silk in the Archbishop's cincture, was the curious history of anti-Catholic prejudice which had poisoned Anglo-Irish relations for centuries and which in England this year came to the surface with remarkable force.

It was eight years since Dickens had published his terrifying evocation in *Barnaby Rudge* (1841) of the anti-Catholic riots, whipped up in 1780 by Lord Aberdeen's fanatical collateral Lord George Gordon. These violent feelings of disquiet were always hovering in the background of English life. They had shaken Oxford for a decade, while John Henry Newman first tried to persuade himself and his followers that it was possible to be a loyal Catholic within the Church of England, and then, when he changed his mind, that it was impossible to be a loyal Catholic outside the Church of Rome. He made the switch in 1845, and four years later, it was hard to know which of the two alternatives the royal pair, the Westminster Establishment, the Anglican bishops and the majority of educated opinion in England found more horrifying.

Then came the bizarre episode of Bishop Wiseman's decision to establish Roman Catholic dioceses in England and Wales. The famines in Ireland and the opportunities for work in the industrial heartlands of England had led to a surge of Irish immigrants, and it was an inevitability that their clergy should have followed them and ministered to them. In London's docklands, where the Irish worked as stevedores and sailors; in Manchester, Birmingham, Leeds, Middlesbrough, a similar story unfolded. Wiseman, whom we met in a previous chapter commissioning Dyce to illustrate missals in Rome, was a learned man who, for example, had read the Karkaphensian Codex in the Vatican Library, the Syriac version of the Old Testament. He was an aesthete and a sybarite, who was said by indulgent friends to 'have his lobster salad side'. The old English Catholics did not like him.

The new Pope Pius IX (elected 1846) told Wiseman that he was considering the creation of English dioceses, set up, as it were, in rivalry to the Anglican ones. Wiseman saw many objections to this scheme, not least the expense of administering such sees, the majority of whose adherents were poor. This was quite apart from the possibility of exciting the 'No popery' gut instincts of the English press and Establishment.

The Year of Revolutions was one in which the Papal States were threatened, Italian nationalism was anti-papal, and it was not the moment to be worrying about the conversion of England, even though this was the year that Augustus Welby Pugin's superb St George's Cathedral was opened in Southwark. In the following year, the Jesuits opened their church in Farm Street in the heart of Mayfair. Sixteen convents of nuns were opened in London, and four male religious houses. Wiseman persuaded himself that the majority of Anglicans, who had been enthusiasts for Newman, would follow him all the way to Rome, rather than dithering in the High Church stance of Newman's friend Dr Pusey. His followers, the 'Puseyites', were seen by Catholics as sham, and by Protestants as Roman fifth columnists. In 1849, the Pope went ahead with the creation of twelve dioceses and the next year he made Wiseman the Cardinal Archbishop of Westminster. On 7 October 1850, Wiseman issued his pastoral letter 'from out of the Flaminian Gate' announcing the new Roman Catholic hierarchy.

It was regarded by the British Government as 'papal aggression', and Victoria and Albert were both caught up in the anti-Catholic frenzy which ensued. Guy Fawkes' Day that year, especially in London, was lively, with many an effigy of Wiseman being burnt. One Anglican clergyman who was unwise enough to go for a walk wearing a cloak was set upon by the rabble until he demonstrated that he was decently clad, beneath the Romish robe, in a good Protestant pair of trousers. In Birkenhead, not only 250 policemen and 700 special constables, but two companies of the 52nd Regiment were needed to still the mob. In sedate Cheltenham, a crowd seized an effigy of the Pope which had been put up in the window of a draper's shop. They carried it to the Catholic chapel, smashed the windows and set fire to the statue.

Lord John Russell's response to the crisis was not to subdue, but to encourage the Protestant extremists. Although his Government was in a state of crisis, he found time to bring before Parliament the Ecclesiastical Titles Act which declared the creation of the new Catholic dioceses to be illegal. Anyone claiming to be the archbishop, bishop or dean of such an illegal see was liable to a fine of £100.[27] Albert saw the advance of Catholicism in England as 'an evil of long standing' which 'will have to be set right by the Protestant body'.[28]

The Queen would retain, long after Albert's death, a greater horror of the Puseyite ritualists within the Church of England than she did of the Catholics themselves,[29] and she did attempt to stay the wilder excesses of Russell's anti-Catholic legislation, asking whether the new laws would not cause embarrassment in the Colonies 'where the Roman Catholic bishoprics are recognised by the Government'. Above all, in John Bull's Other Island, where the Catholic Archbishop of Armagh, Dr Cullen, had already assumed the title of Primate of All Ireland, how could Russell avoid one of two painful choices? Either to appear 'lame' by not prosecuting Cullen, or to cause unnecessary and inevitable resentment by doing so? Despite all these warnings, the legislation did go through, and the Ecclesiastical Titles Act remained for some years one of the strangest laws on the statute book. No attempt was ever made to fine a Roman Catholic for calling himself, for example, the bishop of Salford. Many laws, through the passage of time and the change of mores, become anomalous. Lord John Russell achieved the feat of creating an instantaneous anomaly. The shade of Dean Swift, if aware of these things from his grave in St Patrick's Cathedral, Dublin, must have enjoyed the Lilliputian antics.

1850 would see the birth of Albert and Victoria's seventh child – Arthur, Duke of Connaught. It was a measure of how far they had all come that – the child having been born on the Iron Duke's eighty-first birthday – Albert and Victoria chose Wellington to be one of the godfathers and that the baby took the Duke's name. On their wedding day, Victoria had been so anti-Tory, and so furious at Wellington's mean-spiritedness towards Albert, that she had vowed never so much as to look at him. Now, she and the Duke were great National Institutions, and if Albert did not yet quite enjoy this status, he was on his way towards achieving it. Arthur would always be Victoria's favourite child, and both parents delighted, after the run of girls, in the arrival of another son. Though it was only a decade since his parents had been married, Arthur was born into a very different world from that of 1840.

It was a summer like no other, and marked a great change in their lives, and in national life. The strange drama was dominated by a grotesque incident which showed Palmerston at his most characteristic: if you were of Victoria and Albert's view, Palmerston at his most tricksy, bombastic, dangerous and absurd; if you were the great British public, egged on by sections of the patriotic press, Palmerston at his most swashbuckling and stylish. A Portuguese money-lender by the name of Don Pacifico had his house looted in Athens in what was undoubtedly an anti-Semitic attack. He asked the Greek Government for compensation. Because he had been born in Gibraltar, Don Pacifico asked for the support and protection of the Government in Westminster.

By January, the Greeks had failed to provide adequate compensation to Don Pacifico, and Palmerston, as Foreign Secretary, was able to view the matter as an insult to a British subject. Had he played things by the book, Palmerston would have been obliged to consult France and Russia, as guarantors of Greek independence, about the appropriate course of action. That was not Palmerston's way. He ordered an immediate blockade of the Greek ports by the Royal Navy. In protest, the French withdrew their ambassador from London, a gesture which embarrassed the Establishment, and further enhanced Palmerston's popularity with the country at large.

This was followed by an incident which further demonstrated the gulf between the sanctified protocols of diplomatic life, and the swashbuckling of the Foreign Secretary. The Austrian General Haynau, who had been so powerful a force in suppressing the Italian insurgents against his Empire in 1848, paid a visit to Britain. The public detested him, not least because he had authorized the public flogging of women among the Italian freedom-fighters, or 'rebels' as the Austrians saw them. The press nicknamed him 'General Hyena'. When he was identified by the English crowd, he was pitched into a horse trough at Barclay and Perkins brewery in Southwark – the very brewery which, in the previous century, had belonged to Dr Johnson's friend Henry Thrale. Palmerston issued an apology on behalf of the Foreign Office, but it was a perfunctory one and the Austrians felt the insult keenly. Albert was appalled. 'We in London have in the Haynau demonstration also had a slight foretaste of what an unregulated mass of illiterate people is capable, *le peuple souverain*, which likes to be accuser, witness, judge, and executioner all in one.'[30]

Albert and Victoria spent the summer bombarding the Prime Minister with complaints about the Foreign Secretary, his attitudes and his antics. In particular, they feared the consequences of Britain defending the Danish monarchy in its claims on the Duchy of Holstein. Palmerston was robust in refusing to be influenced by his sovereign, writing to her, 'Is not the Queen requiring that I should be Minister, not indeed for Austria, Russia or France but for the Germanic Confederation? Why should we take up the cudgels for Germany?'[31] He was quite right, of course, Victoria and Albert did indeed hope the British Government would, broadly speaking, support British foreign policy, while attempting to use their influence to bring harmony between Austria and the German Federation; whereas Palmerston's view was that the British Government should pursue British national interests first, foremost and always, and not be shackled by any grand overall view of European *Realpolitik*. For Palmerston, there were no natural allies for Britain; he would tack, this way and that, between the great powers and do whatever he considered suited British interests.

Things came to a head at the beginning of July 1850, when the criticisms of Palmerston, from the royal pair, from the governments of Europe, and from the Tory benches, became so vociferous that the Foreign Secretary felt it necessary to defend himself. John Arthur Roebuck, an extreme Radical MP, proposed a motion 'that the principles on which the foreign policy of Her Majesty's Government had been regulated have been such as were calculated to maintain the honour and dignity of this country'.[32] It was the chance for the Tories, led by Lord Stanley (later the Prime Minister as Earl of Derby), to censure Russell and for all his political enemies to deplore Palmerston. The old rogue delighted in the chance to defend himself. He did so on the night of 25 June, starting at dusk, and speaking for four and three quarter magnificent hours, and ranging from Don Pacifico, to the Austrians, from Schleswig-Holstein to Spain, to Portugal and beyond. Lord John Russell was able to tell the Queen, 'This speech was one of the most masterly ever delivered, going through the details of transactions in the various parts of the world, and appealing from time to time to great principles of justice and freedom. The cheering was frequent and enthusiastic'.[33] A reluctant Albert was obliged to admit, the speech was '*ein Meisterstück*'.[34]

Between Palmerston's triumph in the Commons and the big debate about foreign policy, the Queen suffered yet another assault by a maniac.

Her cousin the Duke of Cambridge was dying, and on the afternoon of the 27th, while Albert was at a meeting of the Royal Commission, she went in an open carriage to see the old man. She was accompanied by 'almost the best of all my ladies' – Lady Frances Jocelyn – who was five years older than the Queen. The two eldest boys, Bertie and Affie, were in the carriage, as was Princess Alice. Colonel Grey and another soldier acted as outriders. The visit to Cambridge House (now 94 Piccadilly, once the Naval and Military Club, called the In and Out Club because of the rather awkward pair of gates, so marked, as one entered the drive) took its course. The old Duke appeared to have rallied. The Queen had a few pleasant words with his heir – her cousin George, with whom she always got on so well – and they made their way out into the courtyard. Because of the awkward arrangement of the 'in' and 'out' gates, the Queen's carriage had to leave the courtyard, on the pavement side of Piccadilly, without the protection of the outrider. Colonel Grey was stuck behind the carriage, still in the courtyard, when a man pushed through the crowd on the pavement.

> A great crowd had collected to see me go out again, & as no one knew that I intended going to Cambridge House, no Police were there. The gate is very narrow, so that the Equerry could not keep close to the carriage, & the crowd got close up to it, which always makes me think more than usually of the possibility of an attempt being made on me. A little in front of the crowd, stood a young gentleman whom I have often seen in the Park, pale, fair, with a fair moustache, with a small stick in his hand. Before I knew where I was, or what had happened, he stepped forward, & I felt myself violently thrown by a blow to the left of the carriage. My impulse had been to throw myself that way, not knowing what was coming next, — till I was caused the moment afterwards by poor Fanny, who was dreadfully frightened, saying 'they have got the man'. My bonnet was crushed, & on putting my hand up to my forehead, I felt an immense bruise in the right side, fortunately well above the temple & eye! The man was instantly caught by the collar, & when I got up in the carriage, having quite recovered myself, & telling the good people who anxiously surrounded me, 'I am not hurt', I saw him being violently pulled about by the people, Poor Fanny was so overcome, that she began to cry. Certainly it is very hard & very horrid, that I, a woman,— a defenceless young woman & surrounded by my Children, should be exposed to insults of this kind, & be unable to go out quietly for a drive. This is by far the most disgraceful & cowardly thing

> that has ever been done; for a man to strike any woman is most brutal & I, as well as everyone else, think this far worse than an attempt to shoot, which, wicked as it is, is at least more comprehensible & more courageous. I ordered the coachman to drive home, & sent off Col: Grey, who was greatly horrified to find Albert, who was riding in the Park with the P^ce^ of Prussia. Went upstairs to my room, to put arnica on my poor head, which was becoming very painful. The Children were much shocked, & poor Bertie turned very red at the time. It is the 2nd time that Alice & Affie have witnessed such an event. Thank God! they are never touched, — that indeed I could not bear with composure. Dearest Albert returned in 20 m. & was dreadfully shocked, annoyed & pained. All my people so distressed, some, quite crying. I went down to meet the P^ce^ of Prussia, who was extremely shocked. We walked about in the garden, talking of the event, & when we came back to the door, Ld John Russell arrived, then Col: Grey came who had been to see the man. His name is Pate, a retired Lieut: of the 10th Hussars, whose only answer, when spoken to, was, that it was a 'slight blow', which I can certainly affirm it was not.[35]

The Queen insisted upon attending the opera that night; they saw Meyerbeer's *Le Prophète*, taking with them Wilhelm, the Crown Prince of Prussia, and Lady Frances Jocelyn, and staying to hobnob after the second act with the stars. One of the sopranos, Madame Viardot, sang the National Anthem and when she came to the second verse and the words 'frustrate their knavish tricks', the crowd erupted into applause, what *Punch* called 'a deafening tumult of love'.[36]

It was indeed a very unpleasant episode. Pate was eventually sentenced to penal servitude in Australia, but within five years he had been pardoned, and returned to England where he lived the life of a 'country gentleman' in the still relatively rural village of Croydon, Surrey. In the ordinary course of things, his extraordinary attack on the Queen's person would have absorbed the national attention for weeks, but these were not ordinary times.

The debate which followed Palmerston's speech took place in the House of Commons during the night of 28 June. In the division which followed, the Government had a majority of forty-six. Among those who had spoken, and voted, against Russell and Palmerston was Peel, who spoke against the deliberate antagonizing of foreign powers.

Peel walked home in the light of dawn. He spent a quiet day, attended a meeting chaired by Prince Albert in the afternoon, and at five went out for a ride. The horse chosen was an eight-year-old which he had purchased fairly recently. A previous owner had got rid of it, because, although suitable for hunting in the Midlands, it had a tendency to kick and buck, and it could not accustom itself to the noises and frustrations of London traffic. Peel's groom advised him not to ride the animal, but he set out nonetheless, up Constitution Hill towards Hyde Park Corner. Near the top, near St George's Hospital, he encountered two young acquaintances, also mounted, the daughters of Lady Dover. Peel's horse began to plunge and kick. Then it swerved violently and threw him over its head. He fell still holding the reins, and the horse collapsed on top of him, striking Peel's back with its knees.

Medical help was sought at the hospital. Peel was in great pain. He had broken a left collar-bone, and probably some ribs. He was taken home and placed on a patent water-mattress on the dining table. Little by little, his condition became worse. A constant stream of visitors came to the house. They included Prince Albert, accompanied by the Crown Prince of Prussia, and Prince George of Cambridge.

When Peel died, on 3 July, the family refused permission for a post-mortem, so the exact extent of his injuries will never be known. The national shock was palpable, indeed without either precedent or subsequent parallel. He was, as G. M. Young remarked, 'the only English statesman for whose death the poor have cried in the streets'.[37] For Albert, it was a loss which was personal and painful in the extreme. He wrote to the Dowager Duchess of Coburg:

> Every day brings us fresh sorrow. Yesterday evening the good Duke of Cambridge died; the family is plunged in grief. The Strelitzes came five hours too late, and found their father already cold. We went with them to-day to see the body. The poor old gentleman slept softly away at the last, after his strength had been quite exhausted by a three-weeks' fever.
>
> Sir Robert Peel is to be buried to-day. The feeling in the country is absolutely not to be described. We have lost our truest friend and trustiest counsellor, the throne its most valiant defender, the country its most open-minded and greatest statesman.[38]

Peel was sixty-two. 'The country mourns over him as over a father,' wrote Victoria.[39]

For Albert, the loss of Peel and the apparent irrepressibility of Palmerston were grievous blows, not only to his heart but to his ambitions. Unable, at this stage, to persuade Russell to dismiss Palmerston for his extraordinarily cavalier attitude to diplomatic protocol, Albert decided to punch beneath the waist. Ten years earlier, when he was newly arrived in Britain, Albert had been told by Stockmar, that listener behind doors and arrases, that manipulator of persons, that Palmerston had, while a guest at Windsor Castle, made his way into the bedroom of one of the Ladies-in-Waiting, Mrs Brand, now Lady Dacre.

Since the royal pair were, in effect, blackmailing Palmerston, he had no choice but to ask for an interview with the Prince. In her version of events, given to Sir Theodore Martin, Victoria printed a lengthy memorandum by Prince Albert of the conversation which passed between him and the Foreign Secretary. He recorded that 'He was very much agitated, shook, and had tears in his eyes, so as quite to move me, who never under any circumstances had known him otherwise than with a bland smile on his face.'[40]

Of course, in the work of 'Sir Theodore Martin', we shall find no mention of bedroom doors being burst open, nor of Ladies-in-Waiting subject to attempted rapes. This is not, however, simply because Sir

Theodore and his distinguished 'silent collaborator' wished to avoid bringing a blush to the cheek of the reader. In the years after Albert died, Queen Victoria came to like Palmerston. He was just the sort of jolly, cynical, Whiggish old roué – coarser than Melbourne, cleverer than some of her own uncles – to whom she had grown fond, and used, during her youth. Moreover, although she would never agree with him about the Schleswig-Holstein Question, she became more jingoistic with every passing year.

In Albert's lifetime, however, Pilgerstein was their prime enemy, and the resurrection of a potential scandal, in the form of Mrs Brand's unpleasant experience at Windsor a decade since, was an indication that, if necessary, Albert was prepared to use dirty tricks. Naturally, there would have been no public scandal, but the Prince believed it would have been possible to use the incident as a weapon against Pilgerstein when trying to persuade the Prime Minister to force his resignation.

From Osborne, in the August of that sad summer, 1850, the Queen sent a strongly worded memorandum to Lord John ('The Queen thinks it best that Lord John Russell should show this letter to Lord Palmerston'[41]) which at least insisted upon the sovereign being informed by her Foreign Secretary about any major protocols or foreign actions. There were to be no more Don Pacificos. The Queen and Albert were, in 1850, at one. She spoke for Albert entirely when she said to their uncle Leopold, 'The state of the Continent is deplorable… I believe that Austria fans the flames at Rome, and that the *whole* movement on the Continent is *anti-Constitutional, anti-Protestant, and anti-English*; and this is so complicated and we have (thanks to Lord Palmerston) contrived to quarrel *so happily*, separately with each that I do not know *how* we are to stand against it all!'[42]

The sincerity of these words was not in doubt, but, behind them, there was an unexpressed fear. For over a year now, Albert had been patiently, and fervently, planning a great enterprise in which Peel was his ally. Already there was strong opposition to it, and, without Peel, there was every danger that it would sink into the sands. This was the Great Exhibition.

FIFTEEN

THE GREAT EXHIBITION OF 1851

THE MAKING OF THE MODERN WORLD

When Turner exhibited his painting *Rain, Steam and Speed* at the Royal Academy Summer Exhibition in 1844, he was depicting something much more than a steam train puffing through apocalyptic cloud. He was painting the Victorians, hurtling towards an unspecified, industrialized, polluted future; like the Israelites of old, in a pillar of fog. Turner himself clearly had a nostalgia for the old world which industry was destroying, but he did not, perhaps, articulate this nostalgia as strongly as did his most fervent collector and admirer, John Ruskin, who – exhibiting 100 of his Turners in 1878 at the Fine Art Society, chiefly watercolour views of landscapes and monuments and churches – made the damning remark that 'instead of Cathedrals, Castles or Abbeys, the Hotel, the Restaurant, the Station, and the Manufactory must, in days to come, be the object of her [England's] artists' worship.'[1] Ruskin, great aesthete and social critic that he was, meant these words to be ultimately dismissive. He deplored factories, both the damage they did to the lives of those compelled by economic circumstances to work in them, and, equally, the ugly artefacts they so carelessly brought forth, caring only for the profit motive.

He was not alone. If there was one matter about which nineteenth-century opinion-formers and ideologues were agreed, it was in their hostility to industry, trade and business. High Tories, such as the MP Colonel Sibthorp, saw the arrival of factories and railroads as destroying the aristocratic past. Karl Marx saw the Industrial Revolution as an ultimate exploitation of the working classes who continued to sustain its achievements and he looked forward to the day when the workers of all lands would unite to overthrow their capitalist oppressors. The Chartists, in a perhaps less ideologically coherent way, also stuck up for the rights of the poor. Dickens lampooned factory-owners and their values, or lack of them, in *Hard Times*, just as in *Dombey and Son* he excoriated the merchants and brokers whose investment skills helped to consolidate the wealth of Victorian England. This essentially romantic vision of Britain, that it was a rural, innocent place, threatened with oblivion by the progress of technology and trade, was carried down throughout the nineteenth century, and became one of the chief reasons why poets, novelists and pundits, whether drawn to the left or the right in politics, professed a real hatred of industrialists and businessmen, and all their products. As D. H. Lawrence would put it in 1923, 'They had a universal desire to take life and down it: these horrible machine people, these iron and coal people.'[2]

There is indeed a paradox in the Victorian success story. Almost all thinking Victorians deplored it, while enjoying the advantages of life in a modern industrialized society. Nor were these advantages wholly material, since with trade, and the growth of a prosperous middle class, came the inevitable movement to extend the suffrage and to allow a closer proximity to democracy than had been dreamed of in any other European country. Britain had become the most prosperous and peaceful country in modern Europe – arguably, the richest country in history. In real terms, the working classes, wretched as was the condition of their housing or the hours of their work, had also become richer than their forebears who had worked as agricultural labourers. A whole new middle class had grown up since the Napoleonic Wars, thanks to the labours of the workers, the ingenuity of the technocrats, the investment skills of the businessmen and merchants. Yet to read of Mr Dombey travelling by train to Birmingham is to see only 'the way of Death… strewn with ashes thickly'.[3] The shrieking monster of a train, the wretched dwellings of the city poor, and the arid mercantilism of Dombey's mind and soul are all leading to something purely negative.

When we turn to the late 1840s, the paradox of Victorian self-hatred by so wide a divergent number of observers is all the starker. For, as Mark Pattison, the sour old Rector of Lincoln College, Oxford, observed, it was a period when, almost overnight, everything changed. After the Year of Revolutions, and after a decade of calamity – Irish famines, Chartist unrest, uncertainty abroad and political convulsions at home – Britain found itself to be a place of peace and plenty. Looking back on theological infighting and churchy feuds which had dominated Oxford in the 1840s, Pattison wrote, 'If any Oxford man had gone to sleep in 1846 and had woke up again in 1850 he would have found himself in a totally new world… In 1850 all this was suddenly changed as if by the wand of a magician.'[4] And what was true of the Oxford Common Rooms was true of the country at large. A modern historian, Boyd Hilton, has written that 'Once fear of the mob had receded, it became possible to come to terms with towns and industry, to recognize that Britain's was an increasingly urban society and that there was no going back… A turning point came in the 1851 census, which revealed that for the first time more Britons were living in towns than in the countryside.'[5]

And there was one figure in public life who appeared to embody that momentous change, one incarnation of optimism: Prince Albert. While Marxists and High Tories, Chartists and Aesthetes could unite in deploring what Mr Dombey saw out of his railway-carriage window – factory chimneys, mills, coke-fumed canals and waggons of factory-made wares drawn over cobbled streets – there was one public figure who had actually journeyed around to visit them, and who saw in the wool-combing machines of Yorkshire, the mass-produced ceramics of Stoke, the shipyards of Glasgow and Newcastle, the steel-works of Sheffield, a dazzling display of varied human ingenuity, and a technological revolution, in which science and trade were married to produce something so glorious that it was not fanciful to see in all this the hand of Almighty God Himself.

It was understandable that Albert, realist as he was, should have viewed the industrialization of Britain with a more benign eye than that of Karl Marx, a poverty-stricken journalist constantly scrounging from his factory-owning chum Engels, or than those of London-based pundits such as Carlyle and Dickens and Ruskin who could conveniently hide from themselves the extent to which the Dombeys and the Gradgrinds were making them and the other Victorians collectively richer than any generation in history. They were not, as Queen Victoria was, the hereditary Duke of Lancaster.

When the Queen came to the throne, the Treasury had made a stab at appropriating the income from the Duchy of Lancaster, which largely consisted in agricultural rents over a wide area of Britain – including farmland on the borders of London, the banks of the Mersey and the pastoral valleys of Carmarthenshire and Glamorgan. The Duchy's Council resisted the Treasury's attempt to appropriate the rents. They thereby changed the sovereign from being an individual who was almost entirely dependent upon the state for her income – Civil List allowances, granted by Prime Ministers and ratified by Parliament – into someone who received an income hugely in excess of anything she could ever spend. For those sleepy little possessions of the Duchy of Lancaster, which in the 1830s had brought in a welcome addition to the Budget, were now coining it in for the Queen. The banks of the Mersey were now filled with warehouses and docks, exporting most of the industrialized produce of the north, and much of the south. The town of Kidwelly and its environs, rather than simply bringing in the modest rents affordable to sheep farmers, was sending men and boys down the coal mines and returning huge personal profits to the Queen. Large areas of Derbyshire, Yorkshire, Staffordshire and Lancashire had similarly been transformed into mining districts, or areas where factories, mills and works, linked by canals, railroads and newly tarmacadamed highways, were building up a colossal fortune for Victoria and Albert. Even the recreation of the wealth-producing masses made the Queen richer. She owned Blackpool, the favoured holiday resort, in 'wakes week', of all the great northern towns. And for the more genteel recreants, she also owned Harrogate – three little villages when she was a girl, but by the time she was having children of her own, one of the most prosperous spa towns in Europe, with a rateable value of over £30,000. As we noted in Chapter Ten, the combination of the Duchy of Lancaster revenue and the inheritance of the miser Neild's bequest made the Queen exceedingly rich.

Albert was, therefore, temperamentally inclined to see the advantages of industrialization and the coming of the Modern Age. He and Victoria would not have been able to finance Osborne and Balmoral out of the Civil List. Without the Duchy, they would still have been spending their summer holidays in the Brighton Pavilion and their rural retreats at Uncle Leopold's house, Claremont. Like many of their nouveaux-riches subjects, they had no reason to complain about the Industrial Revolution.

The event which celebrated the technological and industrial achievements of the previous decades, and which demonstrated the prodigious extent of British industrial wealth, was the Great Exhibition of 1851. Albert's association with the Exhibition was definitive – of it, and of him. The Exhibition was not, primarily, his idea, and he had to be persuaded to give it his backing. When he had done so, however, his influence was a major contribution to its success. And the marriage which it demonstrated, between the Prince and the emergent wealth-producers of the world, was a decisive factor in the stabilizing and strengthening of the monarchy for the future. Old Tories like

Sibthorp would have seen monarchy as belonging unequivocally with the old world which modern industry and a burgeoning bourgeoisie destroyed. Marx and his followers saw the industrial struggle as one which would inevitably destroy thrones and altars. Prince Albert's embrace of industrial technology not only defied both these extremes of view. It led to such an acquisition of wealth by the Royal Commission for the Exhibition of 1851, that he was able to leave behind his one unquestionable legacy: the Albertopolis in South Kensington, a complex of a concert hall, museums and colleges which, to this day, directly benefit from the money raised by the Exhibition. So, in any consideration of Albert's life, the Exhibition must be central.

The idea of exhibitions as a means of promoting arts and crafts did not begin in Britain. Having beheaded those aristocrats who had patronized the arts in their country, the French furniture makers, ceramicists, weavers and silversmiths needed some other means to secure financial backing for their wares. The Marquis d'Avèze, former head of the state-run Gobelins manufacturers of tapestry, organized an exhibition on the Champs de Mars in Paris as a way of reviving the domestic art industry, following the catastrophe of the Terror. The first National '*Exposition*' was repeated in 1802, 1806, 1819, 1823, 1827, 1834, 1839 and 1844. The Exposition of 1849 lasted six months and exhibited 4,532 items. A similar exhibition took place in Berlin in 1844; while being imitative of the French model, it also drew on the medieval German tradition of trade fares at which manufacturers could display their wares.

The Royal Society of Arts in London, housed in a beautiful Adam building in Adelphi since 1835, took note. Albert had been President of the Society since 1843. It was in 1835 that Sir Robert Peel set up a parliamentary committee to inquire 'into the best means of extending a knowledge of the arts and of the principles of design among the people, especially among the manufacturing population, of the country'.[6] The Society, with the backing of Peel, sought ways to improve the standards of design, and there was plenty of room for improvement, as the factories and mills of the Midlands and the north poured forth the cheap furniture and gimcrack metalwork which rightly disgusted the aesthetes.

In 1845, William Fothergill Cooke, the pioneer of telegraphy, made a proposal at the Society's Miscellaneous Committee 'that a National Exhibition of the Products of Industry in Arts, Manufacture and Commerce in connection with this Society be forthwith established'. Cooke himself put up £500 to finance the exhibition. From this small beginning, there grew the idea of a grand exhibition on a national scale. With this in mind, the Society sent a memorandum to the British Association for the Advancement of Science, asking for their co-operation, and informing them of the Society's desire to hold an exhibition which would be representative of the country as a whole.

Cooke was not alone. Liberal politician Charles Wentworth Dilke, London property agent Francis Fuller, engineer John Scott Russell, silk manufacturer Thomas Winkworth were all involved. So too was Henry Cole, that energetic civil servant who had worked at the Post Office with Rowland Hill, designed and pioneered the penny postage stamp, invented Christmas cards, while, under the pseudonym 'Felix Summerley', he designed china and illustrated children's books, while also mastering crafts as various as bookbinding, etching and engraving.

A happy conjunction of talents was forming. In 1845, John Scott Russell became secretary of the Society. A graduate of Glasgow University, Russell was a champion of technological education. He had lectured at the Leith Mechanics Institute, at the British Association and at the Society of Arts in Edinburgh. One of his earliest contributions to the idea of the National Exhibition was that there should be prizes of £50 awarded to the best exhibits to improve 'general taste'. Henry Cole, who bonded quickly with Russell, saw prize-giving as a key factor in the economy of such an exhibition. For such an exhibition to make an impact, it would have to be on a scale which would cost a great deal.

Albert, as President of the Society, watched the evolution of their ideas, but warily. As early as 1846, he had conceded that 'to wed mechanical skill with high art is a task worthy of the Society of Arts and directly in the path of its duty'. Nevertheless, he was loth to give his full and public backing to the planned exhibition. He admired Henry Cole, but not all of his schemes had succeeded. For example, Cole's idea of establishing a National Gallery of Fine Art failed to get enough financial backers and would finally collapse in 1849. Albert was not interested in being associated with an adventurous failure.

The first modest exhibition staged by the Society took place in March 1847 in the great room of the house in the Adelphi. As Francis Fuller, one of the organizers and financial backers, said, 'the public prefer the vulgar, the gaudy, the ugly, even, to the beautiful and perfect. We are persuaded that if artistic manufactures are not appreciated, it is because they are not widely enough known.'[7]

Even the organizers of this first exhibition were amazed by its success. Some 20,000 people crowded into the building to see some 200 exhibits. It was decided to repeat and to expand the experiment. By March 1848, 700 exhibits went on display and 73,000 people viewed them within three weeks.[8] It was at this point that Cole began to develop his idea of a National Gallery of Craft and Art, to be housed perhaps in Trafalgar Square. He tried to enlist Prince Albert's support, and Albert again drew back.

The following year, 1849, a Committee of Management of the Exhibition of British Manufactures was established under the chairmanship of Cole. On 4 June, the Exposition opened in Paris, the grandest of its kind to date, and John Scott Russell, twelve days later, announced the Society's ambition to hold a great British National Exhibition in two years' time – 1851. The Society commissioned Matthew Digby Wyatt, later Slade Professor of Art, to write a report on the Paris exhibition. Other enthusiasts included Herbert Minton, the Stoke-on-Trent potter who had made a great commercial success with 'Felix Summerley's' wares, and Francis Fuller.

It was Fuller who helped change Albert's mind about the whole scheme. In Dover, on his way back from Paris, where he had seen the Exposition, Fuller happened to meet Thomas Cubitt, at that stage working with Prince Albert on the building of Osborne House. Fuller and Cubitt travelled back to London together on the train. In the course of their conversation, Fuller expressed the view 'that we could do a much grander work in London by inviting contributions from every nation; and said that if Prince Albert would take the lead in such a work, he would become a leading light among the nations'.[9] Cubitt, who had reconstructed Buckingham Palace and built Osborne to Albert's specifications, had the ear of the Prince, and he duly reported Fuller's words.

It was a crucial moment in Albert's life. His chairmanship of the Royal Commission on Fine Arts, advising on the decoration of the Palace of Westminster, had shown him to be willing to play a public role, but, interesting as it was, it was hardly vital to the life of the nation. His work as Chancellor of Cambridge had revealed his talents to the academic world, but although he played a pivotal role in the reform of higher education in England, this was at a date when almost no one went to university. The Exhibition offered the chance to put on very public display his conception of the modern monarchical principle. He saw that in the new world which was coming into being, a monarchy should be popular and to some extent populist, at least to the extent of being profoundly involved with the lives of the people. The Society of Arts' decision to improve the standards in industrial design was something which entirely matched Albert's own talents and interests. He was both an aesthete and technologically minded. He was keenly involved with the 'Improvement' of conditions for the working classes, and he was equally fascinated by the speed and skill with which modern technology was transforming Britain into a superpower. So many of his beliefs and interests could be enshrined in this great project, which would not merely exhibit hundreds, perhaps thousands, of human artefacts; it would also advertise to the rest of the world that Britain was the modern world's hub, and the new values, associated with parliament and free trade, were not, as diehards supposed, incompatible with the antique, ceremonious institution of monarchy. They were to be wedded to it.

On 30 June 1849, Albert summoned a meeting at Buckingham Palace, to which he invited his Private Secretary, Colonel Phipps, John Scott Russell, Fuller, Cole and Cubitt. In after time, they all looked back to that meeting as the birth of the Great Exhibition.[10]

With Albert's backing, the organizers of the Exhibition were on their way. They had a little less than two years to realize their ambitions and two fundamental questions faced them from the outset. The first was to generate a number of quite different answers, and it was – What is this Exhibition really *for*? The second, in a sense more urgent, but one which they proved themselves equal to solving, was – How will it be financed?

Albert, from the very first meeting that 30 June, had seen the Exhibition as a primarily international affair. Yes, it had the ambition of promoting good industrial design, and eliminating gimcrack wares; but it was also part of Albert's creed that in wealth-creation, the nations would come together. There would always be others, of whom Palmerston was the most articulate, who believed that the Exhibition was essentially a British propaganda exercise, and the point of it was to show off to the other nations of the world that Britain was in a position of economic and political superiority to everyone else.

The answer to the second question, how it would all be paid for, was investigated in more detail at a second meeting, held at Osborne House on 14 July, when Scott Russell, Cole and Fuller were presented by Prince Albert to Robert Peel, then Leader of the Opposition, and Henry Labouchère, President of the Board of Trade. At this meeting, it was calculated that they needed to raise £100,000 before 1851. Cubitt had calculated that the Exhibition building (wherever it were erected) would cost at least £75,000 and they wanted to hand out £20,000 in prize money.

It was at this meeting that we start to see Albert showing his hand. He had, in the two weeks since the first meeting in Buckingham Palace, decided that it was to be his political project. Henry Cole had been in favour of badgering the Government, before Parliament broke up for the summer recess, to committing itself to a financial backing of the Exhibition. Albert, getting his secretary Colonel Phipps to write to Cole, made it clear that, from now onwards, all negotiations about the Exhibition, between the Commission and the Cabinet, should be conducted by himself.

> The Prince thinks that if the plan is to be matured under his auspices, he must be the Person to treat with the Cabinet Ministers upon it (which indeed is his place as President of the Society of Arts) and that he must be guided by his own discretion (willingly receiving all suggestions and advice) as to the urgency of the time of the decision of the question… H.R.H. is willing to undertake all the negotiations with the Cabinet – but he cannot think that any good would be derived from pressing the subject upon an unwilling or hurried attention, when the alternative of not incurring a responsibility which they had not had time fully to consider would only be a rejection of the proposal.[11]

Albert had averted the first possible impetuous mistake by Cole, namely a premature application to the Cabinet, which would inevitably have been rejected as ill-thought out and extravagant. At the same time, money had to be raised from somewhere. The Society of Arts was not in a position to underwrite the risks. Several potential backers were approached during the summer, and in August, Francis Fuller's father-in-law, George Drew, believed he had found a solution.

He was solicitor to a big firm of public works contractors called Messrs James and George Munday. It was agreed that they would advance a loan for the prizes, and four sixths of the cost of the building, which would, of course, be supplied by Munday themselves.

It was the Commission's first major mistake. The agreement with Munday was made on 23 August and on 7 November the contract was signed. An Executive Committee was formed of whom the Chair was Robert Stephenson MP, son of the inventor of 'The Rocket', the famous locomotive. Other members were Cole, Charles Wentworth Dilke the younger and Francis Fuller, with George Drew acting for Munday. The mistake was committing themselves to Munday, and, at first, appearing to be commissioning Munday to be responsible for the building. The *Patent Journal* would wisely ask, 'The Exhibition of 1851, is it to be made a "Job"? A MUNDAY Exhibition it will be – MUNDAY prizes and MUNDAY profits... In their eagerness to grasp the proffered bonus of £20,000, the Committee would appear to have been willing to make any sort of agreement.'[12]

Luckily, Albert saw that the Exhibition could not possibly be paid for by one firm of building contractors, and the sheer scale of the enterprise became apparent during the autumn of 1849. There was a two-pronged approach. On the one hand, there was a huge meeting in the Egyptian Hall of the Mansion House, in the City of London. Interestingly enough, guided by Phipps, Albert still kept in the background, and did not attend this meeting himself, which assembled between 300 and 400 of the most influential merchants, bankers and politicians, the Governor and Deputy Governor of the Bank of England, the Lord Mayor Elect and the chairmen of many other banks. Next day, 18 October, even the normally churlish *Times* could not fail to be impressed. 'The proposal yesterday submitted, at the suggestion of Prince Albert, to the magnates of the city, is one against which nothing but its grandeur can be objected. An exhibition of the industry of all nations speaks for itself, if it can only be accomplished without signal failure or inequality of execution, and if, as has, indeed, been suggested, it is not engulfed in its own magnitude.'[13] The article became positively lyrical about the place of London in the life of the world: 'This peaceful metropolis is the asylum of the outcast and unfortunate. All parties find refuge here: the Absolutist here meets his Republican foe, and the Imperialist the rebel to whom he is indebted for his exile. We have recently opened our ports to the produce and the ships of all nations. What place so appropriate for the mutual aids and intercourse of peace as this free and open metropolis?'[14]

At the same time, Albert had wisely dispatched Cole and Fuller to the industrial towns and cities. The aim was not merely to collect money from 'backers' but to encourage manufacturers to submit their best wares for competitive display. This was the first really practical stage towards the assembly of the Exhibition. They went to Manchester, Sheffield, Leeds, Bradford, Rochdale, Huddersfield, Kendal, Glasgow, and to the Potteries in Staffordshire. They went to Dublin and to Belfast. They worked with such efficiency and expedition that by the time of the Mansion House meeting in October 1849, the Commission was able to convey to the City merchants some of the extent of the Exhibition's range and remit. There would be wool from Australia and Tibet, there would be spices from the East and hops from Kent, as well as manufactured products from all the great industrial centres of the British Isles.

The organization of an exhibition on the scale envisaged required the detailed co-operation of hundreds of individuals. In almost every area which the Commissioners visited, to awaken both the generosity of donors and the competitive spirit of potential exhibitors (often the same people), committees were set up, just as, in London, the Society of Arts and the Royal Commission would set up many sub-committees to supervise the vital questions: where would the Exhibition be held? How would the expected visitors be accommodated? How would the objections from local residents, and from parliamentary diehards and from hostile journalists be addressed? How would the overseas exhibitors, and their governments, be involved and received? It was an administrative Behemoth, and when one goes through the papers, all carefully preserved by the hand of Prince Albert himself in volume after volume at the Royal Commission for the Exhibition of 1851 Archive (now housed in Imperial College London), you think – by analogy with any comparable thing in modern British life – this enterprise will fail, simply because so many people are involved. By the time the Royal Commission had divided the labours between sub-committees in London, and extended the task to the different regions of Britain to choose their own exhibits and representatives, there were *over three hundred* local committees at work.[15] It could not be possible that so many committees and sub-committees could think creatively.

What astounds us, as we follow the story from inception to opening, is how quickly the Prince and, under him, the Royal Commission were able to think on their feet. They changed tack frequently. When it became clear that one scheme or another was not going to work, they discarded it immediately. A great many things, including, most famously, the choice of design for the Exhibition building, were decided on a last-minute basis. This was a key ingredient in the Great Exhibition's success.

Some of the credit for this must go to the Government Secretary to the Royal Commission, and its unofficial auditor, Stafford Northcote – though, because of family illness, Northcote would take a decreasing role and hand over to Edward Bowring. Northcote is one of the unsung heroes of Victorian Britain; one of the key reasons, incidentally, that Britain did not suffer the tyrannies of France, Spain, Germany or Italy during the twentieth century. Another such hero, also closely involved with the setting-up of the Great Exhibition, was Lyon Playfair. Both Northcote and Playfair learned, during these months, the possibility of creative administration. They were leading lights in the creation of the modern Civil Service. Four years after the Great Exhibition, the Civil Service examination was instituted, and the running of the Government was placed into the hands of independent administrators, chosen on merit and intelligence. Whatever blunders the politicians committed, and however much they might try to grab power to themselves, in Britain there always existed behind the ranks of the politicians themselves the administrators. It was a system which worked triumphantly well until the Civil Service, in our own day, ceased to be recruited by examination only. (The inevitable inefficiencies and minor corruptions began.) Albert was a keen admirer of Northcote, who was a year older than himself, and when Northcote, exhausted by the work of the Commission, tried to resign, it was the Prince himself who persuaded him, despite a potentially dangerous heart condition, to carry on. When one watches Albert at work with Northcote, Playfair and Cole in the work of the Exhibition, one sees that, in a way, it was wrong of Albert's critics (or of Stockmar and Uncle Leopold) to think of him as a frustrated politician. It was truer to say that, when he found his best roles, he was a sort of civil servant in a coronet.

But he was also a public figure, as well as an efficient back-room boy. And this most magnificently manifested itself when he made his speech at the Mansion House on 21 March. This was Albert's first great public appearance as the royal presence on the Commission. It was a huge affair, an enormous banquet at which were assembled all the greatest foreign ambassadors, the mayors and lord mayors of innumerable provincial towns, the Chief Officers of State. The columns of the Egyptian Hall had been decorated with trophies representing the three kingdoms and the Principality of Wales. There were shields representing their most famous raw materials, and groups of stuffed animals backed by laurel. Wales was represented by bunches of leeks, goats' heads and minerals. Ireland by pork and linen. At the far end of the hall, where Albert would sit, high on a throne of royal state, were the colossal figures of Peace and Plenty. Opposite them, Britannia led angels to undo their corded bales of imported goods in an idealized version of the Port of London. Here was a mythologized allegory of free trade which must have delighted the eyes, or at least the heart, of one of the most distinguished guests, Sir Robert Peel himself.

Albert's speech is worth quoting at length because it really is his Manifesto: not just for the Exhibition, but for the entire raison d'être of his public life.

> Nobody... who has paid any attention to the peculiar features of our present era, will doubt for a moment that we are living

> at a period of the most wonderful transition, which tends rapidly to accomplish that great end to which, indeed, all history points – *the realisation of the unity of mankind.* Not a unity which breaks down the limits and levels the peculiar characteristics of the different nations of the earth, but rather a unity, the result and product of those very national varieties and antagonistic qualities.

Sometimes, political speeches, at any period between the French Revolution and our own day, speak of the Brotherhood of Mankind, but without any particular agenda. Albert was not such a man. He genuinely believed that 'the unity of mankind' was a political possibility. At home in Germany, he had believed in an extension of the customs union, the *Zollverein*, to embrace a peaceful federation of German-speaking people, perhaps going on to embrace an entire European Federation. In his view of the peaceful potential of world trade, he thought that the Exhibition would be, not merely a symbol, but a harbinger of what could be brought to pass with sufficient political vision. He even asked Lord John Russell to look into the possibility of establishing a universal world currency. Russell tactfully replied that the relative value of precious metals in the various countries of the world would make such a scheme very difficult to administer.[16]

Albert told the assembled dignitaries at the Mansion House:

> The distances which separated the different nations and parts of the globe are rapidly vanishing before the achievements of modern invention, and we can traverse them with incredible ease; the languages of all nations are known, and their acquirement placed within the reach of everybody; thought is communicated with rapidity, and even by the power, of lightning. On the other hand, the great principle of division of labour, which may be called the moving power of civilisation, is being extended to all branches of science, industry and art.

> Whilst formerly the greatest mental energies strove at universal knowledge, and that knowledge was confined to the few, now they are directed on [*sic*] specialities, and in these again even to the minutest points; but the knowledge acquired becomes at once the property of the community at large.

Albert went on to draw a link between the scientific and technological revolution which had taken place in his lifetime with the second phase of the Industrial Revolution and the development of Western capitalism.

> The products of all quarters of the globe are placed at our disposal, and we have only to choose which is the best and cheapest for our purposes, and the powers of production are intrusted to the stimulus of *competition and capital.*

There could have been many, who were not invited to the Mansion House, who would take issue with these words – for example, the manufacturers and craftspeople of India, who had seen their trade undermined by the British having bought cheap American cotton and chosen to spin and weave it, with cheap labour in Lancashire, and take a huge share of world markets. Such products, which Albert saw as having been 'placed at our disposal', could be seen by others as having been appropriated by manipulating the markets. This is the way that capitalism has worked, and continues to work. Marx and Engels saw the Exhibition as a way of celebrating the way that industrialized nations suck weaker global communities into exploitative patterns of economic dependency.[17]

But for Albert, who was the most eloquent public spokesman for the free-market capitalist ideal, there was something providential about the working of world markets themselves.

> So man is approaching a more complete fulfilment of that great and sacred mission which he has to perform in this

world. His reason being created after the image of God, he has to use it to discover the laws by which the Almighty governs His creation, and by making these laws his standard of action, to conquer nature to his use; himself a divine instrument.

Science discovers these laws of power, motion and transformation; industry applies them to the raw matter, which the earth yields us in abundance, but which becomes valuable only by knowledge. Art teaches us the immutable laws of beauty and symmetry and gives to our productions forms in accordance to them.

Gentlemen, – the Exhibition of 1851 is to give us a true test and a living picture of the point of development at which the whole of mankind has arrived in this great task, and a new starting-point from which all nations will be able to direct their further exertions.[18]

In the late twentieth and early twenty-first centuries, the power of global capitalism in which Prince Albert expressed such faith is often seen, even by political thinkers of the right, as a potentially dangerous, even exploitative, phenomenon. In the early 1850s, although Marx and Engels saw it as exploitative, there were many others who would today be seen as representative of the Left politically, who eagerly shared Albert's faith. The Liverpool Peace Society wrote to the Prince, on 5 April 1850, thanking him for his speech, praising his ideal of 'the realisation of the unity of mankind'. 'Not less striking and beautiful, not less just and true, is the conviction of Your Royal Highness that the blessings which the Almighty has already so bountifully bestowed upon us, *can only be satisfied* in proportion to the help which we are prepared to render to each other, and therefore, also by peace, love and neighbourly assistance, not only between individuals but between the Nations of the Earth.'[19]

The composition of the Royal Commission, which had been announced on 3 January 1850, was intended to be widely representative, but, given the brief – the promotion of free trade, the belief that by abolishing tariffs you promote peace and harmony throughout the world – it would not have enticed any High Tories or anti-capitalist communists to join, even had they been invited. In spring, the *Morning Chronicle* had dubbed the Mansion House banquet 'the inaugural feast of Free Trade'; the organizers liked the title, but the Tory Lord Derby advised them to describe the banquet as something assembled to 'bring into harmonious concord the nations of the world'. The controversial words 'free trade', anathema to Tory ears, were omitted.

The all-male Royal Commission – which included Peel, Labouchère, Gladstone, Cobden and Thomas Bazley, Chairman of the Manchester Chamber of Commerce – constituted, as has been rightly said by a modern historian, 'a microcosm of the various political forces – Whig, Radical, and Peelite – that would soon come to be known as Liberal, bringing together prospective liberals at a time shortly before they came to identify themselves in this way'. The same historian – Jeffrey Auerbach – pointed out that 'in just over a year, they organized a display of 100,000 exhibits in a building (built from ground up in 9 months) that was at the time the largest enclosed space on earth'.[20]

Four tasks faced the Royal Commission in those early months of 1850, and each of them had to be accomplished with speed. First, they had to underwrite the prodigious costs which would amass even before the Exhibition opened, and for this, Playfair, Cole and others kept up a tireless tour of the country, raising funds from local manufacturers. Most of the Commissioners themselves pledged enormous sums which they would have lost had the Exhibition not been a success.

A second, and related point, it was early seen that the commitment to allowing so much of the contract to be with one firm, Munday, had been a mistake. Stafford Northcote persuaded the Commission as early as March to extract themselves from their commitment to Munday, and allow the public to contribute subscriptions. The process of extrication would be protracted and legally complicated. Once they had publicly declared their desire to cancel the Munday connection, however, they enormously broadened their appeal, and subscribers ranged from the highest to the poorest. The Duke of Wellington was persuaded to support the Exhibition,[21] and the newspapers loved the parish constable in Braintree, Essex, who sent a shilling. 'It is all I can afford as I am a working man with a wife and four children, but always ready to every good work.'[22]

Colonel Sibthorp, of course, was unimpressed. He told the House of Commons, 'It was painful to hear the accounts that daily reached him of the measures resorted to, to compel industrious tradesmen and merchants to contribute from their hard earnings to the furtherance of a scheme directly inimical to their dearest interests'. And Sibthorp questioned whether 'this absurd Exhibition should take place at all'.[23]

Having raised the money, however, and disentangled themselves from Munday, the organizers still had two more crucial matters to decide: they had to agree upon a site for the Exhibition, and, having done so, they needed to agree on a design for the Exhibition building.

Where would the Great Exhibition be held? Clearly, it had to be in central London, but at the same time, wherever it took place, there would be grave objections from local residents. The Isle of Dogs was considered, but it was too far away, and, although they did not say so in so many words, the Commission could not relish the prospect of many foreign visitors taking the journey into the East End and seeing the abject poverty with which people lived in the neighbourhood of the Docks. Victoria Park had similar disadvantages, and was too far away from the centre. Wormwood Scrubs and Primrose Hill were considered, but they did not have enough space. A possible site in Regent's Park was abandoned because 'the leases under which the houses in the neighbourhood are held, contain a clear and stringent provision that no new building of any kind shall be erected within the limits of the Park'.[24]

Everything pointed to the use of one of the great parks in central London. When the idea was mooted in the House of Lords, on 19 March 1850, Lord Brougham spoke in the Upper House. 'He should not like either the Green Park or St James's Park choked up with it; but in Hyde Park it certainly must not be.' He went on to quote Mr Wyndham's description in 1808 when it was proposed to build eight houses in Hyde Park that 'the parks were the lungs of this great metropolis'. The noble lord concluded his speech by saying that 'if this building were suffered to be erected in Hyde Park, all the parks would be at an end'.[25]

Hyde Park, however, was the obvious site. It is enormous, and it can be easily accessed from four sides. The Royal Commission knew that the choice of Hyde Park would, however, create an outcry in Parliament and among the public at large. Peel attended a meeting of the Commission at the end of June and advised them how to deal with the inevitable hostility in Parliament. The next day, he was thrown from his horse in Rotten Row, and four days later he was dead. Cole noted in his diary, 'The last words Peel said at the Commission were, "That [Hyde Park] shall be [the] site or none."'[26]

We noted in the last chapter that the death of Peel was a grievous personal loss to Prince Albert, as to the nation. That July was a difficult one. As well as reeling from Peel's death, the nation, and Parliament, was swept up in the Don Pacifico crisis. Nevertheless, inauspicious as this moment was for a battle, the Royal Commission had to be prepared to commit themselves to Hyde Park, and to begin plans for the building. They had, after all, less than a year to go. As Prince Albert said in a letter to Baron Charles Dupin, the leader of Exhibitions in Paris, 'One solitary thing will console us in our very proper grief – the thought that we shall still honour his memory by the assiduous continuation and dignified accomplishment of an enterprise to which he devoted himself with so much zeal; and that the success of our Exhibition will one day be added to all the great accomplishments which survive our illustrious late friend.'[27] Since 1849, Albert had consulted Dupin, who had actual experience of

Left: Roger Fenton's photograph of Queen Victoria and Prince Albert, close to her thirty-fifth birthday in 1854.

Below: Group photograph at Osborne – Albert, Victoria and the family. The grown-up figure in a white dress is the Duchess of Kent, Queen Victoria's mother.

Moving machinery at the Great Exhibition of 1851.

Henry Cole, who helped Prince Albert organize the Great Exhibition.

Joseph Paxton, architect of the Crystal Palace.

Royal Patriotic Building, Wandsworth, built by Prince Albert for orphans of the Crimean War. Note St George, who bears a resemblance to the Prince, slaying the dragon over the central window.

Osborne in the sunshine – Albert's seaside creation.

Above: Prince Albert at Buckingham Palace, May 1854.

Left: Queen Victoria and Prince Albert at Osborne, 1859. By now marriage and life were taking their toll.

Above: Queen Victoria, Prince Albert, Prince Consort, Albert Edward, Prince of Wales, Princess Alice, Prince Philippe, Count of Flanders, Duke of Oporto and King Leopold I of the Belgians, 29 June 1859.

Left: Queen Victoria had this photograph of Prince Albert (taken around 1860) printed multiple times and it was the one she nearly always used in memorializing jewellery.

The last visit to Coburg, 8 October 1860. (*Left to right*) Princess Alice, Princess Frederick of Prussia (Vicky), Prince Frederick of Prussia (Fritz) Ernst I, Duke of Saxe-Coburg-Gotha, Queen Victoria, Prince Albert and Alexandrine, Duchess of Saxe-Coburg-Gotha.

Prince Albert returning from deer stalking.

During his last illness. Albert holding a draft memorandum over the 'Trent' crisis.

14 December 1861.

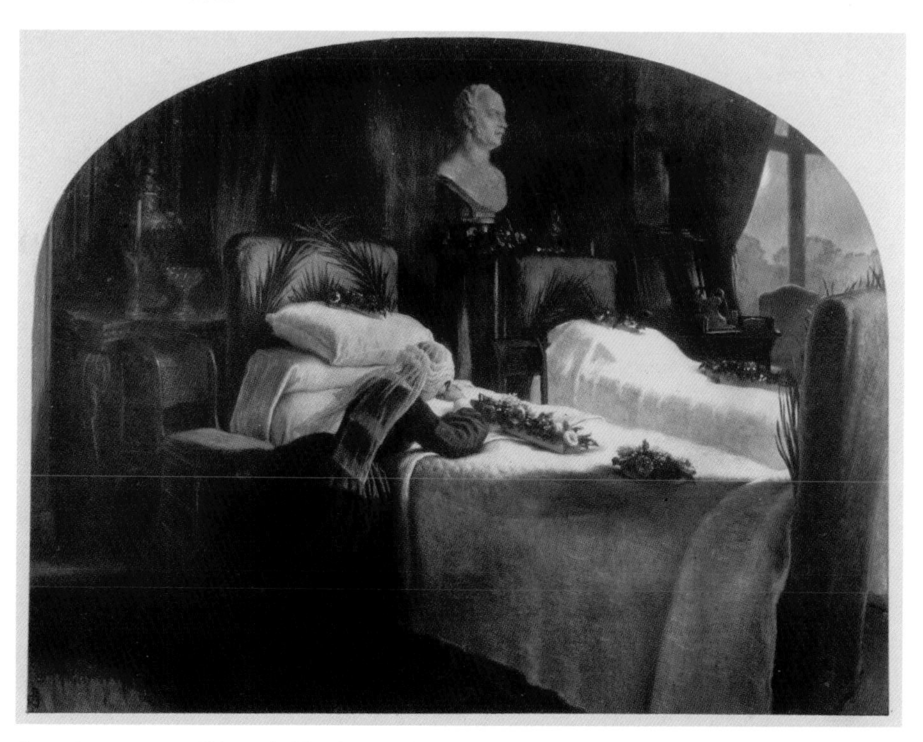

A carbon print of Joseph Noel Paton's *Queen Victoria at Prayer in the Prince Consort's bedroom*. The loss was, and is, irreparable.

April 1863. The group clustered around Prince Albert's bust are: (*left to right*) Princess Helena, Princess Alice and her husband, Prince Ludwig of Hesse, Queen Victoria and Princess Beatrice ('Baby'), the Prince of Wales, the Princess of Wales, showing a photograph of their father to Prince Arthur, and Princess Louise.

putting on international exhibitions in Paris, and as the months of 1850 hurtled by, an expert's advice was more and more needed.

The Times believed that the choice of Hyde Park would be 'the latest in the long continued invasion of John Bull's rights',[28] and Colonel Sibthorp was quick to rise to his feet in the House of Commons and point out that at least ten trees would have to be felled if they went ahead with building in the middle of the park. On 29 June, *The Times* expanded its theme:

> The erection of the building, as proposed, is so deeply offensive to the feelings and wishes of the inhabitants of London, that even the discursive flights of the Parliamentary orators would appear to lose something of their interest by the side of this more homely topic. Dine where you will, go into what drawing room you will, enter into conversation with the first chance stranger, met on a river steamer or in a Kensington omnibus, and on all sides there rises a groan of indignation at the intended pollution of our beautiful park.[29]

Albert's name was so linked to the proposed Exhibition that when it got a bad press, so did he. Lady Lyttelton said that the Exhibition was 'quite universally sneered at and abominated by the *beau monde* and will only increase the contempt for the Prince among all fine folk. But so would anything he does.'[30] *The Times* delighted in reminding the Prince that in linking his name to the proposed 'pollution' of the park, he was only increasing his personal unpopularity.

> The proposed Exhibition is everywhere received as the creation of the PRINCE CONSORT [he was, of course, not officially the Prince Consort yet!]. Should it meet with the success which we trust awaits it, the honour and credit of having originated and carried through so gigantic a scheme,

> will, no doubt, be attributed to PRINCE ALBERT. There is, however, a reverse to the picture. Should the Exhibition not answer popular expectation, or should any unpleasant feelings be aroused in carrying the business through, the PRINCE must be content to have his name associated with an abortive or an unpopular project.

A few days later, the paper was quoting local residents: 'if the PRINCE CONSORT could hear the expressions of deep annoyance which are daily uttered by various gentlemen who have taken houses at high rents in this neighbourhood, and who foresee the certain loss of privacy for themselves and their families as a consequence of his project, we feel sure that his ROYAL HIGHNESS would never consent to the perpetration of such injustice under cover of his name'.[31]

Moreover, *The Times* delighted in pointing out that the unpopularity of the building, and of the proposed Exhibition, went hand in hand with the unpopularity of the Prince.

Albert felt these taunts. He was always thin-skinned about personal criticism. He saw his task, however, clearly, and his attention to detail was unwavering. On 2 July, he was writing to Lord John Russell that he had missed a trick, in replying to the objections of Lord Seymour in Parliament about the potential inconvenience caused to riders in the park.

> I am sorry that Ld. Seymour's scruples prevented your mentioning the Riding in Kensington Gardens in yesterday's debate as it would have a powerful effect. Lord Seymour's objection to the softness of the soil would not hold good for May, June and July, the months during which the riding would take place, and as to the hurdling, the Office of Woods must have quantities in store. If not, they are quickly and cheaply procured, and could even be hired for a couple of Months. There is a quantity of Wattle Hurdles in St James's Park, which ought, on grounds of economy, one day to be changed for Iron

> hurdles, and I am sure will be so. Could Iron hurdles not be used for Kensington and then transferred to St James's?[32]

The plans which the Commission had been considering for the Exhibition building were both expensive and ugly. The Building Committee had been formed a week before the Prince gave his great speech at the Mansion House. It was chaired by William Cubitt, and the other members were Isambard Kingdom Brunel, Robert Stephenson, Charles Barry, T. L. Donaldson, Professor of Architecture at University College, the Duke of Buccleuch and the Earl of Ellesmere. As soon as they allowed it to be known that they were looking for ideas for the Exhibition building, 245 plans were submitted, 38 of them from abroad – from Australia, Holland, Belgium, Hanover, Brunswick, Hamburg and Switzerland. Of the eighteen designs which the committee 'commended', only three were British, which excited predictable scorn from the press. For one reason or another, none of the proposals were so commendable to the Committee that they would actually adopt them. One of the reasons for this was expense. Richard Turner, the Dublin architect who was responsible for the Palm House at Kew and for the glass-house in Regent's Park, made a block model of his design, but the estimated cost was over £300,000, way over the Committee's budget.

That was why, in the end, they came up with the strange botched brick job which rightly excited derision and horror from the public. Had they built it, Colonel Sibthorp would have been right to call the Exhibition 'one of the greatest… absurdities ever known'. The huge brick edifice, which would have been a complete eyesore, was to have been crowned by the ultimate absurdity, a dome, of Brunel's design, three times the size of that adorning St Paul's Cathedral. They believed it would be possible to erect this monster for less than £100,000.

There are so many times in the story of the Great Exhibition where it almost seems plausible to think that Fate was on its side; so many occasions when great mistakes were avoided at the last moment, almost despite the Commissioners' and the Prince's best endeavours.

The choice of architect for the Exhibition building is the ultimate example of this.

It so happened that when the drawings of the new building were made public, the 6th Duke of Devonshire's gardener-architect-agent, Joseph Paxton, was in London. Paxton, the son of a farm labourer from Bedfordshire, was at this date in his late forties. The Duke's gardens at Chatsworth were among the greatest horticultural masterpieces of Europe, boasting such ingenuities as the Emperor Fountain, constructed for a visit by the Russian Emperor which never in fact took place, and the beautiful hothouses, of Paxton's design, the best known of which is the Great Stove. Paxton combined great technological skill with a flawless eye. He also shared with his ducal patron a tremendous sense of style. He thought big. He was an impulsive man with a theatrical imagination. In London for a meeting of the Midland Railway that June, Paxton saw the disastrous plans for the Exhibition building and expressed his reservations to the Chairman of the Committee, John Ellis MP. Ellis respected Paxton and loved his work. He took him at once to see Lord Granville at the Board of Trade who, by great good fortune, was in the middle of a conversation with Henry Cole. These were all men of great dynamism. They believed in getting things done. Cole who, like Paxton, was a driven man, urged him to draw up plans as quickly as possible. Before going back to Derbyshire by train, Paxton walked round Hyde Park to 'step over' the proposed site. On the next Monday he journeyed to North Wales where he saw the floating of the third tube of Stephenson's railway bridge over the Menai Straits. Then on Tuesday he rushed to Derby where he chaired a meeting of the Midland Railway. A pointsman had behaved irregularly, and as his fellow committee members droned on about how to discipline the poor man, Paxton doodled on his blotter. It was to become one of the most famous architectural drawings in history.

For the rest of the week, Paxton laboured at Chatsworth with W. H. Barlow, chief engineer of the Midland Railway, and on 20 June, with rolls of details under his arm, he set off to meet the Building Committee, asking if he could give a lecture on his proposal. By two bits of very good luck, he managed to enlist the support of two highly influential men. The first was Robert Stephenson, who simply happened to be on the same train to London, and who was an invaluable advocate of Paxton's cause with the other members of the Building Committee. The other was the Duke of Wellington. Paxton first called on Lord Brougham, one of the staunchest enemies of the Exhibition, who suggested that he should call on Wellington. It does not seem very likely that Paxton actually did so, even though the Great Duke hugely admired Paxton's work at Chatsworth. He did, however, go to see Granville at the Board of Trade who liked the drawings so much he would not let them go. He informed the Duke of Wellington, who decided to become a supporter not only of the Exhibition, but of Paxton's designs. Though Wellington was the Highest of Tories and the 6th Duke of Devonshire was a Whig, the Iron Duke always admired Devonshire, and the stupendous displays and entertainments which he mounted at Chatsworth. Wellington knew that none of the more imaginative touches, such as the lanterns which hung from all the trees on the occasion of the royal visit, or the engineering feat of the Emperor Fountain, would have been possible without Paxton. The old soldier recognized Paxton's gifts and saw that he was the man to follow his plans through.

The Building Committee were stubborn – Barry in particular. They had their awful plan, and they were somehow wedded to it. Granville, however, arranged for Paxton to sidestep them, and two days later he took the architect, with his drawings, to Buckingham Palace.[33]

It was a cunning move. Paxton had already sent copies of the drawings to the two vital manufacturers who would enable his scheme: to the Smethwick ironmasters Fox, Henderson and Co., and to the Birmingham glass manufacturers, Chance Brothers, who had made the glass for the Great Stove. When detailed plans came to be asked about the viability of his Exhibition building, the technological answers were ready.

Luckily for Paxton, the stubborn Building Committee, thinking to further their cause, published a picture of their proposed eyesore in the *Illustrated London News*. The readership of 200,000 was horrified. It was enough to make Sibthorps of them all.

It was during the week of Peel's death, and the shock which was felt on all sides, that Paxton quietly applied for a patent for his ridge and furrow roof. Even before the Building Committee had adopted Paxton's design, Albert had personally urged Barry and Cubitt to redesign the roof of the proposed building. When Paxton's putative plans were being considered by the Commission, Albert had supported Paxton's longing for a vaulted rather than a flat roof, on aesthetic and practical grounds.[34] Paxton's patent application was successful. Paxton then published his own design in the *Illustrated London News*. It was somehow immediately apparent that this bold venture was right; it was the building they had all been waiting for all along.

It was 1,848 feet long (not, alas, which would have been neat, and as some newspapers claimed 1,851 feet); 456 feet at its broadest point, the transept; 108 feet high at this point. It was to be stupendously simple, and modern. It would be, much of it, prefabricated. It would use no bricks at all. It could be, and indeed in terms of the contract would have to be, temporary, easily demolishable and rebuildable on another site. And it was enormously cheaper than any previous proposal: £88,000, as against the Building Committee's £141,000. And Fox, Henderson only required payment of 40 per cent during construction, with the rest to be paid on completion. The Committee met. While they were deliberating, Samuel Morton Peto, railway entrepreneur and engineer, and a member of the Finance Committee of the Commission, advised the Prince to support it. Peto guaranteed £50,000. By now, the Building Committee were at daggers drawn with one another. Brunel, Cubitt and Stephenson, the engineers, were opposed to Barry, and it did not look as if they would ever agree on a plan. By the end of July, Barry admitted defeat, and Paxton's design was accepted by the Commission. The site was surrounded by hoardings. Paxton, his engineers and workers, had just twenty-two weeks to finish the scheme.[35] By 27 September, Colonel Reid, one of the Executive Committee, wrote to Colonel Phipps to inform the Prince that 'the first of the Iron Columns was erected yesterday, and 18 or 20 will be up tonight. Two ribs which are to surmount the tall elms are finished. The iron columns stand upon concrete instead of being planted in timber and then concrete foundations go down to the gravel, the clay on soft mould being first dug out. Thus the iron columns will stand on firm foundations.'[36]

Albert personally intervened, at a late stage, to take charge of the colouring of the Crystal Palace inside and out. On 5 December, at the thirty-first meeting of the Commissioners, the minutes record that 'His Royal Highness the president, together with Sir Richard Westmacott and Sir Charles Eastlake, were appointed a Committee to examine the specimens of painting on the inside of the Building, and to decide on the mode of colouring to be adopted.' The sash bars were to be white, the ridge piece blue and white, the Paxton gutters left red, but for this red to be of a slightly lighter shade. On every inch of Paxton's Palace, the influence and colouring of Albert's finger was to be seen.

In the event, it took a month longer than planned to erect the 'Crystal Palace', as it was nicknamed by *Punch*, and as it has been known ever since. It was finished in February 1851. Luckily for us, when it was relocated to Sydenham, it was much photographed, so we can see with our own eyes just what a magnificent building it was: six times the size of St Paul's Cathedral. Some two thousand men worked on its construction, its three tiers, its 33,000 iron columns, its 2,224 girders, its 600,000 cubic feet of timber. At 49 × 12 inches, the panes were the largest sheet-glass ever made. As the November storms battered it – causing even the irrepressible Paxton to fear that the doom-mongers might be right, and that the building might be in danger of collapse – it rose, an extraordinary statement of modernity, a thing of light and beauty and optimism and energy. The 6th Duke of Devonshire visited the site every day. Prince Albert visited the workmen, supplying them with gallons of beer. As he wrote to Friedrich Wilhelm IV, King of Prussia,

> Mathematicians have calculated that the Crystal Palace will blow down in the first strong gales; engineers that the galleries would crash in and destroy the visitors; political economists have prophesied a scarcity of food in London owing to the vast concourse of people; doctors that owing to so many races coming into contact with each other the Black Death of the Middle Ages would make an appearance as it did after the Crusades; moralists that England would be infected by all the scourges of the civilised and uncivilised world; theologians that the second Tower of Babel would draw down upon it the vengeance of the offended God.[37]

But the simple fact of the Crystal Palace being so swiftly and so indubitably *there*, the sheer wonder of it, converted all but the diehards. Even *The Times*, even Charles Dickens, came round.

To emphasize the inclusive and nationwide (as well as international) aims of the Exhibition, a banquet had been laid on in York, for 25 October 1850. The host was the Lord Mayor of York, and it set out to be as grand as the Mansion House banquet in March. The meal was designed and supervised by the great Alexis Soyer, and it quickly gained the nickname the 'hundred guinea dish' because of its extravagance.

Prince Albert, who had such a mixed reception from the upper classes, and such an abrasive reception from many Parliamentarians, went down well in the provinces, whether he was meeting academics, inventors or businessmen. The *Daily News* caught, precisely, the reason for Albert's popularity in provincial England, among the aldermen, mill-owners and industrialists. In days gone by, said the paper, royalty would only leave London in order to stay with landed aristocrats.

> Widely different are the objects and purposes with which the present Prince Consort [*sic*] of England visits the first provincial city of the Empire. He comes to it, not at

> second-hand from the mansion of some neighbouring landed lord, but direct from the presence of the sovereign herself, prepared to express as from her own royal mouth the interest she feels in the trade and industry of all her people. Prince Albert, visiting York, becomes its citizen; he takes up his abode with the Lord Mayor and meets face to face and hand to hand the corporation of that important centre.[38]

The other papers made the same point. The *Illustrated London News* said:

> The Prince Consort [*sic*] without any other inducement to labour than the necessity which all healthy-bodied and healthy-minded men feel to have something to do, and the earnestness inspired by a good cause – has for the past year laboured as assiduously as any hard-working professional man in the country. The working classes, ordinarily so called, begin to see that they are not the only workmen, and that compulsory idleness is a greater curse than labour.[39]

The *Spectator* said:

> Without obtruding any outlandish innovation upon the country into which he had wedded, Prince Albert has certainly taken no small share in promoting a new spirit, which corrects many narrow notions of the English people. A pupil of German philosophy, he lends an exaltation of purpose and a breadth of feeling to his treatment of practical affairs, which exalts the common and reunites the useful to the beautiful. At the same time, in a mood suited to his temperament, and certainly well-suited to his peculiar position, he awaits to capture the prevailing idea spontaneously arising among his English countrymen and rather to impart additional vitality to the English idea than to confuse it with strange associates.[40]

The speech he made at the York banquet was a masterpiece. He made his defence of the Exhibition, and of the royal involvement with the scheme, into a panegyric for Peel. Since Peel had now been canonized by the public, Albert made those who had mocked or doubted the Exhibition seem as though they would despise the national hero, the only statesman in history for whom the poor wept in the streets of London when he died. He was also able to represent himself, not as a 'student of German philosophy' with his head in the clouds, but as a practical Englishman, the friend of Peel, the common-sense son of the calico-manufacturer of Tamworth. He also seemed to be expressing some sympathy with those who had hesitated (as Peel had done) before endorsing the ideals of the Great Exhibition.

> Warmly attached to his institutions, and revering the bequest left to him by the industry, wisdom, and piety of his forefathers, the Englishman attaches little value to any theoretical scheme. (Cheers.) It will attract his attention only after having been for some time placed before him; it must be thoroughly investigated and discussed before he will entertain it. Should it be an empty theory it will fall to the ground during this time of probation. Should it survive this trial it will be on account of the practical qualities contained in it; but its adoption to the end will entirely depend upon its harmonizing with the national feeling, the historic development of the country, and the peculiar nature of her institutions. (Loud cheers.)[41]

When he left York, exhausted, Albert was entitled to the self-congratulation which enlivened his note to Colonel Phipps – 'Everything at York went off remarkably well – people much pleased, journey quick, my stomach deranged from hurry, nervousness, and M. Soyer.'[42]

By now, the wind was in the Exhibition's sails; the fact that they had only months to go before its opening was a spur to organizers and contributors. British exhibits, and potential exhibitors, were on the whole of high quality, as Lyon Playfair was in a good position to observe. In June, he had reported from the Potteries, 'one dessert service, preparing at an actual outlay of £800, is one of the most beautiful things I have ever seen'.[43]

Lieutenant-Colonel J. A. Lloyd, the Commissioner charged with liaising between local committees and London, wrote ecstatically from Oldham that the machinery he had inspected was 'magnificent'. 'The wealth-producers of the Midlands and the North had found that their champion is a German Prince.'[44] Where Oldham, Burnley and Manchester led, Bolton, Macclesfield, Stockport and Sheffield followed, not merely subscribing money but constructing machines for display. Nor was it forgotten that these machines and mills and mines, though the technological achievement of clever engineers and inventors, needed the men and women who operated them, streaming into many a 'works' by cobbled streets before dawn. They too were to be welcomed to the Great Exhibition, and Alexander Redgrave, a specially drafted civil servant from the Home Office, drew up a memorandum to make arrangements with a railroad company to make a register of possible lodgings in London where visitors could affordably stay.[45]

Not that every part of the country showed the vim and enterprise of the northern manufacturing towns. When Playfair went to Penzance in September, he was appalled that 'instead of finding Cornwall in an active state of preparation, nothing at all has been done'. He set off to Wales, where he found beautiful woollen stuff being woven. 'The better kinds rank highly among the woollen dresses of our ladies and the commoner sorts are equally excellent in their way.' By the time he went back to Plymouth at the beginning of December, he was pleased to report, 'I have left Cornwall in capital heart and in good working order', though it is not entirely clear what this meant.[46]

Foreign exhibitors were no less ardent than Playfair to prepare their exhibits. As early as August 1850, the Persians had already assembled enough exhibits to fill their allotted 1,000 feet of space, manufactured goods, raw produce and minerals, as Colonel Phipps was told by Bowring at the Board of Trade, who also wanted the Prince to know that English residents in China were helping to ensure that the Chinese collections would be ready on time – even though it was feared 'it would be quite impossible to get the Chinese to move at all'.[47]

Most of the Persian exhibits came, not directly from Tehran or Isfahan, but from London dealers or English collectors. Fabrics and rugs and clothing predominated, but there were also khorapan daggers with ivory carved handles, papier-mâché work-boxes, carved chessmen, and carved panels depicting Persian history. Articles from the collection of Mr W. F. Mills included smoking materials and a series of portraits of the Shahs.[48] For Europeans, the Exhibition was full of political significance. Baron Christian Karl von Bunsen, the Prussian Minister in London, was a close friend and ally of Prince Albert. He wanted a co-ordinated *Zollverein* display at the Exhibition, that is to say, a united German collection of exhibits. This was at a time when 'Germany' was still a patchwork of thirty-nine states loosely linked together by trade agreements and the like, though united in a common language. To Albert and Bunsen's distress, Bavaria and Württemberg were not so keen on Prussia, as they saw it, organizing their contributions to the Exhibition. Hesse-Darmstadt and Frankfurt also opted out of Bunsen's idea of a united German exhibition-space. It was only twenty years away from the united German Empire, formed under a Bismarckian Prussia, but their unwillingness to co-operate in 1850 showed how far the cause, and the method, of German unification had to travel. Equally, the fact that, by the time of the 1862 Exhibition in London, Prussia had managed to draw in all the German states except Austria, the Hanseatic towns and the Mecklenburgs, was a sign of how far, in a short twelve years, the power of Prussia had solidified.[49]

The German and French preparations for the Exhibition demonstrated clearly how much it was seen as an attempt to spread a liberal political agenda, especially in France.[50] They could look across the English Channel, however, with mixed feelings. The Industrial Revolution, which the Exhibition was celebrating, had undoubtedly made Britain the richest and most productive country in Europe. Britain had 6,621 miles of railroad track, France, a much bigger country, a mere 1,869 and Germany 3,639. Britain had seen enormous population explosions in the manufacturing cities. Manchester had swollen, in the previous fifty years, from 20,000 to 335,000; Birmingham from 15,000 to 242,000. On the other hand, with the political radicalism and liberalism which these statistics encouraged, there were also the chilling costs in the quality of human life: the Sheffield knife-grinders, whose life-expectancy was thirty-two years. More than half the men employed in the industrial towns died in their thirties or forties, usually of infectious diseases.[51]

Pessimists, Tories, Marxists and backwoodsmen could look at the exhibits, and the countries preparing them, with a sense that things could only get worse. Optimists, and those with an eye to the future, could see everyone, even the sickly working classes, being, in real economic terms, better off. And there were, for the prophetically inclined, signs of a future in which even Britain, with its colossal resources and enterprise at this date, would be outstripped. One of the categories exhibited would be Raw Materials. The size and quality of the American coal reserves were for the first time to be exhibited to the world. And, when it reached London, the gigantic tea-service, made entirely out of Californian gold, would tell, for those with eyes to see, a story of the future. Sheer mineral wealth, however, was not the only thing on display. The Great Exhibition would be an eloquent example of the phenomenon neatly summarized by one modern economist from Chicago: 'Science didn't make the modern world. Technology did, in the hands of newly liberated and honored instrument makers and tinkerers.'[52]

SIXTEEN

THE GREAT EXHIBITION

'SO VAST, SO GLORIOUS'

With the arrival of 1851 itself, the building complete, the opening ceremony weeks rather than months away, there were many problems to be overcome. The King of Prussia wrote to Prince Albert with anguish about the likelihood of terrorist attacks.

When the Exhibition opened, W. M. Thackeray published an Ode in *The Times*, clunking but true, which contained the verse:

> From Rhine and Danube, Rhone and Seine
> As rivers from their sources gush
> The swelling floods of nations rush
> And seaward pour.
> From coast to coast in friendly chain
> With countless ships we bridge the strain
> And angry Ocean separates
> Europe no more.

The King of Prussia's fear, shared by police throughout the Continent and in London, was that, together with the bales of cargo, the examples of industrial ingenuity, the inventors and the tourists, would be travelling the communist desperadoes, intent upon murdering his brother and his nephew, the heirs to the Prussian throne. This was what the German secret police had told him. 'It is time that these godless villains, the crème and general mass of whom London, grâce à vos vielles lois liberals, is full of, would be able to make a fine gesture in their fashion if they caused harm to the Prince of Prussia and his son. P.S. Countless hordes of desperate proletarians well-organised and under the leadership of blood-red criminals, are on their way to London now.'[1]

Albert, and the Foreign Office, took soundings from Lord Normanby, the British Minister in Paris, to see what the *on dit* was among French secret agents. Normanby, in some haste, clearly, for his syntax was wobbly, reported on 'the projects of the Revolutionary Party, which has no doubt a most extended European organization, gives warnings of various characters, as to the use of which it is intended to be made the promiscuous assemblage of foreigners to be collected within the next few months in England'. Normanby's spies told him that the communists were moving around in the populous regions of Britain, 'to pervert the spirit of the population by Socialist emissaries, who are to circulate about by the conveyances established for the purpose of carrying the district population to the Exhibition. For the purpose of corrupting the Manufacturing population, the agents seem to have been selected are Germans rather than French, from their generally speaking the English Language better, and with less accent'.[2]

Terrorist attacks and communists seducing the proletarian Exhibition-goers were only some of the calamities which needed to be faced. Prince Albert, who shared with the Duke of Wellington an eye for the minor detail, the practicalities of life, had been inspecting the Refreshment Rooms, and, in February 1851, he addressed the danger of fire.

An unmistakably Albertian question was sent to Lord Granville. 'What provisions had been made to answer the apparently contradictory requirements of the Refreshment Rooms: that there should be a plentiful supply of hot food, and at the same time a minimal danger of fire breaking out?' Granville replied, somewhat tautologically:

> The meat is to be cooked away from the Building and brought in cold – the potatoes to be steamed by steam which will be given to the Contractors. It has been arranged to send boys under the flooring to clear out all the shavings and combustible matter. The water will be laid on, so as to be applicable directly to every inch of the building, so that with tolerable watching there ought to be no danger of fire. It would be impossible on account of time and expense, to cover the boarding with metal.

Granville added, loftily, but – by missing out a negative, inaccurately – 'I do not know why the flooring was not iron in the first instance, but then I am an Ironmaster.'[3]

Many historians, in their surveys of the Exhibition, have noted that modern sensibilities respond with mixed emotions to the celebrated De La Rue Patent Envelope Machine, a device which drew huge crowds and was one of the most popular exhibits. One contemporary described it as having 'a series of the most beautiful mechanical movements it is possible to conceive'. It cut and folded, gummed, forwarded and delivered thousands of envelopes per hour. It was worked by two little boys. We should perhaps feel admiration at its ingenuity, and discomfort that it was being operated by child labour.[4] The little boys, underneath the very floors of the Refreshment Rooms gathering scalding-hot wood-shavings, as the patrons tucked into meat and potatoes above their heads, remind us that we are in that foreign land, the nineteenth century. Neither Granville, Grey, nor Albert himself raised any objection on the grounds of 'human rights'.

An even more glaring reminder, however, that we are in that very different world of the nineteenth century is found in the fact that so many Europeans looked to England, aghast at its hyper-energy, which so contrasted with the lassitude of those at home. Even the Germans, stereotypically addicted to work and efficiency in twenty-first-century British eyes, deplored their own lethargy compared with that of the British,[5] while the King of Portugal, Albert's cousin Ferdinand, wrote despondently that 'no one here is ready for this exhibition, as they are in other countries... You can't imagine what an effort it takes to persuade some of these people to send ANYTHING.'[6]

Then there was the question of protocols. Right up to the day of the opening ceremony itself, the Corps Diplomatique was causing ripples of confusion to the Court of St James's. Naturally, the foreign ambassadors were all invited to the ceremony. It was assumed that one or another of these distinguished gentlemen would make a speech in honour of the Queen on behalf of the Corps. This task fell to the lot of the Belgian Minister, perhaps because, for some reason, King Leopold, whose finger was normally in every pie, would not be attending. To the idea of being represented by the Belgians, the Russian Minister took profound exception. Why should a Belgian Minister speak for the Russian Minister and Court? Baron von Bunsen, the Prussian Minister, tried to pour oil on troubled waters by explaining that the etiquette of the occasion must be decreed by the Court of St James's, and that the Russians had no place to be raising objections.[7]

Albert, on top of his general supervision of the Exhibition, took a detailed interest in the British exhibits. Among those which he had specifically commissioned were tables made from that most august of British minerals, Derbyshire Bluejohn. The Bakewell Committee wrote from that county to state that 'thanks are eminently due to His Royal Highness for the highly successful attempt which His Royal Highness has made to elevate and improve the taste displayed in the inlaying of Derbyshire marbles, by the admirable designs furnished by His Royal Highness through Mr Gruner of London.' Two tables were manufactured, by Mr Woodruff of Bakewell, following Grüner and Albert's specifications.[8]

Such exhibits could cause no controversy, but this was not universally the case. Some appeared in danger of provoking downright hostility, possibly riots. Here again, the Belgians seem to have been the unwitting cause of trouble. Contemporaneous with the planning and establishment of the Great Exhibition had been Bishop Wiseman's controversial decision to set up a rival Roman Catholic hierarchy of bishops in England, parallel to their Anglican equivalents. Lord John Russell, by drafting the Ecclesiastical Titles Act, had declared the Roman measure illegal. It was with some innocence of all this, and its implications, that the *Gothaisches Tageblatt* listed, among the ceramics from the Henneberg und Comp. Porzellan-Fabrik, such as a fine fruit vase and a double-handed drinking-beaker, 'a porcelain picture of the Pope'.[9] The Belgian Commissioners for the 1851 Exhibition, being closer to Britain, were more aware of the potential offensiveness of such objects in Protestant London. A '*brodeur de Bruxelles*', M. Van Halle, claimed acquaintance with Wiseman, who had recently been elevated to the College of Cardinals. His promised exhibit to the Exhibition was a set of figures depicting the Pope and Twelve Cardinals, all sporting exquisite vestments, and lacy *broderie* of M. Halle's design. While admitting the impressiveness of these figures, especially that of Cardinal Wiseman, the Secretary to the Commission, M. Bombay, recommended that only the embroidery should be displayed – while not wishing to seem intolerant, M. Bombay suggested, '*Les vêtements, oui; les poupées, non.*'[10]

Way in advance of the Exhibition, Protestants were objecting to the Medieval Court in the Crystal Palace, planned and executed largely by Augustus Welby Pugin. 'I saw Pugin this morning,' Granville assured Grey on 20 March, 'and found him perfectly ready to do anything we wished.'[11]

As the very day approached, however, the organizers were faced with a decision which it seems, in retrospect, almost incredible they had not confronted before. The Queen was to open the Exhibition. The musicians had been trained and hired. The Archbishop of Canterbury had been enlisted to say a prayer. The Corps Diplomatique would – whether or not any of them made a courteous address to Her Majesty – be amassed in their court dress. But no one else. This Great Exhibition, which had been designed to declare the unity and amity of the entire human race, would be opened, as it were, in secret. No ordinary member of the public would be allowed into the Crystal Palace until the royal party had been safely spirited away.

It is understandable that – given the warnings from the police forces of Britain, France and Germany – they feared that an assassination attempt would be made. After all, by now, there had been several attempts on the Queen's life. If she were to mingle with hundreds, perhaps thousands, of people in the Crystal Palace, there could be no way of guaranteeing her safety. On 19 April, less than two weeks before the opening, Albert was telling Lord Granville that while he and the Queen were 'prepared to listen and give their best consideration to any practical and feasible plan for opening the Exhibition which might be more popular than the present arrangement', he was terrified of the risk. 'That the Queen could move about in a moving Crowd of from 10–15,000 persons in a Building so filled with Obstructions is a manifest impossibility'.[12] The very next day, however, the Queen herself realized that the howling critics in the press were right. This was not a moment for caution or cowardice. She decreed that season ticket-holders be admitted to witness the opening ceremony.

Though Albert had considered this option 'a manifest impossibility', it was in fact not merely a brave choice, but one entirely in keeping with Albert's wish to make the monarchy more accessible, and more observable. Both of them knew from experience, however, that they were taking a risk, and there was surely no other monarch in Europe who would have appeared, at this date, in such close proximity to crowds.

Winterhalter's *The First of May 1851* draws, even more than his Holy Family depiction of them in the drawing room at Osborne, upon the iconography of Renaissance Christian art. It is a magnificent painting, and its political message could not be clearer. In the foreground kneels the Duke of Wellington. His birthday was 1 May, the opening day of the Exhibition, and it was also the birthday of the third of the royal princes, Arthur, Duke of Connaught. The child holds out lilies of the valley, a traditional good luck charm. It is an Adoration of the Magi, with only one magus. Wellington represents the old order; he was the Last Great Englishman, the Tory of Tories, the Victor of Waterloo. He holds up a jewel-encrusted casket. Prince Arthur later heard rumours that it was a casket intended to be opened on his twenty-first birthday, and that it was filled with priceless treasure. This rumour, his mother told him in 1871, 'is utterly without any foundation... Dear Papa & Winterhalter wished it to represent an Event, like Rubens and Paul Veronese did, periods of History – without any exact fact... It only shows how wrong in fact it is not to paint things as they really are.'[13] Not that the Prince went without presents from his godfather. The Queen's journal recorded 'the visit of the good old Duke on this his eighty-second birthday, to his little godson, our dear little Boy. He came to us both at five, and gave him a golden cup and some toys, which he had himself chosen, and Arthur gave him a nosegay.'[14] So there was some truth in Winterhalter's allegory.

The event which the painting symbolizes is clear enough, for the Crystal Palace is seen in the background. Winterhalter shows us that the old order kneels before the young pair because of the Great Exhibition. It is a demonstration of the new order, wealth created by trade and ingenuity. The Queen, wearing the George III fringe tiara,[15] made from brilliants cut from stones belonging to her grandfather, looks downward with a demure, shy, mysterious smile. She is the Madonna of the picture. Prince Albert, standing behind her, is no St Joseph, however; he dominates the picture. Like the Iron Duke, he is wearing the uniform of a British field marshal. He is holding a paper in his hands, presumably a plan or programme of the Exhibition. He gazes confidently out of the canvas towards a great future. He is truly the master of the situation. He had taken a gamble, of associating himself with the Great Exhibition. He had been reminded by a carping press that he thereby risked associating the monarchy with a failure. But from the very first blast of the Hallelujah Chorus, which burst forth after the Queen had declared the Exhibition open, it was clear that a stupendous success was about to be demonstrated.

In all the millions of words in Queen Victoria's journals, which do not eschew hyperbole, there are no more ecstatic pages than her description of 1 May 1851. She described the immense crowds in Green Park and Hyde Park, their approach to the immense edifice from which hung flags from all the nations upon earth: 'The glimpse of the transept through the iron gates, the waving palms, flowers, statues, myriads of people filling the galleries and seats around, with the flourish of trumpets as we entered, gave us a sensation which I can never forget… The sight, as we came to the middle where the steps and chair (which I did <u>not</u> sit on) were placed, with the beautiful Crystal fountain just in front of it, was magical – so vast, so glorious, so touching'.[16] She did not exaggerate.

> The tremendous cheers, the joy expressed in every face, the immensity of the building, the mixture of palms, flowers, trees, statues, fountains – the organ (with 200 instruments and 600 voices, which sounded like nothing), and my beloved husband the author of this 'Peace Festival', which united the industry of all nations of the earth – all this was moving indeed, and it was and is a day to live forever. God bless my dearest Albert, God bless my dearest country, which has shown itself so great today![17]

The fact that Victoria and Albert had walked among their people without the slightest danger was a source of pride to the political classes.[18] It could not have happened in any other country of the world.

Of course, when we, of the twenty-first century, imagine going round the Crystal Palace, we know that we should see something very different from what confronted the eyes of the *six million* visitors in the summer of 1851. We know, for example, that the Exhibition did not herald an era of universal peace. Within three years of the Exhibition closing, Britain and France would be fighting a war against Russia. There are many, moreover, who could question Cobden and Prince Albert's belief that international trade inevitably led to peace. Such commentators would say that, on the contrary, trade rivalries were as much a cause of war as the old credal and territorial rivalries which led nation to take up war against nation in the past.

Likewise, the Indian section of the Exhibition, which for most Victorian visitors was a source of wonder and pride, might, were we to visit it from our era, very well occasion embarrassment or anger. One of the most popular of all the exhibits was the Koh-i-Noor diamond, which was in a 6-foot-high cage, locked to the floor, and surrounded by other jewels. It was lit by gaslight from below. Weighing 186.5 carats, it was an object to be gazed upon for its own sake, quite apart from its significance. For the Victorian imperialist, it was not just a great jewel, but a symbol of British power. It was, in fact, loot taken from Lahore by British forces, and was the rightful property of the Maharajah Duleep Singh, as, in grown life, he was brave enough to point out. The Queen herself always felt embarrassed by 'owning' this object, and had a close affection for the Maharajah, whose portrait by Winterhalter shows him as a boy of fifteen, to whom she first showed the jewel in 1854. The Marquess of Dalhousie, who took possession of the jewel in Lahore and personally presented it to the Queen in 1850, expressed the hope she might 'long live to wear the Koh-i-Noor as a trophy of the glory and strength of Your Majesty's Empire in the East'.

(Prince Albert was disappointed by the jewel, incidentally, feeling it had been badly cut and was so lacking in brilliance that, after the Great Exhibition, he entrusted Sebastian Garrard to re-cut it, reducing it to a little over 105 carats.[19])

Whatever thoughts the imaginary twenty-first-century visitor to the Crystal Palace were to entertain about the British Empire, these thoughts could not be Victorian ones. 'It was all so unimaginably different/And all so long ago,' as a poet of the 1930s wrote about the Ancient Greeks.

And what would such a visitor make of the exhibits themselves? Of the Water Closet, Patent, by Jonathan Downton, of Commercial Road, Limehouse, the portable Cleansing Machine, from Deane, Dray and Deane, Wheel (model of) and Windmill (model of) from William Fitt of Ponder's End, of the Fire Engine, made entirely of Galvanized Iron, worked by 9 Men and will deliver 44 gallons a minute, of the many steam engines, for marine and land use, of the Pile Driver submitted by the Electric Telegraph Office, Lothbury, of the Oscillating Cylinder ingeniously supplied by H. J. Day of 27 Lower Northampton Street, Clerkenwell, every one of whose machines was of interest to HRH Prince Albert?[20]

John Ruskin scornfully said that the Crystal Palace had only one thought behind it, to build 'a greenhouse larger than ever greenhouse was built before'.[21] Many of the exhibits shared, with the building in which they were displayed, a record-breaking quality which was seemingly pointless. The penknife with eighty blades, an object balancing on an encrusted Renaissance-style pediment, was made by the great Sheffield stainless steel company, Joseph Rodgers and Sons. They had sent it in to make a sensation, and it is an object which is still remarkable, merely for having so many blades. A sort of metallic Struwelpeter, rather than being the best, still less the most practical, object they made. The firm of Jennens and Bettridge of Belgrave Square, London, and of Birmingham, exhibited a papier-mâché piano.[22] Prince Albert himself, as well as his justly celebrated model houses constructed of hollow bricks, affordable and decent dwellings for the working classes, chose as an exhibit a garden seat made entirely out of coal. There were tables and statues made of zinc on display in the Crystal Palace. No wonder the young William Morris found the whole Exhibition 'wonderfully ugly'.[23]

Posterity has, on the whole, sided with Morris. By the time of the Festival of Britain in 1951, the organizers of that event seemed to be guided by the principle of separation from the ideals and aesthetic of their Victorian predecessors. Their aesthetic was spare, austere, Scandinavian-socialist and, surely, a bit bland. Their guiding principles were anti-imperialist, anti-triumphalist. Numb from the recent wars, the men and women of 1951 looked forward to a pared-down, economical, modest Britain, rebuilt on a Swedish or even a Finnish model. Nikolaus Pevsner's book about the Great Exhibition, written to coincide with the 1951 Festival, was a sneering indictment of the factory-made, ersatz 'antique', over-decorated Victorian triumph. Pevsner was a professor of architecture who had travelled throughout Europe, and been able to see the glories of European achievement. So too, in his own day, had been John Ruskin, who was in a position enjoyed by very few of his contemporaries.

Neither Ruskin in 1851 nor Pevsner in 1951 were quite able to imagine what it was like to be one of the six million visitors who had not enjoyed their own personal advantages. One of the most eloquent commentaries on the Great Exhibition was George Cruikshank's cartoons of London and Manchester in 1851. The streets of the northern manufacturing hub and commercial centre were deserted. The world and his wife had gone to the Crystal Palace.

In January 1851, Paxton had infuriated Prince Albert by writing to *The Times* and suggesting that there should be days when the Exhibition was open *free of charge* to the poor. This upset the elaborate hierarchy of costings, which the Commissioners had devised in order to recoup their costs. On 25 January, Paxton realized he had jumped the gun, and wrote penitently to Colonel Grey, asking him to convey to the Prince 'the sincere regret I feel that any act of mine should give His Royal Highness annoyance, nothing could be further from my thoughts, when I wrote the letter in question, on the contrary, I thought the Royal Commission would be rather glad to see what turn public opinion would take in any Scheme, that might require a Parliamentary Grant to carry it out; and feeling myself independent to act, I wrote the letter.'[24]

The Commissioners decided to issue a number of season tickets at three guineas apiece. Admittance on the second and third days of the Exhibition was to be £1. The fourth day's admission was five shillings, and this would continue until 22 May when the price would drop to one shilling for the first four days of the week, rising to half a crown on Fridays and five shillings on Saturday afternoons.[25] The shilling days were understandably popular. This would have been easily affordable for working-class visitors. An agricultural labourer would at this date have been paid £20 per year; a maid of all work, ten years later, would cost, according to Mrs Beeton, between £14 and £20 per year. Lower-middle-class incomes varied between £300 and £1,000 per annum.

It was therefore possible for almost everyone in the country to afford entry on at least one of the days. A little under a quarter of the entire population (twenty-seven million living in the UK at this date) made the journey. The pessimists who predicted widespread crime or disorder were proved wrong. (It was typical of Albert that he noticed what had slipped the observation of the other Commissioners, that the exhibitors themselves might be admitted to the Exhibition, and he arranged that, bringing two friends each, every exhibitor should be allowed to visit the Exhibition when it had closed on specific days.[26])

What the visitors saw was an abundance of evidence of how the world had changed and was changing. Ruskin and Morris (and later Pevsner) might wince at the elaborate gas lamps and brackets, designed by Winfield of Birmingham. They might feel understandable dismay at the thought of some self-made mill-owner or northern alderman having his food fetched from the extraordinarily lumpy neo-Jacobethan 'Kenilworth Buffet'. And, true, if the aim of the Exhibition had been, at least in part, to improve the quality of industrial design, it could not be viewed as an unqualified success.

But it would be to miss the point. Part of the excitement of the Exhibition was its incontinence, its variety, its juxtaposition of the graceful statuary of Puttinati of Milan with the giant rubber boat built by Goodyear. You would come out of the gallery exhibiting Scottish vegetable produce, gasp at the Gobelins tapestries in the French room, and then find yourself confronted by some of the most ingenious machines ever devised, such as Nasmyth's Steam Hammer, capable of making steel girders of a size never before seen, but so delicate that you could use it to crack a boiled egg; or the Jacquard loom, developed in France, and capable of producing the most elaborate woven patterns. The fact that so many people could read about the Exhibition in so many far-flung lands and so soon after the correspondent had filed his article was in part attributable to the ingenuity of Herbert Ingram's vertical press, on which *The Times* was printed and which was displayed in the Exhibition.

Once settled at Osborne in August, Albert began drawing up a detailed memorandum of how to spend the profits of the Great Exhibition. And it is here that we see him laying the groundwork for his most lasting achievement: namely, the establishment of the so-called Albertopolis in Kensington.

Most historical judgement is debatable; which is why, from generation to generation, the same stories, the same biographies, can receive such a variety of treatments. The Great Exhibition itself is a case in point: was it as truly internationalist, in Albert's idealistic vision, as he believed, or was it based on colonialist and exploitative ideas? Was it internationalist or nationalist? Did it glorify free trade at the expense of exploiting the working class, and child labour? All these questions will remain forever open, endlessly discussed from different viewpoints. Likewise, Prince Albert's contribution to the history of the British monarchy: did he save it, or by creating 'the

Royal Family' as an icon of bourgeois rectitude did he store up trouble for future generations of the family, whose private lives were held up to a merciless scrutiny, and whose inability to match his monogamous and saintly behaviour made them look dissolute or negligent of public duty?

And then again, we could debate endlessly the question of Prince Albert in Europe, Prince Albert as the mind behind the dynastic marriages into which his children and grandchildren were guided. To my mind, he cannot be blamed for dreaming that Prussia and Russia would one day become more liberal, more like England; but one can see how the opposite argument would go: that his lack of realism about Europe actually made things worse: that he put his eldest and most beloved daughter through hell by making her the ambassador for all his enlightened values in the increasingly militaristic and autocratic Germany of Bismarck.

So there are many ways of viewing Albert's achievements and that is why he will deserve the scrutiny of historians for generations to come. One thing, however, cannot be disputed. That is that he was the driving force behind the creation of the Albertopolis: what would become the Royal Albert Hall, the Victoria and Albert Museum, Imperial College, and the other colleges nearby devoted to craft and needlework and music; as well as the Natural History Museum and the Science Museum. Of course he did not establish all these wonderful institutions single-handed, but it was his vision which enabled the multi-faceted, multi-talented Victorian founders who brought to birth all this cultural glory.

The Great Exhibition, which so many pundits had predicted would be a disaster, left a profit of £186,000. Albert devoted August, while he was down at Osborne, to drawing up a memorandum. First, he proposed that '20 to 30 acres of ground nearly opposite the Crystal Palace on the other side of Kensigton [*sic*] Road called Kensington Gore (including Soyer's Symposium) are to be purchased at this moment for about £50,000'.[27] The Royal Commission would eventually receive its Supplemental Charter on 2 December 1851, which enabled it to have charge of the surplus from the Exhibition. It had by then become clear that they could buy a much wider extent of land. The contractor John Kelk was put in charge of the negotiations, which involved them buying 130 acres at a cost of some £400,000, underwritten, but not financed, by the Treasury.

As had been the case during the planning of the Great Exhibition, the Prince's eye retained its double vision, of the bigger picture, and of the minutiae. March 1856 found Colonel Grey, on Prince Albert's instructions, writing to Bowring. Albert had been down Gore Lane in Kensington with a map and marked with a cross the trees which would need to be removed before a building programme began. 'Of the cherry trees, he thinks you should only leave about 1 in 3'.[28] When the South Kensington Museum came to be built, the Prince gave advice about how it should be decorated, favouring 'distemper for the walls and screens'.[29] He favoured painting the walls of the museum building 'as a tent with stripes, which HRH thinks might be the prettiest'.[30] As the building neared completion and he visited the undecorated galleries, he added, 'merely as a suggestion', that the roof should not be painted the same colour as the walls; and he paid close attention to how the pictures, when hung in the gallery, would be lighted.[31]

The story of how the Albertopolis developed lies slightly outside the scope of this book, since Albert would die in 1861, and so many of the great buildings and institutions in Kensington came into being after this date. Henry Cole and Colonel Henry Scott saw through the building of the Royal Albert Hall. Henry Cole was the first director of the South Kensington Museum, opened on 22 June 1857 by Queen Victoria, which was rebuilt as the 'V and A', designed by Aston Webb, in 1899–1909. The Natural History Museum, designed by Alfred Waterhouse, opened in 1881, and was largely the inspiration of the great zoologist Sir Richard Owen. Imperial College, today one of the foremost scientific universities in the world, grew out of Prince Albert's patronage of the Royal College of Chemistry in 1845. Later came the Royal School of Mines and the City and Guilds Colleges.

All these institutions, together with the Patent Museum, the Royal School of Art Needlework, the National Training School for Music, which became the Royal College of Music in 1883, and the Royal College of Organists, are manifestations of the rich and varied talents of the Victorians. Albert was their midwife, not their creator. It was his common sense and his overall vision which prevented the surplus funds of the Exhibition being frittered away in a whole variety of disparate schemes.

In one of his ambitions, however, he was unsuccessful. Early on in his plans for South Kensington, he got wind of the scheme, supported by some in Parliament, to move the National Gallery from Trafalgar Square to be the centrepiece of the new complex of museums, colleges and galleries.

He had two motives. One was a genuine concern for the paintings, which he believed to be in danger from the high levels of pollution in central London which was so often black with CO_2-fuelled Dickensian smogs. Another motive, and perhaps the chief one, was his desire to control. To some temperaments, there is something pleasing about London's lack of Napoleonic Order. The British Museum, with its many treasures, is in Bloomsbury, housed since the 1830s and 1840s in Robert Smirke's superb Greek Revival building. The National Gallery (founded in 1824 after George IV and Sir Francis Beaumont persuaded the Government to purchase thirty-eight stupendous paintings by Raphael, Rembrandt and Van Dyck) is in the less impressive building (1832–8) by William Wilkins. Albert's tidy mind would have loved them all to be in the same giant campus, in what he optimistically supposed would remain forever the unpolluted country air of Kensington.

For six years after the closure of the Great Exhibition, Albert continued to harangue Prime Ministers and Chancellors of the Exchequer with the idea that his Albertopolis would be incomplete without the National Gallery. For a while, there was a thought that Kensington Palace itself might be converted into the Gallery, though Albert and Victoria rejected this as 'impossible in every view'.[32] When

The Times, unsurprisingly, opposed the idea of moving the Gallery to Kensington, Albert told Lord John Russell that the Queen was 'indignant... and trusts the Government will not allow itself to be beaten upon the National Gallery question by a knot of persons who seem to delight in nothing but making mischief and showing their own importance'.[33]

Albert refused to recognize that the Government was not, as it happened, committed to moving the National Gallery. By June 1857, a Royal Commission, headed by Lord Broughton, with H. H. Milman, Michael Faraday, C. R. Cockerell and George Richmond, was set up to decide the question. They summoned as witnesses Eastlake, Cooke, Mulready, Ruskin and J. F. Lewis, and the broad consensus was that the pictures, though in danger of pollution, could be protected by being framed in glass. Their final report was that 'considered architecturally, the site of Trafalgar Square stands, by common consent, without a rival'. The chief argument for moving – the environmental one – had to be weighed against the fact that building and traffic in Kensington was becoming as dense as in central London, and that it would not be long before pollution attacked the once-rural retreat of the Queen's girlhood. 'Whether it will continue to be so when the neighbouring land to the south-west and west shall be covered with buildings and become part of the metropolis may by some be doubted.'[34]

Albert was beside himself with rage at the findings of the Commission. Indeed, he dismissed them as 'no report at all, not being unanimous'. (They in fact voted three to one.) They were 'Gentlemen who opine that, although Kensington Gore has many important advantages over any other site, they prefer Trafalgar Square as more <u>accessible</u>'.[35] There was contempt in his underlining of the word.

His failure to move the National Gallery – for which we must surely be grateful – does not, however, diminish the achievement of the Great Exhibition and his huge contribution to its creative aftermath.

As for the dear old Crystal Palace itself... Albert watched its demolition with regret. To Lord Derby, he wrote, 'One cannot help regretting that so fine an edifice should be doomed to destruction, but must console oneself with the consideration that it was intended only for a temporary purpose which it has admirably fulfilled.'[36]

It was demolished but not destroyed. It was re-erected and enlarged near the River Thames at Sydenham, where for eighty years it was used as a concert hall, theatre, menagerie and exhibition space. Outside it were fountains, supplied, in Paxton's typical manner, by two huge water towers. There were regular firework displays. Fire was to be part of its destiny. Flames destroyed the great Alhambra Court in the 1860s. But Francis Kilvert, the diarist of rural life in Wales during the 1870s, gave us a glimpse of the Crystal Palace in January 1870, when, during a visit to friends at Mitcham, he attended a pantomime, *Whittington and His Cat*. He and his friends were 'very much amused'.[37]

The innocent suburban philistinism which the young parson enjoyed in 1870 was, perhaps, far from Prince Albert in spirit, but the Crystal Palace, with its raucous and varied life, was, long after his death, part of his legacy. Then, just at a point in history when the British monarchy was perhaps entering its gravest crisis – that of the Abdication of King Edward VIII – the Crystal Palace burnt down: on 30 November 1936. The conflagration could be seen right on the other side of London, from Hampstead Heath and Primrose Hill. Those who watched it burn, and who were about to hear of King Edward's intentions to marry an American divorcee, saw much of what Prince Albert stood for go up in flames. All that survived of the exhibits were some strange plaster prehistoric animals, themselves as obsolete in that 'low dishonest decade' as the Victorians themselves.

SEVENTEEN

STORMY WEATHER

It was a bitterly cold September at Balmoral. Even the Queen, with her fondness for the cold, felt it to be uncomfortable. Then, on Thursday, 16 September 1852, she told her journal, 'We were startled this morning at 7, by a letter from Col: Phipps, enclosing a telegram with the report, from the 6th Edition of "the Sun", of the death of the Duke of Wellington, the day before yesterday. We however did not at all believe it. Would to God that we had been right, & that this day had not been cruelly saddened in the afternoon!'

Like many of Victoria's set-piece journal entries, the grand narrative of a national loss is set within the context of something which was, on the surface, domestically trivial. When she went for her afternoon walk, she realized she was missing the watch which the late Duke had given her. Panic ensued and a gillie named Mackenzie was sent back to the Castle to search for the watch, which, inevitably, she had left behind by accident.

Victoria had the qualities of the almost-novelist, and her diary often conveys the way that, when the mind is concerned with a great matter, it concentrates for protection upon a small. Interspersed with her usual descriptions of all the scenery they passed, 'We walked along the Loch, the road being excellent & having been widened to admit of a carriage, — to the Alt na Dearg, a small burn & falls, & very fine & rapid. Up this a winding path has been made, admirably done, — which we rode up, but some little bits are rather steep for riding. The burn falls over red granite rocks, etc etc', and anxiety about the loss of the watch, there was the sense that Britain had lost a giant. When Mackenzie rejoined the party and assured the Queen that her watch was safe at home, he also brought the latest delivery of letters,

> amongst which was one from Ld Derby, which I tore open. Alas! it contained the confirmation of the fatal news, that Britain's pride, her glory, her hero, one of the greatest men, she ever produced, was no more! What a great & irreparable loss!... One cannot think of this country without 'the Duke', our immortal hero! In him centred almost every earthly honour, a subject could possess, his position was the highest a subject ever had, above all Party, — looked up to by all, — revered by the whole nation, the trusted friend of the Sovereign!... To Albert, he invariably showed such kindness & such confidence. There will be few dry eyes in the country. — — We hastened home on foot, to the head of the Loch, & then rode back to Alt na Guithasach in a very heavy shower. Our whole enjoyment spoilt. A gloom cast over everything! Albert dreadfully sad... Talked much of all the arrangements, which must result from the sad event.[1]

Albert's comment in a letter to Stockmar about the Great Duke was self-revealing: 'He is a fresh illustration of the truth, that to achieve great results, and to do great deeds, a certain one-sidedness is essential. That feature of his character, to set the fulfilment of duty before all other considerations, and in fulfilling it to fear neither death nor the devil, we ought all of us certainly to be able to imitate, if only we set our minds to the task.'[2] Both the Queen and the Prince managed to blot from their memory the Duke's early opposition to their marriage, his guileful refusal to squash rumours that Albert had been a Catholic, his part in the parliamentary decision to reduce Albert's allowance, and their wish to exclude Wellington from their wedding. The 1840s saw a blossoming of admiration and friendship on both sides. In 1845, Victoria had invited herself and Prince Albert to stay at the Duke's seat, Stratfield Saye. It was where Albert, according to Florence Nightingale, learned to 'miss at billiards'. Both Albert and Victoria had been delighted by the Duke's attentiveness. He fetched her for dinner, helped her to more pudding, and took her for informal gentle chats in his library while Albert was with Anson and the gentlemen in the billiard room. Wellington, ever so practical and brilliant at all aspects of military administration, formed the highest opinion of the Prince, and decided that Albert would be the best person to succeed him as Commander-in-Chief of the British Army.

The reason was partly political. Wellington, a high old Tory, believed a royal personage should command the army, 'as with the daily growth of the democratic power the executive got weaker and weaker, and it was of the utmost importance to the stability of the Throne and Constitution, that the command of the army should remain in the hands of the Sovereign and not fall into those of the House of Commons'.[3]

This would become the hottest of potatoes during the first Liberal administration of W. E. Gladstone, from 1868 onwards; by this time, the C-in-C was Victoria's great friend and cousin, George, Duke of Cambridge. She and Cambridge both shared Wellington's view of things, whereas Gladstone and the Liberals very decidedly did not. The Secretary of State for War, Cardwell, pushed through reforms of the army, abolishing flogging, for example, and attempting to introduce a more efficient way of training officers than merely selling commissions in the grander regiments to rich young aristocrats. These differences all lay in the future. When Wellington died, however, Albert was wistful at having, rather reluctantly, turned down the Duke's suggestion that he should be in charge of the army. They lived in dangerous times. Although a British reader might grow impatient with Albert's constant belief that British painters, musicians, scientists, universities, etc. were all inferior to their German equivalents, it was unfortunately true that the Prussians trained their officers in a modern, professional manner. A war loomed, two years after

Wellington's death, in which the incompetence of the British was to be advertised to the world, particularly in all the matters at which Albert, and Wellington, would have been adept – supplies, medicine, transport of the troops, and so on. It is hard not to believe that many lives would have been saved in the Crimea had Albert, rather than Lord Hardinge, succeeded Wellington as C-in-C.

Be that as it may, there was nothing fake in Albert and Victoria's grief-stricken reaction to the Duke of Wellington's death, and, indeed, it was felt by almost everyone.

Greville, so sparing in his praise of human beings, wrote that Wellington was 'beyond all doubt a very great man – the only great man of the present time – and comparable, in point of greatness, to the most eminent of those who had lived before him'. And Greville added, having listed the Duke's great qualities – the simplicity, the lack of vanity or personal ambition – 'It was in this dispassionate unselfishness, and sense of duty and moral obligation, that he was so superior to Napoleon Bonaparte, who, with more genius and fertility of invention, was the slave of his own passions, unacquainted with moral restraint, indifferent to the wellbeing and happiness of his fellow-creatures; and who in pursuit of any objects at which his mind grasped, trampled underfoot without remorse or pity all divine and human laws'.[4] 'Here,' Albert wrote to his stepmother, 'we live under laws, and not under despotism, like the blessed Continent.'[5]

As Britain prepared to bury the Great Duke with, as the Poet Laureate said, 'a Nation's lamentation', it was with just such thoughts of governance and *Realpolitik* at the back of everyone's mind. Wellington had not defeated Napoleon single-handed, but his military genius had set the seal on Napoleon's defeat. In the generation since, the other nations of Europe had been convulsed with revolutions and indecisions about how they should be governed. Victoria and Albert, consumed as they often were by egoism and self-regard, were not megalomaniacs. They would have agreed with Greville that they were not 'great' in the way that Wellington or Napoleon had been deserving of this epithet. They had, however, the humility to see that a constitutional monarchy made life less hellish for those who lived under it than many of the political alternatives. Their patient collaboration with the representative politicians in both Houses was a crucial part of the story. Both of them felt extreme frustration when confronted by politicians, above all Palmerston, whose policies they deplored. But they both helped, in their different ways, to further the benign purpose to which they felt called, helping Britain be a better place to live than France, Germany or Italy. Marx, in his best joke which is repeated so often as now to be spoiled, compared the rise to power of the two Napoleons – the first Emperor and his preposterous nephew. It was Victoria and Albert's destiny to live with the 'farce' of Louis Napoleon; but in burying the Great Duke, they recalled the years when Europe lived through 'tragedy'.

Albert took charge of the funeral arrangements including the design of the enormous bronze carriage, made from melted-down French field guns from Waterloo. The original sketch design was made by the artist Richard Redgrave, who resented not being given more credit.[6] Henry Cole took the drawing and asked the German architect Gottfried Semper to adapt it. The result, further tweaked and 'improved', looked like what it was: a funeral carriage designed by a committee. It may still be seen at Stratfield Saye, the seat of the Dukes of Wellington. It was a funeral such as the century had never seen, taking place over two months after the Duke's death, on 21 November. The old man had been embalmed, to give time for the Poet Laureate to write his great funeral Ode, and for the capital to prepare for the funeral in London. A million and a half people lined the streets to witness it, a huge procession, from Chelsea Hospital, where the body had lain in state, to St Paul's on Ludgate Hill. After the dignitaries of Church and State in their carriages, the marching bands, the officers carrying the Duke's standard, the eighty-three Chelsea Pensioners in their scarlet coats, the mourning coaches of the heralds, the representatives of the Tower, the East India Company, the Cinque Ports, and so on, Clarenceux King of Arms walked, carrying the ducal coronet on a cushion. Then came the vast funeral carriage,

21 feet long and 12 feet wide, weighing 18 tons, and decorated with trophies and heraldic achievements.

As it wobbled along Pall Mall, drawn by twelve horses, it sank into the mud by the Duke of York's statue. Sixty strong men were required to drag it out with ropes. Then, as it teetered down the Strand, it came to the next obstacle, Temple Bar, where the structure had to be lowered by machinery for it to squeeze through. The mechanism for transferring the coffin to the bier at the west door of the cathedral failed to work for over an hour. Even if you did not share Greville's view that the car was 'tawdry, cumbrous and vulgar', or Carlyle's, that the design was the 'abominably ugliest', the logistic failures of the procession were unworthy of Wellington, a 'details man' who would have planned everything so punctiliously. Wellington's twentieth-century biographer Lady Longford wrote of the funeral, 'nothing was wanting; except the simplicity which had been the hallmark of the hero'.[7] For many of the witnesses, including the Queen, who was reduced to tears by the sight, the most moving thing was seeing Wellington's groom, John Mears, leading his riderless horse behind the cortège, the reversed, eponymous, boots hanging from the saddle.

It was understandable that Albert, only thirty years old when the Duke had proposed it, felt unequal to following in Wellington's footsteps. Lord Hardinge became the Commander-in-Chief, a jobbing staff officer who had impressed Wellington during the Peninsular War in the early days at Torres Vedras, nearly forty years earlier. He had since done three years in India, and seen respectable service in the Sikh war, but in 1852, he was an elderly sixty-seven. The *Dictionary of National Biography* fairly tells us, 'age was telling on him, and a feeling of loyalty to his departed chief rendered him unwilling to disturb routine arrangements that had been sanctioned by Wellington'.[8] In asking Albert to take over as C-in-C Wellington had spotted a young man who would not have shrunk from making necessary reforms, and whose willingness to 'disturb routine arrangements' had made itself known at Cambridge, and in consequence in all the universities – Oxford, Trinity Dublin and in Scotland; in the

Royal Household; in the world of the arts; and in industry. Albert could not see an arrangement without wishing to disturb and improve it, and as he advanced through his thirties, with an energy which was far from tireless – his work made him ill and tetchy – he turned his attention seemingly to everything: to everything, that is, except what was making his wife so very unhappy. Far from putting his feet up, after the exhaustion of the Great Exhibition, he had merely thrown himself into the multiplicity of all his activities. And perhaps it was this, the knowledge that as Commander-in-Chief of the British Army he would simply not have had time to be managing everything else in the three kingdoms as well, that ultimately persuaded him to decline the old Duke's suggestion.

When Victoria called him 'a terrible man of business', in a letter to Uncle Leopold of February 1852, she was not accusing him of business incompetence, but saying that he was addicted to work, and could not, seemingly, relax: for even family life was a relentless programme of education for the children.

Over the door of Schloss Friedenstein, at Gotha, where Albert spent so much of his childhood, are engraved the words '*Friede Ernehret Unfriede Verzehret*', 'Peace nourishes, conflict eats you up'. When his own marriage was under strain, he felt the truth of the words. It had begun, in 1840, as a passionate love affair – on her side at least – punctuated by storms. Perhaps she even felt that the loving reconciliations, after their tempests, made the squabbles possess something approaching sweetness. He had never felt this. He complained very early to Stockmar that she was too passionate for him. By the time they had been married twelve or thirteen years, the balance of things had shifted. The love affair punctuated by storms was now an everlasting battle punctuated by moments of relative calm. Albert, with his busy schedules, found her histrionics ate most terribly into the work timetable. 'I try my best to be patient,' he told her testily in May 1853, 'but I feel the dreadful waste of most precious time, and of energies which should be turned to the use of others in "mere and long recriminations". I feel they give you relief when you have made yourself unhappy by exciting your own feelings, and then I am anxious to further them, but this requires uninterrupted time, and a mind quite disengaged from other thoughts; this unfortunately I have not often.' He was candid enough to tell her that he no longer had the time for her unhappy displays.

This particular row took place at Osborne. It was just a month after Leopold, their eighth child, had been born. From this perspective of time, we can see that she was clearly suffering from postnatal depression. Albert saw merely intolerable histrionics. It was three years since the birth of the last child, Arthur. Four would pass before the birth of another.

The storm, as they nearly always did, arose out of nothing. Albert and Victoria were sitting together turning over the leaves of a collection of prints. 'I complained of your turning several times from inattention the wrong leave [*sic*],' Albert punctiliously recalled, 'in a book which was marked by me as a Register... This miserable trifle produced the distressing scene'.[9]

There was an outburst. Victoria, feeling herself at fault, as she turned over the wrong print, accused Albert of being unfeeling. He tried to point out the 'groundlessness and injustice of the accusations'. Presumably his self-defence was accompanied by a little laugh, because she then ratcheted up the row and said he was 'turning what you say into ridicule'. He then retreated into silence. He claimed this was because he believed trying to 'reason with a person in a state of excitement' would make things worse. He 'turned a deaf ear to your attacks'. She began to shout at him. It was at this point of the row that he felt 'I have no choice but to leave you when I see the conversation' taking this turn. 'I leave the room and retire to my own in order to give you time to recover yourself, then you follow me to renew the dispute and to <u>have it all out</u>'. By this stage of a row, he confessed, 'it is I believe out of human power, certainly out of mine', to make things better. He wrote to her: 'I try to forget such scenes as quick [*sic*] as possible and to return to our ordinary state of cordiality and unity, but this even is a special grievance and construed as a want of love for you'. Of course, by writing down the different stages of the dispute, Albert was demonstrating that he could not 'forget such scenes', even though he claimed he did 'deeply pitty [*sic*] you for the suffering you undergo'.[10]

One of the things which makes these fragmentary recollections of marital hell so especially painful to read, over a hundred and fifty years since they were scribbled out, is that it is not really a two-sided row. The rows took the form of Victoria screaming for anything up to a day, but she was screaming that she wanted love and understanding, while Albert ran from the room and wrote down why he found her behaviour intolerable. There would then be an uneasy period of sulky silence between the pair. Then, however, there was an agreement between the two of them that it was she who had been at fault. The reproaches which she cried out in her rage would be repented of. The Angel was by now completely in control. In the book which she kept entitled *Remarks, Conversations, Reflections*, she reproached herself for her lack of control. She was so blindly in love with him, up to, and beyond, his death, that it never crossed her mind, except in the heat of passion, to reproach him for his coldness, his need to control and his lack of sympathy. This would be the pattern of their lives together for the next eight years, until Albert's death.

On the surface, when things went well, there existed what he called 'our ordinary state of cordiality'.[11] For her, though, there was no such state. He was her Angel. She did not want cordiality. She wanted ecstasy. She wanted an opera. She referred to his birthday in 1857 as 'that day of all others which is blessed as having given birth to my beloved and adored Husband, the <u>purest</u>, the <u>greatest</u>, and the <u>best</u> of human beings, and like of whom, at his age and in his position, has never been seen. So beautiful too!'[12] She was in love, physically and emotionally, and his absorption in business, his tendency to snub or to imply superior intelligence or knowledge were deeply wounding when she was in a vulnerable state.

It was not possible, either by mutual discussion or by any outside 'marriage guidance', for them to find a way out of these dark pathways, for the truth is that both of them were thin-skinned, vulnerable people; both carried burdens from childhood, and both had 'unresolved issues'. Victoria's short fuse at least allowed her to let off steam; whereas Albert, who tried to use cold reasonableness as a weapon, could really only find consolation in work. The overwork, to which he was addicted, exacerbated tiredness, 'rheumatic' pains in his joints, of which he began to complain from now onwards, and intestinal troubles, which were to afflict him for the rest of his days.

Louis Napoleon Bonaparte was the third son of Napoleon I's brother Louis. His rise to power, after the tempests of 1848 and the exile of Louis-Philippe, demonstrated remarkable political deftness. Scrupulous he was not. By being all things to all men, he managed to be elected as President in 1851 by 5,434,226 votes against his rival Cavaignac. By December 1851, he had managed to disband the National Assembly and to conduct the coup d'état labelled farcical by Karl Marx. France once again had an Emperor, and a highly bellicose War Minister in Saint-Arnaud. There were many in Britain who now actually dreaded invasion.

Not so Palmerston, the Foreign Secretary, who believed that Napoleon had prevented something much worse – a counter-coup by Louis-Philippe. He breezily told the French Ambassador in London how highly he approved of Napoleon III's actions. The British Ambassador in Paris, Lord Normanby, was appalled. To Albert and Victoria's great delight, it was felt that this time Pilgerstein had gone too far. Lord John Russell demanded his resignation. And so their great political enemy was removed from the scene.

It was Stockmar who fed Albert's obsession with Palmerston. In 1853, he wrote an enormous memorandum on Palmerston for Albert to absorb.

> The Element he lives in is quarrel. He is clever. But it is the cleverness of a child, who has not the intellect enough to estimate the relative value of objects or morality enough to prefer his duty to the personal gratification of vanity and insolence.
>
> England has twice, against Louis XIV and against Napoleon, preserved Europe from despotism. Her task in 1848 was to have preserved the continent against anarchy. She failed in doing this by Palmerston's total incapacity. He sacrificed every alliance on the continent to his impatience, spite and short-sightedness. He trampled on the weak and insulted the strong.
>
> His reckless immorality, his overweening conceit in his political knowledge of affairs and his statesmanship led him to interfere on the continent in a way destructive to fine and sound English policy and to the true interests of European order and civilization.[13]

And so on. For pages. 'Flashy'. 'Crude'. 'Bamboozling'. All epithets, no doubt, which could be justified. What neither Stockmar nor Albert could bear to add was another one, which was the political reality in Britain between 1848 and 1868 – 'indispensable'.

After the abolition of the Corn Laws, British politics was in a state of flux for over twenty years. No one party or interest commanded enough support to be able to form an administration without coalition. Russell's Government staggered on for only two months after the replacement of Palmerston at the Foreign Office by Earl Granville. Lord John was replaced by the fifty-three-year-old Earl of Derby, a figure whom neither Victoria, Albert nor Stockmar could really understand.

As befitted one whose grandfather had instituted, on Epsom Downs, the greatest flat race in the history of the world, he spent much of his time at the races. (In April 1855, the Earl of Malmesbury, who would serve as Derby's Foreign Secretary, fondly recalled that

Derby had spent so much of the week at Newmarket that he had not bothered to follow the peace negotiations at Vienna that concluded the end of the Crimean War.) As well as having a passion for the turf, and being one of the richest landowners in the north of England (Cheshire and Lancashire; one of Lord Derby's nicknames was 'King of Lancashire'), Derby had published a translation of Homer. Behind his facetious insouciance he concealed fierce political ambition, but the parliamentary numbers simply could not stack up to keep him in office for long. His first spell as Prime Minister lasted just ten months. Having destroyed Peel, and their own party, by voting against free trade throughout the 1840s, the Tories in the Commons now declared themselves in favour of it. They had a Commons minority, so when the Whigs moved a vote of censure against them, accusing them of supporting free trade for purely opportunistic motives, no one knew how the vote would go. Palmerston, that old Whiggish friend of Radicals, swung it for the Tories. 'We are here an assembly of gentlemen, and we who are gentlemen on this side of the House should remember that we are dealing with gentlemen on the other side.'[14]

Derby, who was, of course, in the Lords (Palmerston's viscountcy being Irish, not English, he sat in the Commons), knew that he depended on Palmerston. In fact he spent much of that year away from the Commons, suffering from gout and, at sixty-eight, appearing much aged. When the Derby Government's Budget was debated, however, he returned to his old friends, the Peelites and the Whigs, and voted the Tories out of office. This was in part, it was admitted, because Lady Palmerston was tired of having him hanging about at home and they were 'beginning to feel the pinch without a ministerial salary'.[15] When Aberdeen was asked by the Queen to form a coalition, she knew that Palmerston would have to be a member of it. She held out against his being allowed back to the Foreign Office, and so he was made Home Secretary. Hence the paradoxical situation that Palmerston, the most gunboat-happy Foreign Secretary in history, was at the Home Office as the country moved towards war; and it was peace-loving, hesitant Aberdeen who was the unlikely warlord who took it upon the country to resist the Russian invasion of Turkey.

Not that it seemed like that when Aberdeen first took office in 1852. He and Palmerston were agreed that they should prepare for war, but both assumed that the war, when it came, would be with France. As the events relentlessly unfolded, which led finally to the declaration of war on Russia on 28 March 1854, Albert was constantly in correspondence with Stockmar, with Russell (who was serving as Foreign Secretary and waiting eagerly for Aberdeen to fall, and return the premiership to him and the Liberals/Whigs) and with many others. Few wars in history have taken longer to break out. When it actually happened, it seemed inevitable, but Albert, Victoria and Aberdeen – and his Peelite allies in the Cabinet, James Graham (First Lord of the Admiralty) and Gladstone (Chancellor of the Exchequer) – were hoping for a negotiated peace as the crisis lurched from one thing to the next – to Napoleon III taking possession of the Holy Places in Palestine for the Roman Catholics, and the Russians retaliating by extracting from the Sultans the promise of privilege to all Orthodox Christians within the Ottoman Empire; to the sinking of a Turkish squadron at Sinope in the Black Sea, to the various positionings of the British Fleet, both in the Dardanelles and in the Black Sea. Palmerston had urged from the first that the Royal Navy should in effect police the Black Sea, and threaten to sink or seize any Russian vessel which threatened the Turks. Had they managed to limit the war to naval engagements, it might have been shorter, and involved less catastrophic loss of life.

Albert's need to be involved led him to constant political involvement which was regarded by many, in the Government and in the press, as interference.

A typical memorandum, penned at Balmoral, recorded:

> I had a long interview with Sir James Graham this morning, and told him that Lord Aberdeen's last letter to the Queen made us very uneasy. It was evident that Lord Aberdeen was,

> against his better judgement, consenting to a course of policy which he inwardly condemned, that his desire to maintain unanimity at the Cabinet led to concessions, which by degrees altered the whole character of the policy, while he held out no hope of being permanently able to secure agreement. I described the Queen's position as a very painful one. Here were decisions taken by the Cabinet, perhaps even acted upon, involving the most momentous consequences, without her previous concurrence or even the means for her to judge of the propriety or impropriety of course to be adopted, with evidence that the Minister, in whose judgement the Queen placed her chief reliance, disapproved of it.[16]

This was a perfectly fair statement of the facts. What was in question was the propriety of Albert making such statements and interventions. It is probably true to say that this period, in the run-up to the Crimean War, represented the low point in Albert's personal popularity with the British public, and with the press, who, thanks to the highly popular and unscrupulous Palmerston and others, were kept fully abreast of Albert's need to interfere.

Political popularity is as elusive a thing to explain as sex appeal. Palmerston, who had both in abundance, said things which you would not, at first perhaps, imagine to be popular, still less populist. During the election of 1852 he told his Tiverton constituents why he did not want an extension of the franchise, and why he deplored the ballot box. 'To go sneaking to the ballot box, and poking in a piece of paper, looking round to see that no one could read it, is a course which is unconstitutional and unworthy of the character of straightforward and honest Englishmen'. Ballot voting was not, in fact, introduced until 1872. When the Peelites, and Russell, brought in a measure to extend the franchise to the urban working classes, and to small urban property-owners, Palmerston resigned. You would not think this was a way to wow the would-be electorate. Palmerston, however, was wildly popular with a now war-hungry press. Before issuing a bland statement of correction, he allowed the newspapers to believe that he had in fact resigned because of the 'massacre' of Sinope, and the failure of Aberdeen to retaliate against the Russians. Palmerston was now fixed in the public mind as the one Cabinet Minister with the guts and spirit for a war. He issued a short statement that he had not in fact resigned because of foreign policy. By then, he had withdrawn his resignation and resumed his seat in the Cabinet. He had also allowed the press to believe that he had been hounded out of office by the Prince. He had made sure that his letter of resignation was published in the newspapers and he had leaked the fact that he had been disgusted by Aberdeen's refusal to help the gallant Turks.[17]

The public were at this stage unwilling to attack the Queen herself, but they had no hesitation in attacking her husband. Moreover, even those who were friends and supporters of Albert, and most especially Aberdeen, found that the newspapers, malicious and unreliable as they were, did raise serious constitutional questions. They said that Prince Albert was in regular correspondence with foreign powers. This was true. They said that he had no constitutional right to be present when the Queen was in consultation with her ministers, and this was true. *The Times* questioned whether the Queen's reliance upon the Prince was politically healthy.

On 5 January 1854, the Queen wrote to Aberdeen that 'a systematic & most infamous Attack appears in the Mor. Herald and the Standard'.[18] She saw some 'design' in this. Aberdeen was sympathetic, but he was compelled to admit that the position of the Prince was 'anomalous'. It was a perfectly reasonable thing to say. At no point in history, either before or since, unless one chooses the unhappy analogy of King Philip II of Spain when married to Mary Tudor, has a reigning monarch's marital consort sought to exercise such influence on the conduct of the British Government.

Victoria was so infuriated, and intimidated, by the press attacks that she told Aberdeen she was not in a position to take part in the State Opening of Parliament. Aberdeen told her that her absence would be 'most unfortunate', but her fears had been justified. The crowds hissed her as she emerged from her carriage and entered the Palace of Westminster. Albert's unpopularity hurt Victoria deeply, and she urged Aberdeen to formalize her husband's status by giving him the title of Prince Consort. It was hardly a judicious moment to be raising such concerns. Not only did he have a war to avert, and then to fight. Aberdeen's own personal reputation was at rock bottom, with virulent personal attacks appearing almost daily in the newspapers. Disraeli, in *The Press*, the weekly he had founded with Lord Stanley, anonymously chided the Prime Minister: 'His mind, his education, his prejudices are all of the Kremlin school.' This was because he believed the aims of the war – to prevent Russia from obtaining control of Constantinople and the Straits – could have been achieved by peaceful negotiation. Albert agreed with him. Hence there arose such absurdities as the, for a while, widely believed rumour that Aberdeen and Albert had been imprisoned in the Tower of London for High Treason. (A crowd gathered, two months before the actual outbreak of war, hopeful of seeing Aberdeen and the Prince led off to the Bloody Tower in chains.)

Albert's view of the crisis was that it was 'the just punishment of Heaven on the Emperor for <u>the embroilment which he has brought upon Europe</u> by his wilfulness and obstinacy'. It would, Albert was convinced, never have happened 'if there were a <u>Germany</u> and a <u>German</u> sovereign in Berlin'.[19] Once war had broken out, Albert, however, played as active a role as he was allowed. It would have been unimaginable had he not done so. The allied forces of France and Britain, 58,000 strong, sailed from Varna on 20 September 1854, crossing the Black Sea and landing at Old Fort, near Eupatoria, a few days later. The autumn saw allied victory at the Battle of Alma; there followed a gruelling campaign, hindered by failures of communication, widespread cholera killing more British troops than the Russians ever did, and the pyrrhic victories, achieved in ghastly conditions, of Balaklava and Inkerman. The Queen's cousin, George, Duke of Cambridge, was so appalled after Inkerman by the sickness and bad management that he retreated to Malta and never returned to the Crimea. 'In the great world, the star Mars continues to blaze in bloody splendour in the sky,' Albert wrote to his stepmother. 'God grant that the unhappy Sevastapol may soon fall and peace be made as a result'.[20] When he heard of Albert's rather know-all attitude to tactics in the war, the Duke of Cambridge burst out, 'Why doesn't he come himself and have a try? He wouldn't stay 24 hours I know, the fine gentleman.'[21]

Albert, that September, had spent the inside of a week at Boulogne with Napoleon III. They spent a whole day reviewing the French army at Saint-Omer, who were divided into two infantry divisions of about eight thousand men each. The main purpose of Albert's visit, however, was not military, but political. The British wanted a closer assessment of Napoleon III, and the French, likewise, were quizzical to find themselves allies of the British in this war.

Albert's memorandum, written as soon as he returned, conveys his impression of Napoleon, his Government and his entourage, in sixteen closely printed pages of Sir Theodore Martin's biography. They discussed everything, from the character of Aberdeen – Napoleon candidly admitting that his distrust and dislike were deep – to the best method of raising taxes; from the Schleswig-Holstein Question – 'about which he confessed to the same ignorance which is common with English statesmen'[22] – to the future of Poland. While they spoke, the Emperor chain-smoked – the new fad – cigarettes ('not being able to understand my not joining him'[23]). Albert felt that the atmosphere was more that of the officers' mess than a Court. About the war, in which their two countries had so precipitately been engaged, Napoleon was candid enough to admit that he had been, from a military point of view, unprepared for it. Albert was much more cautious about admitting his many fears and reservations about the inadequacy of the British military. 'The Emperor was almost the only person among the French at Boulogne who had any hope of the success of the expedition against Sevastopol, and the astonishment was great that our whole party of English officers were so sanguine about it.'[24]

In general, however, this was a successful diplomatic meeting. The surprising entente was now guaranteed. Albert invited the Emperor to visit Britain, and believed, with complete accuracy, that Victoria would like the Empress Eugénie, to whom Napoleon was evidently devoted. Albert, unsurprisingly, found that Napoleon's 'general education appeared to me very deficient… He was, however, remarkably modest in acknowledging these defects'.[25] Napoleon, like most people who met Prince Albert, was a little overawed by the range of his knowledge and his willingness to display it. '*Lorsqu'on a su apprécier les connaissances variées et le jugement élévé du Prince, on revient d'auprès de lui plus instruit et plus apte à faire le bien.*'[26] ('When one has learnt to appreciate the varied areas of knowledge and the elevated judgement of the Prince, one can only come away from his presence better informed, and better qualified to do the right thing.') It is a good assessment of Albert: much hangs on the conditional with which it begins. Not everyone saw his point.

The Crimean War was the first in history to be reported by war correspondents while it was actually happening. It was also the first war to be photographed. No military operation can be conducted without blunders, sickness and horror. Even Wellington's successful campaigns in India, the Iberian Peninsula and Belgium had been punctuated by mishaps, but no *Times* reporter was on hand to witness them and report them immediately in the columns of the press. William Howard Russell's dispatches to *The Times* were revealing to the British that their armies were badly organized, and poorly led. His account of the Charge of the Light Brigade at Balaklava inspired Tennyson to write his splendid Ode, but it also revealed to the world that the cavalry officers, all aristocrats who could not distinguish between riding to hounds and conducting a cavalry charge, were catastrophically incompetent.

The war had the effect of advertising to the truly powerful in Europe, above all to Bismarck, that Great Britain had no military capacity to deter a major aggressor. When the Schleswig-Holstein Question came to the point of war a decade later, the Prussians, trained and armed to the utmost, knew that the British were not in a position to threaten military intervention. They therefore had the relatively easy task of defeating Denmark, establishing Prussian control of those Baltic duchies and beginning the relentless march towards Prussian supremacy over Europe, the unresolved consequences of which would, with blood, iron and fire, be evident for the next eighty years. The British could conduct colonial wars against Asian or African people heavily outgunned; but the Crimean War had been a geopolitical disaster for them, even though the British and French 'victory' was dressed up as something to celebrate, and Alma Streets and Inkerman Squares were jerry-built all over England, as loyal Englishwomen knitted their families cardigans and balaclava helmets.

The Treaty of Paris, which finally brought the war to a conclusion after long and complicated negotiations, was signed on 30 March 1856. By then, the British appetite for war had evaporated, and the political landscape at home had changed dramatically. Aberdeen's conduct of the war had been lacklustre and incompetent. A motion in the Commons, inquiring into the condition of the army outside Sevastopol, led to a humiliating defeat for the Government – a vote of 305 versus 148. Aberdeen had to go. When he resigned as Prime Minister at the beginning of 1855, the Queen wept. She tried to persuade Lord Derby to form a government. The figures he could command in the House simply made it impossible. Lord Lansdown was equally and evidently unable to do as she wished. Lord John Russell, much as he yearned to resume power, knew that he was not in a position to do so. There was only one man who had the energy, the popularity and the sheer oomph for the task: it was their old enemy Pilgerstein. 'I am, for the moment,' Palmerston wrote to his brother on 15 February, '*l'inévitable*.'[27]

It was true. A journalist would write: 'England said: Try Palmerston. The country looked on in hope, beginning to breathe more freely... Palmerston was the man to whom the business of war could be committed and in whose hands the name of England was safe.'[28]

Strangely enough, although this was the political outcome which Stockmar and Albert had dreaded more than any other, Victoria very quickly came to believe that it was true. Political truth is different from other truth. Palmerston was actually able to deliver a no better 'deal' for Britain in the peace negotiations than Aberdeen would have done, but he made people feel better. Presentation mattered more than ever. That was one of the lessons of the war. Albert and Victoria could see it.

The casualties of the war, considering the fact that it had (relative, for example, to the Napoleonic Wars, or the Franco-Prussian War of 1870) comparatively few battles, were dreadful. The allies lost some 70,000 in battle, the Russians 128,700; total losses – for, as is well known, disease was rife – were 252,000 allies (45,000 British) and 256,000 Russians.[29] Many, with no thick coats in a deadly winter, perished from exposure. However the figures are calculated, many families were bereft.

Albert's response was to found an orphanage. The result, which may be seen in the London Borough of Wandsworth to this day, is the gigantic Royal Victoria Patriotic Building. Albert raised the huge sum of £1.5 million for the widows and orphans, of which £35,000 were spent on the monster-Gothic building, designed by architect Major Rhode Hawkins. It is a French château owing something to Scotch Baronial, and the life of the orphans housed there must have made them envy the nearby prison which had been erected in 1849. There were no fireplaces in the orphans' rooms, and they worked hard, running a laundry. (Water had to be pumped up to the ugly brick towers.) Sexual abuse by the chaplain led to the death of at least one of the little girls, and the orphanage eventually fizzled from use. In the First World War, the Royal Victoria Patriotic Building was a hospital. In the Second, it was 'the London Reception Centre' where aliens, potential enemy agents, were 'interrogated'. It then followed the predictable path of an unloved, ugly Victorian building. It was requisitioned for use as a school. It fell into such disrepair that there was talk of demolition. The Victorian Society campaigned for its preservation. It was purchased for £1 in 1980 by a property developer and is now used for a variety of purposes, including a restaurant, flats and, perhaps inevitably, 'an arts centre'. Despite all such efforts to inject optimistic life into the place, a pall of sorrow hangs over it, as it juts from the trees on the edge of Wandsworth Common, as though haunted, not merely by the Prince's good intentions, but by the misery which had begun to possess his private life by the time he had collected the money for the enterprise and persuaded the often furious and habitually postnatally depressed Queen to lay the foundation stone in 1857.[30] The statue of St George killing the dragon, in the niche over the entrance, bears a presumably deliberate and unmistakable resemblance to the Prince, serving as a reminder, not only of his charitable good intentions, but of the demons he had to slay in his remaining five years of life.

Much of the hard work which consumed Albert's time – minute considerations of the state of the army, for example, fine-comb analysis of the international situation – would perhaps come to nothing. He and Victoria, however, understood in their own way what was clear to Palmerston, that political success is not simply a matter of competence but of publicity. The monarchy, and the role it was to play at home and in Europe, was not to be defined in terms of its narrowly constitutional role but as an emblem. Victoria and Albert had always guarded their private and family life as sacred, while they performed their public duties. They had also paradoxically come to recognize that the two could not easily be separated. The greatest public roles they could in fact perform were not nakedly political. They were to provide a figurehead, an iconography of family life. And, through their children, they were in a position to use dynastic marriage to forge and shape the future state of Europe.

Albert's wish to influence events in Europe was, thus, intimately bound up with his control over the family, and over his wife. In November 1855, he congratulated her: 'The four weeks of unbroken success in the hard struggle for self-control cannot fail to strengthen your self-confidence, and thus to make each future triumph easier.'[31] Victoria for her part noted, with approval, the iron, schoolmasterly control which Albert exercised over the family. 'The younger children who [*sic*] he constantly kept in order if they eat badly or untidily, saying, "Manierlich essen, nicht so grosse, Hücke", and used to say to Vicky when a girl, ever, "Elegant" when she did not eat elegantly. He could not bear bad manners and always dealt out his dear reprimands to the juveniles, & a word from him was instantly obeyed.'[32] The pertness of Vicky, aged thirteen, was an especial worry to him. 'You are now standing on the threshold of the youthful period of your life and that makes me very, very anxious for you. For if you give the impression to others that you delight in giving them pain you will engender resentment and contempt.' She replied with a grovelling apology: 'I want so very much to tell you how extremely sorry I am to have behaved in the way I have done lately.'[33] Ideas of family hierarchy, in the novels of Samuel Butler or the satires of the Bloomsbury Set, for example, have perhaps skewed our perspective of how fathers and children, husbands and wives, 'ought' to behave. There are still many cultures in the world where it would be seen as a husband and father's duty to control, and to be the master of his house.

There were differences between Victoria and Albert, however, in their attitudes to child-rearing. When the Queen placed the Prince of Wales under the charge of Lord Clarendon, in 1855, to tell him what to do and how to behave, Albert confided in Clarendon that the Queen's severe way of treating her children was 'very injudicious' and would cause trouble later, especially with Bertie. Clarendon told Prince Albert that he and his wife had reared six children. 'Now we have never used severity in any shape or way, never in their lives had occasion to punish any of them, and we have found this mode of bringing them up entirely successful.'[34] Clarendon's example shows us that not all Victorian families were as disciplinarian as others. Nevertheless, Vicky, like her mother, completely idolized Prince Albert, and did not resent his high standards. Not in the least. The late Victorian and Edwardian reactions against the old-fashioned disciplines of family life have perhaps made the severity with which the royal children were reared seem crueller than they did at the time. It is hard to get these things right. Albert was disloyal to Victoria in his implication that it was always she who urged fierceness towards the children; her own memory, that he was a martinet insisting on good behaviour at all times, suggests that he did not need his wife's prompting to wish to impose order on the nursery. These children were not, in his eyes, simply children. They were being trained up for a purpose, namely to be constitutional monarchs and the parents of more constitutional monarchs, spread across Europe. The children were going to be Albert's way of imposing decency on the whole continent – neither the anarchy of radicalism or socialism, nor the irrationality and unsustainability of autocracy such as was practised in Russia.

In 1854, Albert would write to his old tutor Florschütz that he was 'always especially glad to receive your good wishes on my ageing birthday, for they remind me so vividly of the days of my childhood'.[35] He described to Florschütz the carefully staged tributes brought to him by the children. 'Vicky recited the whole scene of Shylock before the Court in the *Merchant of Venice*, presented her father with an essay in French on the life of Francois Premier, and played the piano. Her drawings were particularly good.' Bertie had been called upon to recite the all-too-apt scene in *Henry IV Part One* in which, supposing his father to be dead, Prince Hal placed the crown on his own head. He also recited a monologue from Schiller's *Wilhelm Tell* ('*Durch diese hohle Gasse muss er kommen*'; 'Through this narrow alley must he come') and wrote an essay on the Fate of Ulysses. Affie recited the Fable of Vulpis and Corvus, produced two essays in German on the foundation of Wartburg and of the monastery of Reinhardt-Brunner. 'The little ones did their part likewise'. There was then a photo-opportunity. Dr Becker, Albert's librarian, 'took a photo – Osborne – in which you will see us altogether'.[36]

The first photographs ever taken of British royalty were of Prince Albert, on 7 March 1842, at Brighton. The photographer was William Constable. In that month, Albert also visited Richard Beard's studio in Parliament Street, where a daguerreotype was taken. Both Victoria and Albert were early enthusiasts for photography. Partly because the agents of the pioneer photographers Daguerre and Fox Talbot rigidly enforced patent restrictions on other photographic methods being pioneered, photography did not become popular at once. In 1851, Frederick Scott Archer pioneered a collodion process without patent restraints. Fox Talbot fought this in the courts, but in 1854, the year that Dr Becker took the family portrait at Osborne, legal restrictions were lifted. By then the collodion process had been refined and enhanced, and there was a rapid increase in the growth of photographic studios.

Victoria and Albert became patrons of the Photographic Society soon after its inception in 1853, and in 1855, Prince Albert would contribute £50 towards a study of how to prevent fading in photographs.

Both the Queen and the Prince commissioned and collected photographs from the early days. Some of the greatest nineteenth-century photographs are in the Royal Collection, work by such masters as Julia Margaret Cameron (their eccentric Isle of Wight neighbour) and John Jabez Edwin Mayall. Roger Fenton, working in the 1850s, most famously worked in the Crimea, but he also took unforgettable portraits of the royal children. For the first time in history, we see, not Holbein drawings or Velázquez versions of royal faces, but the people themselves. We see Vicky aged nearly sixteen, her left arm casually lolling on the shoulder of the slightly-out-of-focus, Edith-Sitwellesque melancholy little figure of her thirteen-year-old sister Alice; we see Lenchen and Louise, ten and eight, stiff as dolls in their huge, puffed-out Highland tartan skirts, staring truculently at the lens; we see little Affie at Balmoral, the socks beneath his kilt slightly ruckled, looking downwards at the grass on which he is lying. They are as vivid as Fenton's Crimean soldiers, and once we have seen them, they never quite go back between the pages of a book. As sharply as the words of their mother's diary, they remind us that the extraordinary enterprise of a constitutional monarchy was a human enterprise. Fenton's views taken of Windsor, Great Park and Castle, show the place more or less unchanged, as it may be seen today. Fenton's images of the family are the first in a series of which come down, once more, to our own times: the depiction of the Royal Family by photographers being both something which they have, ever since Prince Albert's day, orchestrated and arranged, and also profoundly dreaded when it became predatory or intrusive.

Victoria and Albert – but most especially Albert, who was really the mind behind the scheme – were in the process of making the British monarchy new. Photography would play an essential, dangerous part in this. In the first instance, the royal pair's interest in photography was purely private. They rejoiced, as, in time, would most families in the developed world, in seeing images of themselves and their children, swiftly, accurately and mysteriously produced in the fraction of time it would take Landseer or Winterhalter to achieve less vivid results. It was not until May 1857 that Albert opened the Manchester Art Treasures exhibition. This was a groundbreaking exhibition which in many ways was an extension of Albert's greatest work, the establishment of the museums in South Kensington. By taking paintings and photographs from the Royal Collection to Manchester, Albert intended them to be enjoyed, not by the cognoscenti so much as the working classes. It was an almost Ruskinian or Arnoldian mission to bring culture to the masses. The photograph of Albert was taken specially for the exhibition by Lake Price, and it was shown again in February 1858 at the South Kensington Museum. It was hung in pride of place, and it established the centrality of Albert to the cultural mission. It advertised the Royal Family as the standard bearers of modern culture and inventiveness.

It was the Queen's turn, in May 1858, to have a photograph of herself publicly exhibited at the London Photographic Society's fifth annual exhibition. This was the exhibition which also included one of the most emblematic photographs ever taken of Victoria and Albert with their dynasty – Leonida Caldesi's group portrait of them all in the sunken Italian gardens of Osborne. The brilliant thing about the picture is that it does not look posed. It looks just pre-posed. Baby Beatrice sits on the Queen's lap. Albert is gazing, a little testily, in the direction of his wife. Of the eleven individuals captured, only Affie looks at the lens. Everyone else turns away 'quite leisurely'.

In 1854, Disdéri patented photographic *cartes de visite*, and having caught on to the commercial possibilities of a phenomenon as life-changing as iPhones in our day, people were soon collecting *cartes de visite* and assembling albums, either of their own family or of celebrities. The 1850s saw the commercialization of photography and August 1860 would see the publication of Mayall's *Royal Album*, fourteen *carte-de-visite* portraits of Albert, Victoria and their children. John Plunkett, author of *Queen Victoria: First Media Monarch*, was surely right to say that this moment 'heralded the beginning of a turbulent relationship between mass culture, photography, and the royal family'.[37] After only a few days on sale, the wholesalers had already ordered 60,000 sets. Nor were they cheap. One photographic retailer, Charles Asprey of 166 Bond Street, advertised the *Royal Album* at four guineas. From now onwards, the Royal Family had become, however little they really wanted this, public property. Albert had made of himself, his wife and his nine children not merely pieces in a dynastic game, and not merely a group of people who happened to be related to the Head of State. They were themselves the selling point of constitutional monarchy. No sane person would have held up the private lives of George IV or William IV as role models to be followed by a respectable bourgeoisie. Albert, the Angel, had done so. The children who stared so broodingly at the photographer's lens were in awe of their father's intelligence, his command, his unspotted virtue. Not all of them, however, would be able to raise themselves to the Angel's exacting example.

EIGHTEEN

THE PRINCE CONSORT AT LAST

AT THE BEGINNING of 1856, the Queen's doctor, Sir James Clark, expressed the deepest misgivings to his diary: 'Queen's health has not improved. She had complained of feeling weaker than usual more especially of late, & has been at times frequently low and nervous. I have never seen her weaker than this day under her ordinary state. I feel at times uneasy regarding the Queen's mind, unless she is kept quiet... The time will come when she will be in danger'. Victoria had told Clark that if she had another baby she feared she would die. (She was seven months pregnant when he wrote these words.) The doctor added, 'I too have my fears but they are more for the effect on the mind than on the health. Much, very much, depends on the Prince's management. I must again have some conversation with him on the subject. My position is a difficult one, not that my path is not clear. My difficulty is to impress the Prince with the necessity of keeping the Queen free from all mental irritation.'[1]

Their last child, Beatrice, whom we encountered as an old woman in the Second World War in the opening pages of this book, was born on 14 April 1856. 'Nine is certainly a formidable number and makes me indeed a patriarch,' Albert noted.[2] Victoria had been plunged into her usual postnatal depression; by now, she was totally dependent on Albert to help her with the political role she was supposed to fulfil. Yet, while she called him 'my Master', and he addressed her as 'my

Child', constitutionally he was simply her husband. The politicians had been reluctant to allow him so much as the title of Prince Consort. Although Victoria badgered successive Prime Ministers about this matter, it never seemed quite pressing enough to be brought to the top of any premier's in-tray; and the politicians of all parties, precisely because they could see how hungry for power Albert was, were happy to leave him as a constitutional anomaly.

If it embarrassed and irritated Albert to be so regarded, his misfortune was that he was following an anomalous precedent. The previous three Queens Regnant were Mary Tudor, Mary Stuart and Anne. Mary Tudor was married to the King of Spain who technically reigned over England as King Escort, but who was seldom resident. William of Orange, intensely ambitious and obsessed by the Protestant succession, insisted upon being the King Escort, and indeed reigned after the death of his wife Mary. Anne Stuart, however, was married to the altogether less forceful figure of Prince George of Denmark, a weak character with inadequate English who, it was considered, would have made a poor fist of being King. He was therefore passed over, and, upon her death – her immediate heir being her Roman Catholic nephew James – the succession was passed to the Hanoverian cousins.[3] The Victorian politicians could point to Prince George of Denmark and say that his example showed there was no necessity to make Albert a King Consort.

The differences between the two men could not have been more marked, of course, and the reasons for withholding such authority, in both cases, were almost precisely opposite to one another. George of Denmark was not made King Consort because of his palpable unsuitability; Albert was not granted such a role because his suitability had been demonstrated all too clearly, and neither Palmerston nor Derby, the two party chiefs who were most concerned with the matter as the 1850s drew to a close, wanted a royal personage to usurp their role at the helm of government.

In May 1856, the Queen sent a memorandum to the Lord Chancellor, Lord Cranworth, which showed every sign of having been punctiliously dictated by Albert himself. Most revealingly, the copy of the memorandum is preserved at Windsor among the Stockmar papers.

> The present position is this: that while every British subject down to the Knight Bachelor, Doctor and Esquire has a rank and position by Law, the Queen's husband alone has one by favour – and by his wife's favour, who may grant it or not!… The only legal position in Europe according to International Law which the Husband of the Queen of England enjoys is that of a younger brother of the Duke of Saxe Coburg.
>
> The question has often been discussed by me with different Prime Ministers and Lords Chancellor, who have invariably entirely agreed with me but the wish to wait for a good moment to bring the matter before Parliament has caused one year after another to elapse without anything being done. If I become now more anxious to have it settled it is in order that it should be before our children are grown up, that it might not appear to be done in order to guard their father's position against them personally, which could not fail to produce a painful impression upon their minds.[4]

For as long as Albert had no status beyond that of 'husband' and 'younger brother of a foreign Duke', he was already in the embarrassing position of being below his own son, the Prince of Wales, in orders of precedence. While Bertie was a minor this was not very likely to be a matter of public display. Soon enough, however, Bertie would have grown from being the fifteen-year-old which he was when this memorandum was written, to being a young man. The 'painful impression' which the memorandum fearfully envisaged being formed in the minds of Bertie and his siblings was as nothing to the 'painful impression' which it caused in the bosom of their father. There can be no doubt that this awkwardness helped to complicate relations between Bertie and his parents. The non-royal majority of the human race find it easy to forget that the eldest son and heir's entire raison d'être is to inherit; and that this is another way of saying that his raison d'être is for his parent to die. This cannot enhance cordial feelings between monarch and heir apparent. The embarrassing fact that Bertie took precedence over his father in all matters of ceremonial protocol could only exacerbate the hostility of his parents. For this reason alone, it was desirable that some official title be given to Prince Albert.

An obvious answer would be, as happened in the case of Elizabeth II, to make the Queen's husband into a British Duke, with precedence over the other Dukes, royal and non-royal. The trouble with this solution was that it would give the possessor of such a title an executive power, the right to sit, and to vote, in the House of Lords. Whereas the mid-twentieth-century Duke of Edinburgh could be firmly told that he must not exercise such a role, it was by no means certain, in the minds of Palmerston, for example, that Albert would not turn up and listen or, worse, take part in debates. And it is hard to believe he would not have exercised his right to a parliamentary vote. The modern historian David Cannadine was surely right to say that Albert 'wanted to be a sovereign who governed as well as reigned',[5] and Palmerston, Derby and the rest, whatever else divided them, were united in not wishing this to happen. Derby warned the Queen and the Prince that the process would have to come before the Cabinet and then be discussed in Parliament, and this inevitably risked a hostile press,[6] especially since there was no way of persuading the Cabinet to make Albert King Consort, and the diminished title of Prince Consort would have to be cobbled together as a compromise. The Privy Council approved the measure and, on 25 June 1857, it was formally ratified, and his title was included in the liturgy. *The Times* refused to be impressed. 'In spite of the poet there is much in a name, and if there be increased homage rendered to the new title on the banks of the Spree, or the Danube, the English people will be happy to sanction and adopt it,' their leader-writer sneered.[7]

The compromise was, however, agreed. It is difficult to doubt that, whatever the fears entertained by the politicians about allowing Albert too much power, they recognized not only his competence but also the fact that he was, superficially, a much more stable figure than the Queen herself. An all-male Establishment could not be expected to understand the swoops and soarings of her moods. We, who consider ourselves wiser than our ancestors in such matters, can diagnose postnatal depression and a highly volatile personality existent in a fundamentally sensible woman. Sir James Clark, together with the other physicians and the entire political class, could not forget the spectre of her insane grandfather, but although in her wildest fits of grief for Albert she herself felt she was losing her reason, Victoria was not by any definition a madwoman. Maddening, sometimes, but not mad. For our part, we can also see, with hindsight, that she had one of the most fundamental requirements in a Head of State, robust physical health. Albert, we can see, but his contemporaries of course could not – he was still only in his late thirties, after all – was frail, with only a little time left to live.

The only sure method of birth control, as her physicians told this highly sexed woman, was abstinence, and if this were the method employed to guarantee the end to the procession of babies, it could only have added to the Queen's misery. She made no secret of her passion for her husband and for what she called 'fun in bed'. In bed, hitherto, she had been sure of his love, even if, during the hours of day, he was so often distracted and busy. She felt increasingly that he was a cold fish who was withholding love. He, in turn, found it necessary not merely to check and monitor her moods, and her loss of 'control', but also to criticize her behaviour as a mother. This letter, of 13 October 1856, was written in German:

> It is indeed a pity that you find no consolation in the company of your children. The root of the difficulty lies in the mistaken notion that the function of a mother is to be always correcting, scolding, ordering them about, and organizing their activities.

> It is not possible to be on happy, friendly terms with people you have just been scolding, for it upsets scolder and scolded alike. I say, 'scold', as the harshest term which applies – minor corrections, of course, matter less, but the stronger word makes the principle clearer.[8]

Their different attitudes towards their children only heightened feelings of tension. Vicky, born 1840, was the first to be lined up in Albert's dynastic plan. When she was only fourteen, menstruation began, and the Queen, the reverse of squeamish when physical matters were discussed, hastened to inform the Queen of Prussia. The child had 'developed amazingly of late', and was now capable of bearing liberal-minded Albertian heirs to the Prussian throne. It was during the summer of 1855, when the Crimean War was still dragging on, that twenty-three-year-old Fritz, the Crown Prince Friedrich Wilhelm, paid a visit to Balmoral. He and the child walked up the slopes of Craig-na-Ban, he picked her a sprig of white heather, and they had their first kiss. That evening, she ran into her mother's room, 'very much agitated', to say that she loved him. It was agreed that they should become engaged, but that they must wait until Vicky was seventeen.[9]

Both parents had misgivings. Albert dreaded losing the member of the family with whom he had most in common. Victoria loved Vicky, but was disconcerted by her cleverness, jealous of Albert's affection for her, and open-eyed about the dullness, and essential weakness, of Fritz. She felt they should consult Stockmar before allowing the match to proceed. Over this great matter, as over so many small ones, there were the inevitable marital squabbles. One evening, a year after he had written to her to congratulate her on her self-control, there was a huge outburst. They were reading a letter together, from Fritz. Victoria made one of her disparaging remarks, and Albert asked, 'What makes you so bitter?' He must have been experienced enough to know that this was a dangerous remark, and it produced a major storm.

He wrote one of his longest letters of rebuke, revealing that, in this round of the fight, the gloves were off:

It is my duty to keep calm, and I mean to do so, but unkindness or ingratitude towards others makes me angry, like any other kind of immorality. Fritz is prepared to devote his whole life to our child – whom you are thankful to be rid of – and because of that you turn against him: Stockmar, who has shown us nothing but kindness for as long as we remember, is suddenly asked, old and ill as he is, to drag himself over here, and his coming is taken as an offence. This is not a question of bickering, but of attitudes of mind which will agree as little as oil and water, & it is no wonder that our conversations on the subject cannot end harmoniously. I am trying to keep out of your way until your better feelings have returned & you have regained control of yourself. You, on the other hand, want to talk it all over with me again, though it has all got to work out as you have planned beforehand. That makes my task almost impossible. I cannot admit you are right, and yet I have to see you suffer, and you make a parade of your suffering before my eyes so as to make me feel your reproaches still more keenly, as much as to say: 'see, this is all <u>your</u> work'. The injustice of this is not calculated to encourage me any more towards reconciliation – indeed I felt it would be wrong to attempt it. I don't want to prove myself in the right, or to set up as being perfect, but only to see you happy, which even if I did not love you, would ensure my own happiness. Neither will I play the game of Greatheart and <u>forgive</u>, that is not at all how I feel, but I am ready to ignore all that has happened, and to take a new departure as the sailors say [*sic*?!] although I shall probably be accused of being unfeeling) and try in future to avoid everything which might make you unhappy, & your state of mind worse. <u>Cure</u> it I cannot. You alone can do that. I only wish I could.[10]

While the Head of State and her consort were at loggerheads, the worst crisis in Anglo-Indian relations took its bloody course. 'Oh!' the Queen sighed, 'when I think of my own sorrows... and reflect on the fearful and appalling horrors which have taken place in India... How unbearably small does every suffering of ours appear!'[11]

Whether or not the British Empire came about as a fit of absence of mind, it undoubtedly developed from a haphazard combination of mercantile takeover and overt military action into an armed imperialism which was intentional and organized. The Queen had never been in any doubt that British power could only be sustained in Asia by force. Like most Britons, until our own generation, when so much more has been unearthed about the essential violence of the East India Company's activities from the beginnings, she would have maintained there was a distinction between peaceful commerce and military conquest; but since at least the time of Clive a century earlier, the British military presence in India was fixed.

To call the disturbances which began at Barrackpore in January 1857 a 'Mutiny' is, of course, to accept the imperialist programme: that is, that the inhabitants of the country where the violence erupted were 'rebelling' against their legitimate rulers. In this book, the word is only used for shorthand – meaning the (relatively localized) massacres and reprisals which dominated 1857, largely quelled by September 1857, but continuing in a more desultory way in Oudh until the end of 1858. The supposed 'cause' of the Mutiny – the belief of the Sepoy soldiers that the cartridges of the newly introduced Enfield rifles, which had to be bitten before insertion into the gun, had been greased with the forbidden fats of pigs and cows – was symptomatic of the much larger British insensitivity to India, its religions and customs. Appalling violence was perpetrated on both sides – European women and children hacked to death in the prison at Cawnpore, revenge killings by the British no less horrible, Lucknow, Delhi and elsewhere eventually brought back from 'rebel' control.

After the Mutiny, it became apparent that the ramshackle way in which India had been occupied by the British should be formalized, a

Viceroy appointed as a representative of the Crown, and the Imperial nature of the enterprise made explicit. Central to the whole affair was reorganization of the Indian army. While implying no defence of the imperialist idea, it is surely legitimate to say that, had the reorganization been incompetently done, there would have been future massacres and reprisals. The administration of 'British India' or 'the Raj', conducted by only around a thousand members of the Indian Civil Service, is surely a tribute to the relative benignity and efficiency of the military presence, no serious acts of violence occurring after 1858 until the Amritsar massacre of April 1919. This was in no small part owing to royal intervention.

Albert had been the Duke of Wellington's choice as Commander-in-Chief of the army. As we have seen, the post went to Lord Hardinge, who resigned on 15 July 1856 to be replaced by Victoria's cousin George, Duke of Cambridge. Hardinge was an old man, Cambridge was thirty-seven. He had seen active service in the Crimea, had his horse shot from under him at Inkerman and was mentioned in dispatches at the Alma. His own approach to army reform, which would differ so markedly from that proposed by the Liberal Edward Cardwell in the late 1860s, belongs outside the scope of Albert's life. As far as India was concerned, it was Cambridge who urged the amalgamation of the troops of the East India Company with the army of the Crown, giving him general control of the Indian army from London. Albert heartily endorsed this crucial change. There can be little doubt that George and Albert helped Victoria draft the long memorandum sent to Palmerston in July 1857 when the Mutiny was at its height.

> The present position of the Queen's army is a pitiable one. The Queen has just seen, in the camp at Aldershot, regiments which, after eighteen years' foreign service in most trying climates, have come back to England to be sent out after seven months to the Crimea. Having passed through this destructive campaign, they have not been home for a year before they go to India for perhaps twenty years! This is most

> cruel and unfair to the gallant men who devote their services to the country, and the Government is in duty and humanity bound to alleviate their position.[12]

Wellington had always cared passionately about the lot of the common soldier – in the Peninsular campaigns, for example, testing on his own shoulders the weight of the kit-bags and equipment each man was required to bear. In Alan Bennett's *The History Boys*, the inspirational teacher, speaking to his pupils about Hardy's poem 'Gunner Hodge', suggests that it was only in the Boer War, or possibly the Zulu wars, that the 'cannon-fodder' began to have names and 'the names of the dead were inscribed on war memorials. Before this, soldiers…[*sic*] private soldiers anyway, were all unknown soldiers, and, so far from being revered, there was a firm in the nineteenth century in Yorkshire of course, which swept up their bones from the battlefields of Europe in order to grind them into fertiliser'.[13]

The humanization began earlier, though, in the Crimean War, partly as a result of William Howard Russell's reporting in *The Times* of ordinary soldiers' suffering, partly through the just fame brought by that war to Florence Nightingale and her nursing endeavours at Scutari; but also because of Royal Sympathy.

Stockmar had thought of modern monarchy in largely constitutional terms, sometimes giving Albert good advice, but sometimes filling his protégé's head with unrealistic hopes about the extent of his political influence or control. There were, however, other guides, including Victoria herself. She had learned from her aunt by marriage, Queen Adelaide, the value of personal engagement between sovereign and people. This was deepened by the creation of the Victoria Cross, the highest possible commemoration for gallantry, which was awarded to soldiers of all ranks. Royal personages tend to have a fixation on decorations, orders, medals, uniforms and the ceremonial trappings of power, Albert being no exception. It was, however, Victoria, as they pored over the different designs for the medal, who thought up the perfect and unforgettable legend. 'The motto would be better "For

Valour" than "For the Brave", as this would lead to the inference that only those are deemed brave who [*sic*] have got the Victoria Cross.'[14]

Victoria was insistent that the medal should recognize neither birth nor class. It was gazetted on 5 February 1856,[15] and the first ceremony at which the recipients received the VC was held in Hyde Park on 26 June 1857. The award was backdated to include soldiers of conspicuous valour who had served in the Crimea – there were 111 of those, of whom the Queen invested personally 62.[16] Ever since, the holders of the VC have been held in awe, by comrades and public alike. The close link between the monarch and the 'ordinary' servicemen is something which remains to this day. It may be said to have begun with the 'sailor king', William IV, but Victoria and Albert, with their much closer sense of the way in which constitutional monarchy works, forged the link. To this day, it is one of the most palpable and observable phenomena of the contemporary armed forces, something often overlooked by theoretical republicans. Victoria, who prided herself on being a soldier's daughter, understandably blinding herself to the extreme unpopularity of the Duke of Kent in the army, would retain this sense, and this link, to her dying day, even when other aspects of her public duties, during widowhood, appeared to be neglected.

Foreign affairs preoccupied Albert's attention for much of the time during this period. It could hardly be otherwise: war in the Crimea had not solved any of the problems of the supposed 'sickness' of the Ottoman Empire. The rivalries between France and Russia were still being played out in Turkey, and in the Near East and the Mediterranean and in the Balkans. The aftermath of the Indian troubles and the setting up of the 'Raj' were a lasting source of anxiety. Tectonic shifts were taking place in the balance of power in Europe – the rise of Prussian military and international strength, the 'struggle for mastery' between Russia, the Prussians and France – which would have momentous consequences for the future of the world. Clouds, a little larger than a man's hand, were forming in the United States, with the Southern States increasingly alienated by the political sensibilities and commercial prosperities of those in the north.

It was a time when Britain might have hoped for strong or united government at home. Although there were men of prodigious political intelligence and energy in both Houses, the party system itself was, not merely in flux, but, to all appearances, in a state of dissolution. Sir James Graham believed there 'is not one man in the House of Commons who has ten followers, neither Gladstone nor Disraeli, not Palmerston'. The 'total destruction of parties and of party ties and connexions' appeared imminent.[17] It was no surprise that, in such circumstances, the monarchy, which had no executive power but a duty to advise, should be more than usually involved in the executive process, and Albert at this period became almost a one-man Civil Service and Diplomatic Service, writing thousands of words a week to a succession of Home Secretaries, Foreign Secretaries, Colonial Secretaries and Secretaries for War, whether (as from February 1855–8) they were serving under the premiership of Palmerston, or (from February 1858–9) under Lord Derby. Moreover, these politicians, unlike Peel, did not see the necessity to woo the electorate by constant appearance in 'the regions'.

On 30 April 1857, for example, when the last of George III's surviving children, the Duchess of Gloucester, died aged eighty-one, protocol would have decreed that Albert went into mourning and cancelled all engagements. He did not do so, heading north on 5 May to open the Manchester Art Treasures exhibition. (It was here, as mentioned in the last chapter, that photographs of the Royal Family were first put on public display.[18]) 'No country invests a larger amount of capital in works of art of all kinds than England,' he had told the exhibition organizers.[19] And so he left Buckingham Palace at six in the morning, changed trains at Cheadle, reached Stockport by eleven, and spent the rest of the day being shown round Manchester by the Mayor, Mr Watts. At half past seven that evening, as he wrote to the Queen, 'I am half undressed, to put myself in plain clothes for dinner, and half dead too with the day's fatigues.'[20]

Such duties were never neglected. In the same month, June 1857, that the South Kensington Museum was opened he was presiding over the first meeting of the Conference on National Education. At this date, it was estimated that of 4,908,696 children in Britain, 2,861,848 received no education of any kind. Albert insisted to the meeting that, although new schools, public and private, had been built at an unprecedented rate in Britain, the fundamental 'evil',[21] his word, would not be addressed without a shift in public opinion. 'Public opinion is the powerful lever which in these days moves a people for good and for evil, and to public opinion we must therefore appeal if we would achieve any lasting and beneficial result.'[22]

Very many of the politicians in both administrations, both Palmerston's and Derby's, failed to grasp this fact. It was a chaotic few years in public life. November 1857, for example, saw a major financial crisis, with a run on the banks, the Bank of England interest rate rising from 5.5 per cent to 9 per cent and its gold reserves reduced to a little over £7 million. 'Bad times are approaching. Bankruptcies are spreading, thousands of artisans are turned into the streets through the consequent stoppage of works. Want and discontent are very visibly on the increase in manufacturing districts.'[23] Albert could say these things because, unlike so many politicians, he had seen them in his peregrinations through the English regions and his constant attention to all the details of the crises in education, poor relief, public health and so on. In such a time, while politicians in the narrower parliamentary sense were unable to provide firm government, the part played in public life by royalty, political in the broader sense, was vital. Victoria and Albert had become something which was difficult to define, but which was much stronger than a term such as 'figurehead' would denote.

This was especially the case as far as foreign policy was concerned, where the British relationship with France, and the personal relationship between the royal pair and the Bonapartes, was inextricably entwined.

When, in 1870, Prussia triumphed over France in war, and Napoleon III was dethroned and disgraced, the Queen of England felt 'these news take one's breath away'.[24] Her eldest child, Vicky, was, after all, by now the Crown Princess of Prussia. Nevertheless, the Queen warned her daughter, lest it should be misinterpreted by diplomats and politicians, that she had sent a personal message of sympathy to the Empress Eugénie, and wanted her son-in-law Fritz to be aware of it. To Vicky herself, the Queen said, 'I know you will feel for the poor Emperor and Empress and think of '55 and '57'.[25]

Those had been the years of the closest bond between the Bonapartes and the British Royal Family, as the Crimean War was brought to an end, in which they had found themselves to be unlikely comrades in arms. The relationship had been a complicated one. Palmerston, as seen in the last chapter, had breezily saluted Napoleon's coup d'état, and never made any secret of his fondness for the faintly farcical Emperor. More cautious British opinion regretted Bonaparte's buccaneering spirit. The Crimean War had achieved nothing beyond the unnecessary loss of life, the further decline of the Ottomans and a strengthening of Russian hostility to the Western powers. Napoleon's interference in northern Italy looked like antagonizing Austria. He would eventually meet his Nemesis by antagonizing Bismarck's Prussia. The British military and political Establishment never much trusted Napoleon, nor would they have been surprised, at any juncture, had he decided to emulate his uncle and wage war on Britain. Albert was acutely aware of all these things. For Victoria, also aware, her feelings were complicated by a genuine fondness for both Bonapartes, especially for Eugénie, and when she thought 'of '55 and '57' she thought of happy times together, popping across the Channel in their royal yachts to visit one another at Osborne and Cherbourg.

Yet, as she recalled in 1870, Albert had never felt the personal warmth which she extended towards their French friends. The atmosphere at Court, and the whole 'feel' of Paris, that dreaded city where his mother had died so young, was abhorrent to the Prince Consort. 'The system of corruption, immorality and *gaspillage* [waste] was dreadful. Nothing annoyed dear Papa more than the abject court paid to the Emperor and the way in which we were forced to flatter and humour him, which was short-sighted policy and spoilt him,' wrote Vicky.[26]

Nevertheless, the Queen was right to remember that, whatever Albert's fastidious distaste for Bonaparte's private morals, there had existed a friendship. True, the state visit to France paid by the Queen and her husband in 1855 had delighted Victoria more than Albert, even if, in retrospect, they had both regretted taking with them the impressionable Bertie, Prince of Wales. It was his first visit of many to what, in this regard, Victoria came to see as 'that horrid Paris'.[27] As her secretary Sir Charles Grey was later to observe, however, many of the things which she hated in Bertie, his self-indulgence, his greed, his frivolity, were qualities which she had herself. The great ball in Versailles given in her honour, the garlands of flowers, the gleaming chandeliers, the roar of fireworks, the groaning tables, the flowing champagne thrilled her, made her feel like the frivolous Whiggish girl who had flirted with Lord M. She joked with Napoleon that she would come back to Paris incognito the next year, take a cab from the station, and knock on the door of the Tuileries, begging for some dinner.[28]

Albert, whose constitution simply did not allow him to guzzle and scoff, could not play these social games. When he was chatting with scientists, or northern inventors, or Cambridge dons, he was natural and unpompous. He could not, however, play the silly courtly games which the French Court expected of him, and he never lost his awareness that Victoria and Napoleon were Heads of State, that they should be aware that upon their relationship depended the destinies of millions of their fellow Europeans.

From 6 to 10 August 1857, Napoleon and Eugénie came for a return visit to Britain; not a state visit to London with grandees and fountains of Veuve Clicquot, but a private holiday at what Victoria called 'poor dear unpretentious Osborne'. The weather was awful, as they trudged round Albert's model farm and made a visit to Carisbrooke Castle, where the Royal Martyr had been, for a while, imprisoned after the Civil Wars. (And escaped.) They sailed up and down the Solent in the rain. The Empress Eugénie eschewed the breakfast room, where the windows were wide open despite the squalls in the garden outside, and came downstairs having breakfasted in her room, chastely dressed in simple cambric and muslin dresses – clothes in marked contrast to her dazzling outfits when on public display in Paris.

Albert, meanwhile, was using the time for diplomatic talks with the Emperor. Though Albert never saw the point of Palmerston, the Prime Minister saw the virtue of Albert, who had mastered all the details of the case, and all the diplomatic implications, using the Emperor's visit to Osborne for semi-formal talks. As Albert complained, 'this place... is meant for a haven of rest, but is now selected by all sorts of exalted personages as the place to pay us visits, a total *bouleversement* of its original purpose'.[29]

A crisis, not for the first or last time, had blown up in the Balkans. Two principalities, Moldavia and Wallachia, had achieved independence from Turkey in the Treaty of Paris in 1856. Napoleon III wished them to be united, and to become French dependencies. Russia supported them in this, presumably because it hoped to use the united new country – what would eventually become Romania – against the Turks. Britain – even the Francophile Prime Minister Palmerston – was against the union.[30] Moldavia had voted, in elections, for union; Wallachia not. France was in favour of enforcing the union, willy-nilly, and for the international community to back this. There was, just about, justification for this position, since the Paris Treaty allowed for the possibility of both provinces being governed by the same ruler or Domnitor – which indeed would happen in 1862, when the modern country which we call Romania was born. In 1857, however, the powers were divided. Austria was opposed to the union because it strengthened the hand of France and of Russia. Britain was unconvinced by the Napoleonic position. It was Albert who persuaded Napoleon, while he was on the Isle of Wight, to annul the elections in Moldavia, which some observers felt had been fraudulently conducted. This was on condition that fair and open elections would be held again in both provinces. Neither the Austrian Emperor nor the British Prime Minister could restrain the desire for Romanian national identity. Albert, however, had held the pass, and he had demonstrated that royal personages had their diplomatic uses in the unfolding European story. Palmerston was impressed. 'The Prince,' he told the Queen, 'can say many things we cannot.' 'Very naturally,' was Victoria's proud reply.[31]

While the men, in one part of the house, tried to settle the future of Romania, as the English rain beat against the panes, the women were forming what was to be a lasting bond. Sixteen-year-old Vicky, the prospect of herself becoming a royal consort growing ever nearer, was especially entranced with the Empress, and, at Eugénie's request, plaited her a straw bonnet. 'She is my *beau idéal* of a woman and I am quite enchanted by her,' Vicky gushed.[32] Eugénie, for her part, tactfully but firmly told Vicky that she must improve her wardrobe before she became the Prussian Crown Princess. 'I'm telling you this because I want you to sparkle in Berlin.'[33]

To cement the friendship, Victoria and Albert took the royal yacht across the Channel to Cherbourg that August to visit the Imperial couple in Normandy – country which reminded Victoria of Devon.[34] It was only a flying visit, but they took all the children, with the full entourage of governesses and nursery maids. The children were in heaven, but for the parents, and their firstborn, there was a haunting sense that their family unity was about to be disrupted, and that everything they were now doing together, with Vicky, would be for the last time.

When they went north to Balmoral, Albert noted, 'Vicky suffers under the feeling that every spot she visits she has to greet for the last time as home. The "Johanna sagt euch Lebewohl" of the Maid of Orleans comes frequently to my mind. It has been my lot to go through the same experiences.'[35] The reference is to Schiller's play in which Joan of Arc, before setting out to see the Dauphin and rescue France, leaves behind her father's farm. Neither Vicky nor Albert could have quite foreseen the roasting which the Prussians had in store for her. Albert, perhaps revealingly, misquoted the Schiller line, since, surveying the scenes of her childhood, the Maid in fact says not merely 'Joan bids you farewell' but 'an eternal farewell' ('*ewig Lebewohl*').

It was all the more reason that they should enjoy their last Scottish holiday together as a family. Victoria chose this occasion to invite herself to Lord Aberdeen's house – Haddo. It was partly a gesture of gratitude to him for having procured them Balmoral after the death of his brother. It was also a little signal to Palmerston that she and Albert missed the Aberdeen style – diffident and nuanced. Aberdeen was shy about his house, which was, as the Queen tartly observed, 'plain'.[36] The extent of Aberdeen's lands and properties, however, was staggering, and he impressed the Queen by making his entire tenantry, over six hundred people, line the route as the royal party approached, as if he were an old Highland laird in pre-Hanoverian days. The countryside was the reverse of Highland, and the Queen noted, as her landau, containing herself and Vicky and Alice, drew near to Haddo, that the landscape was 'frightful, the hills totally disappearing & nothing but stone walls'.[37] The royal party, however, were touched by Aberdeen having erected triumphal arches for them along the dull way, and, while finding the house 'small' and the 'air very different from the dear Highlands & one sadly missed the mountains', she appreciated his hospitality and noticed that the dining room ('very narrow, reminding me somewhat of the one at Frogmore') contained 'some fine pictures'.[38]

That winter, as they agonized about the impending departure of their daughter to Berlin, and about the unfolding horror of news from India, something happened which wrenched Albert back into his childhood and deprived him of the only female companion of his childhood, who was roughly his age. This was his, and Victoria's, first cousin Victoire, daughter of Prince Ferdinand of Saxe-Coburg-Gotha and belonging to the Catholic branch of the family. She was married to Louis, the second son of King Louis-Philippe, the Duc de Nemours, and since the revolution of 1848 she and her husband had lived at Claremont, scene of the fateful death of Princess Charlotte in childbed in 1817.

Forty years later, history repeated itself, though the baby in this case, a healthy girl called Blanche, survived. The death was not only a great shock to Albert and to her other relatives. It also tore away yet another part of his childhood. With the many calls on his time, he did not see very much of his cousin, but her presence in England was a comforting token of old Coburg.

'Two days ago,' he told his stepmother, 'I hurried to Claremont immediately on receiving the woeful news, and I was there yesterday with Victoria to join our tears with those of the mourners. Close by the room in which the corpse of the Mother is still laid, lies her little baby in rosy slumber, and shall never know her mother; a few days ago they were one!'[39]

It was a stunning blow, a reminder of how quickly and easily death comes up and surprises us, regardless of our age; but a reminder, too, that they had been so fortunate – with nine children whose mother was, physically at least, unscathed by the experience of childbirth. (They knew nothing, at this juncture, of Prince Leopold's haemophilia and its dire implications.) The health of the baby, the deadness of the corpse, the desolation of Nemours – who would, as it happens, live until 1896 – were all stark shocks, putting their own marital squabbles into perspective. The year, like its predecessors, had been punctuated by some fairly awful rows, often precipitated by the Queen's handling of the children. 'You are quite mistaken,' Albert told her, 'if you think I am not concerned to maintain your maternal authority with the children. On the contrary, it is my constant care to safeguard it and to preserve the warmth of the children's feeling for their Mother, and that is just the reason why I have felt it my duty to warn you of the rocks on which all our efforts are wrecked. I admit it was an error of judgement to speak to you yesterday about Alice's weeping, for I ought to have remembered the state of your nerves, but I really did not think they were shaken to the extent they have since shown themselves to be, and there was nothing in what I said to excite a healthy person to such an outbreak. You yourself will not reproach me with harshness, & did not even do so at the time. You were only deeply wounded & excited and I did not manage to keep myself under control. You asked me to promise "not to scold you again before your children". To that I willingly agree – what you call scolding [*schelten*] I would call simply the expression of a difference of opinion'.[40]

This had clearly been a humdinger of a row, and the older children in particular had witnessed it. Albert's perpetual complaint, to and about the Queen, was that she was too harsh with the children, and asked him to join in. At the same time, it would be a complete mistake to imagine them always at odds, and there was real sincerity, surely, in his olive branches. 'It should not be difficult to live happily in love and harmony. I don't care about making professions. Let us turn over a new leaf & show by our deeds tomorrow what yesterday & today might have been and shall be.'[41]

One matter on which they were in complete harmony was the question of the jewels which they believed to be theirs by right, but to which Victoria's odious uncle the King of Hanover laid claim.

1857 was the year when a sixteen-year-old wrangle about royal jewels was resolved, much to Victoria and Albert's annoyance. Both of them loved jewels, and the part played in royal lives by precious stones is not to be underestimated. Witness the queues at the Tower of London to see the Crown Jewels today. Most royal marriages, across the entire continent of Europe, have involved substantial presentation of jewels. They possess a symbolic value, naturally. For their wearers, too, they bring a sense of continuity with the past, and a personal security.

Since the Civil Wars, the English royal houses had not been especially rich in bling. The Stuarts had taken a certain amount into their varied exiles, and it had not returned. In 1761, jewels worth

£50,000 were presented by George III to Queen Charlotte. He could not personally have afforded this prodigious sum, and the funds were supplied by Parliament. They were always, however, known as the Hanoverian Crown Jewels. They included five pearl necklaces which had belonged to Queen Anne, and another which was said to be the finest rope of pearls in Europe. There were other stupendous items including Indian treasures presented to Queen Charlotte by the Nawab of Arcot.

Victoria had always worn the 'Hanoverian' jewels, and was especially fond of the string of pearls. She could not become the Queen of Hanover, because a female succession was not allowed. Her much-hated uncle the Duke of Cumberland took the throne, and claimed possession of the loot. Victoria resisted, and from the early 1840s the dispute chuntered disagreeably along. The King of Hanover died in 1851 and thereafter the Hanoverian claims were renewed. There was more delay. Victoria continued to wear the disputed jewels. Then a Commission was set up and, on 5 September 1857, Sir Tite Barnacle eventually decided that the Hanoverian Crown Jewels, though purchased, for the most part, with money voted by an English Parliament, belonged to Hanover. Victoria was obliged to surrender them. Despite the loss, Victoria was never short of jewels. As well as the Koh-i-Noor, appropriated in 1850, she bought a lot. That said, it was a miserable end to the year. By then, however, Albert and Victoria had other things on their minds. By the time the jewels were handed to the Hanoverian Minister in London, Count Kielmansegge, on 23 January 1858, the marriage of their firstborn was almost upon them.

NINETEEN

KING IN ALL BUT NAME

As the wedding approached, the emotional realities of the situation sharpened like the light of a cruelly cold dawn. Albert's life had hinged on separation. The first had been the unexplained departure of his mother from his life. The second, when, an emotionally vulnerable twenty-year-old, he was separated from his father's house and his Fatherland. Now, at thirty-nine, he was to lose the person in the world to whom he was closest.

To say that he was her hero is to understate. They were besottedly fond of one another. 'Every time she hears the glorious first monologue in Goethe's *Faust*, she must think of "Darling Papa". Everything that is good, great, clever and beautiful, whether in books or in real life reminds her of him, as a reflection of his perfect character!'[1] Those familiar with *Faust Part One* – and the monologue is something most German-speaking students would have by heart – might be surprised by the analogy. Faust, true, like Prince Albert, has made himself the master of every subject in the world; but the monologue expresses nihilistic despair at the futility of such studies unless they could be translated into power. He says that he wishes to abandon the life of abstract study altogether

> That I may detect the inmost force
> Which binds the world, and guides its course.

Vicky was quite intelligent enough to realize that for her father, as indeed for herself, the pursuit of learning was intimately connected with the desire to change the world. Old Carlyle, sitting in his Chelsea attic, would spend twenty-five years writing the life of Frederick the Great, the Enlightened Despot whose palace at Potsdam made such a profound impression on Albert. Now Vicky was marrying the man who would be Frederick's eventual successor as King of Prussia; but the nineteenth century was not the Age of Enlightenment. The political programme had moved on, and the frustrating thing, almost killingly frustrating for Albert, is that he would only have been able to push forward all the reforms he believed necessary for Europe with the one weapon he lacked; the weapon which, in the kind of constitutional monarchy of which he approved, any monarch would always lack: absolute power. Albert's life was a frustrated attempt to round the circle: to force people to become, in his sense, liberal; to build the Good City. His failure is only partly attributable to the shifting balance of power and the rise of Bismarck. Albert had many great qualities, but he was not by temperament (as Bertie would be) a diplomat. He was not an extrovert, he was not conversational. He was unsuited to the exercise of 'soft power'. His failure would be softened, partly by death, but chiefly by the huge successes he clocked up along the way in the fields of art, craft, industry. He immeasurably enriched the culture, and Britain still reaps the benefit, with its Albertopolis in South Kensington, its array of museums and the great concert hall erected in his memory. More than that, he definitely spread what Arnold called Sweetness and Light in the nineteenth century. If Arnold was right, and he surely was, that Britain as a new industrialized power was in danger of being run by the Philistines, Albert was the champion of the anti-Philistines in art, music, schemes of public welfare, encouragement of science and enlightened technology. He failed, however, as a European diplomat. It was not in his power, or anyone's power, to temper Napoleon III's military and political ambitions; nor to neuter the growing power of Russia; nor to influence in a positive direction the imperialistic strengths, fatefully to be pitted against one another, of Russia and Austria. The dynastic marriages of his children, which he wanted to use as bargaining chips in this strategy, would be to no avail. His own inevitable failure was laying the foundation for his daughter's failure, the consequences of which were truly catastrophic.

Albert told Stockmar that such was the enthusiasm, among ordinary Germans, for the match, they had found 'in Vicky herself answer to their expectations'.[2] In preparation for her new life, Albert had set his daughter the task of translating a pamphlet, published in the previous year – *Karl August und die Deutsche Politik*. This encapsulates the misapprehension he was under: that it would be possible to recreate, in the mid-to-late nineteenth century, the enlightened culture of the early decades. Karl August, Grand Duke of Weimar, had been the man who lured Goethe there in 1775. The *Goethezeit* was the glory age of German culture, coinciding with the lifetimes of Kant, Schiller, Beethoven, Fichte, Hegel and countless others. Spiritually speaking, Albert himself belonged to these times, but there could have been few pamphlets of less relevance to the political reality which Vicky was to encounter in Berlin in 1858.

Just as his own Coburg minders had prepared him for his destiny by rigorous programmes of education, so Albert had forced his highly intelligent firstborn to read, study and to prepare. The seventeen-year-old girl was to be the ambassador, in the Prussian Court and capital, for the values which Albert had espoused and nurtured for the whole of her life: an alliance between royalty and a representative Parliament; an enlightened acknowledgement of the role played by the Crown in the cultural and spiritual life of a nation. Vicky was to be the Great Exhibition made flesh; the Society for the Improvement of the Working Classes translated into German; the friend of poets, novelists and scientists; the patient, behind-the-arras adviser to all government departments, whether in the field of armaments, colonial administration, urban planning, drainage, museums, university education… He had prepared her so well. As he led her into the Chapel Royal at St James's Palace, on 25 January 1858 – the very place where he had been married, eighteen years earlier – he was making a political statement to the governments of Europe. Victoria remembered, in the chapel, 'myself kneeling by my Angel's side… when I also felt that joy and confidence when one gives oneself for life to the man of one's heart and choice. It is a great and elevating sensation.'[3]

Disraeli, who was present at the bridal ball in Buckingham Palace, said that there were as many princes present as at the Congress of Vienna.[4] Greville said that the wedding was celebrated with 'amazing éclat'. The invention of the steam engine and the extension of railroads permitted Londoners to see crowned heads from as far away as Berlin, Coburg and Naples. The British public had never seen anything like it. Their feelings about the German Prince and his extended European family were predictably mixed. The bridegroom, Prince Frederick William of Prussia, as they called him, looked splendid in his helmet and uniform, though as matters transpired, he would show himself to be a weak character, entirely unable to liberalize the increasingly militaristic, bellicose, iron and blood world of Bismarckian Germany, and in any case, not a 'liberal', in the English sense, at all. Prussia in the 1850s was determined never to repeat what they regarded as a mistake, the few months of a liberal constitution cobbled together in 1848, and swept away in the same year by a staunchly conservative regime. Prussia was destined to become less and less Albertian in complexion, and the young 'Englishwoman', as she would always be stigmatized in the German press, was going like a sacrificial lamb into the Teutonic jaws. Fritz, however, would always be a loving husband. The Cockney crowds shouted out at him, as he trotted past them, 'If you don't treat her well, we will have her back!' They should have been shouting it at his entourage, and at his father, the Prussian King's brother, destined to become the German Emperor in 1870 and to declare himself governing by Divine Right; and at the Prussian Court, ministers and journalists, who were not there, but who would make Vicky's life so unhappy.

Hindsight enables us to see that Victoria and Albert's dynastic ambitions were to be wholly unsuccessful. We can see that neither in Berlin, nor in Darmstadt (where Alice was to become Grand Duchess), nor in the next generation when Alice's daughter Alicky became the Russian Empress, was there any glimmering hope of Prussia or Russia adopting Albertian principles. None at all. We know that, after 1870, the European powers, locked in an unstoppable arms race, were marching towards 1914, the cataclysm of world war and revolutions which would, for most Europeans, bring monarchy and all its stabilizing influence to an infernal, bloody end. Our knowledge makes the heartbreak of the young players in the drama – script by Prince Albert, directed by Prince Albert – all the more poignant.

After the wedding and the celebrations came the terrible moment when Albert and the two elder boys, Bertie and Affie, accompanied Vicky to the waiting royal yacht at Gravesend. 'My beloved Papa,' she wrote, 'the pain of parting from you yesterday was greater than I can describe; I thought my heart was going to break when you shut the cabin door and were gone – that cruel moment which I had been dreading even to think of for 2 years and a-half were past – it was more painful than I have ever pictured it to myself... To you, dear Papa, I owe most in this world. I shall never forget the advice it has been my privilege to hear from you at different times, I treasure your words up in my heart, they will have with God's help, an influence on the whole of my life.'[5]

This was undoubtedly true, and in nearly all her letters from Germany, she alluded to 'the large beloved Photograph' which she had of her sire and hero opposite her bed which she kissed every day; she also said that she dreamed of 'dear Papa' and 'thinks she hears his voice'.[6]

It spoke volumes that Vicky wrote to each of her parents in different languages: to her father in German, and to her mother in English. With her mother, Vicky's relationship had always been more complicated. The Queen would never have spelt out to herself the obvious fact that Vicky and Albert, on an intellectual plane, had more in common with one another than she ever quite had with her husband. Nor would she ever have expressed jealousy of the intimate love Albert felt for his firstborn. Instead, the Queen, while recognizing the good qualities in her son-in-law, found it surprising that Vicky 'is so excessively fond of him! I feel that none of our daughters will ever find such a Husband as Albert – that is hopeless, but there are others who are far superior in talents to Fritz... All this I feel – & it makes me a little bitter & angry – when I ought to be grateful and thankful only – as my beloved Angel says, May I find strength & assistance from On High.'[7]

One thing for which posterity can be grateful, however, is that Victoria and Vicky, once they were separated, began a correspondence which ended only in death. Their frequent letters to one another are one of the most invaluable of historical sources. They reveal that, whatever the awkwardnesses and complexities of any relationship between Queen Victoria and another human being, the two women were deeply fond of one another. And these highly distinctive documents, so revealing of the characters of both women, never fell into the hands of Princess Beatrice, so here we can read the voice of Queen Victoria uncensored. Within only a few weeks, and by the time Vicky had sent her mother, by special messenger, a delicious 'pumpernickel and tower-shaped cake',[8] the two women had exchanged thousands of words. The Queen said at one point that Vicky was more like a sister than a daughter to her, and the outpourings from the pen of the mother to the exile contain some of the most candid of all that candid writer's observations: as, for example, when she said how much she envied her daughter not being a Queen Regnant. 'Though dear Papa, God knows, does everything – it is a reversal of the right order of things which distresses me much and which no one, but such a perfection such an angel as he is – could bear and carry through.'[9] Likewise typical, in its candour and lack of tact, is the letter in which the mother tells her first child how much she envies her being married without, at that very early stage, the incumbrance of having children. 'They are an awful plague and anxiety, for which they show one so little gratitude very often! What made me so miserable was – to have the two first years of my married life utterly spoilt by this occupation!'[10]

It was not long before Albert was planning a visit to the *Heimat* to see his daughter, and, as the Queen wrote to Vicky, 'believe me, my dearest child, that the greatest proof of my love and affection for you is, that I have encouraged him in it – though you will know how miserable and (from my isolated position) lost I am without my master.'[11]

In the early weeks of Vicky's absence, Albert poured out letters to her, covering a wide range of subjects, from literature to politics. He was also self-revelatory to her, in a way which, in any surviving letters at least, he never quite had been with his wife. When she had been gone only a few weeks – in April 1858 – he wrote:

> That you should sometimes be oppressed by homesickness is most natural. This feeling, which I know right well, will be sure to increase with the sadness which the reviving spring and the quickening of all nature that comes with it, always develop in the heart. It is a painful yearning, which may exist quite independently of, and simultaneously with, complete contentment and complete happiness. I explain this hard-to-be-comprehended mental phenomenon thus. The identity of the individual is, so to speak, interrupted; and a kind of Dualism springs up by reason of this, that the *I which has been*, with all its impressions, remembrances, experiences, feelings, which were also those of youth, is attached to a particular spot, with its local and personal associations, and appears to be what may be called *the new I* like a vestment of the soul which has been lost, from which nevertheless *the new I* cannot disconnect itself, because its identity is in fact continuous. Hence the painful struggle, I might almost say, spasm of the soul.[12]

Has he not anticipated here in one paragraph some of the Bergsonian ideas of the 'self' which permeated and inspired Proust's

A la recherche? Or if that is too – to use a word with which Albert begged to be excused for his reverie – abstruse, does it not explain his very consciously adopted device for 'getting through', not merely his public duties, but also the pains of married life? That is, by self-consciously adopting 'Dualism', he was sometimes 'not there'; the busy encourager of chambers of commerce in English northern towns; or the silently exasperated husband, unable to understand another explosion of emotional rage from his wife, being another 'I'; elsewhere, wandering the Thuringian forests with a gun under his arm and a volume of Schiller in his pocket? Vicky, by contrast, never mastered the 'dualist' knack, and was always, naïvely and sometimes a little clumsily, '*die Engländerin*', so far as her Prussian Court and the Prussian press were concerned.

When Vicky left home, Albert, never in very good health, suffered from sleeplessness and his old digestive troubles. 'I have been very unwell a night and a day', was a typical record, 'and am only slowly recovering. I was attacked by a most violent bilious fit, which has left me very weak'.[13] He was determined, nonetheless, to visit Vicky in her new world.

It was thirteen years since Albert had visited Coburg, and one of his aims, when he went back to Germany to visit Vicky, was to reconnect with his earlier 'I'. He left as soon as he tactfully could after the Queen's birthday, which fell on 24 May. Three days later, he crossed for Germany, by way of Dover and Ostend, accompanied by his personal equerry, Colonel Henry Ponsonby, and by his German librarian, Dr Becker. Thanks to the speed of the express train from Frankfurt, they were able to find themselves in Coburg on 29 May. They heard Ernst's composition 'Santa Chiara' performed in the theatre. Albert confided in his wife, to whom he wrote several times a day, that he had 'struggled manfully to keep drowsiness at bay' during his brother's concert. He dined with Ernst and Alexandrine at half past three in the Schloss Ehrenburg, and then walked in the Hofgarten, and climbed the steep hill to the old castle, the Veste, noting a novelty – a new church being built on the edge of the park

– a Catholic church. 'Right in front of the palace,' he noted, with presumably mixed feelings.

It was a hot June, as, lost in his own boyhood, he revisited the childhood scenes. 'The Rosenau was truly lovely today'. He picked flowers for the Queen there. Then he visited the Kalenberg, admiring the substantial rebuilding which Ernst had undertaken. All too soon, it was over, and he travelled to Gotha, laying a forget-me-not on his grandmother's grave: the mother of his mother.[14] There was little enough time left for the flying visit to Vicky in Berlin.

Albert was struck by how ill King Wilhelm looked, pot-bellied but skinny, and by how much weight he had lost, even since Vicky's wedding of a few months earlier. Nonetheless, the King was standing there to greet him, in full dress uniform with helmet and sword, when Albert arrived. The two-to-three-day visit allowed Albert to see how happy his daughter was with Fritz. They had only lately moved out of Berlin to the summer palace of Babelsberg at Potsdam, with its green lawns sloping down to the River Havel. Albert was not to know it, at this early juncture, but his daughter was three or four weeks pregnant. She and her husband were in cramped quarters directly over the palace kitchens, and the heat was unbearable. Nevertheless, when Albert returned to London, he was able to give the Queen a good account of it all. The fine way in which he spoke of his reunion with Vicky, and of the young Princess's marital happiness, provoked a desire in the Queen to be emotionally competitive, and she wrote to Vicky establishing that she was closer to Albert than Vicky was to either man.

> I must... repeat to you what you say about your feelings towards your husband are only those which I have ever felt and shall ever feel! But I cannot ever think or admit that anyone can be as blessed as I am with such a husband and such a perfection as a husband; for Papa has been and is everything to me. I had led a very unhappy life as a child – I had no scope for my very violent feelings of affection – had no brothers or sisters

> to live with – never had had a father – from my unfortunate circumstances was not on a comfortable or at all intimate or confidential footing with my mother (so different from you to me) – much as I love her now – and did not know what a happy domestic life was! All this is the complete contrast to your happy childhood and home. Consequently I owe everything to dearest Papa. He was my father, my protector, my guide and adviser in all and everything, my mother (I might almost say) as well as my husband.[15]

This letter gives some indication of the emotional burden Albert as a husband was expected to carry, and it must be remembered in the context of their marital storms. It is not for us, the men and women of posterity, to patronize Queen Victoria, but within the terms of the marriage, the frank avowal that he was everything to her goes some way to explain why she treated him, not only as her lover and her political adviser and her friend, but also as so many other things as well. So, while genuinely worshipping him, she could rail against him as an uncertain child rails against its parent. 'I am accused of want of feeling, hard-heartedness, injustice, hatred, jealousy, distress, &c &c I do my duty towards you even though it means that life is embittered by "scenes" when it should be governed by love and harmony,' he wrote wearily after one of the storms. 'I look upon this with patience as a test which has to be undergone, but you hurt me desperately, and at the same time do not help yourself.'[16] The handwriting in which these words are written is scrawled and exhausted, so very different from the copy-book neatness of his handwriting as a young man.

Despite his feeling ill and tired, Albert arranged that he and his wife should revisit Germany. They planned to make a simple, private, visit to Babelsberg, but this was not allowed. The old King insisted that, both in Potsdam and in the capital itself, it should be a state visit to Berlin in the height of summer.

The royal yacht left Gravesend for Antwerp on 10 August. A full entourage was present, including Sir James Clark, the Queen's physician. Seeing the trip through Clark's wholly English eyes, we are able to recognize how very different Prussia was from the England which Vicky had left behind. At every dinner and formal occasion, and throughout the streets of Berlin and Potsdam, there were uniforms. The whole place was like a military barracks. 'The whole country seems occupied in playing at soldiers'.[17] At the same time, Clark was aware of the intellectual impressiveness of the German scene, visiting Alexander Humboldt in his extreme old age, and taking the Prince Consort to 'see some electrical experiments with Electric fish from the Coast of Africa'. The man conducting these experiments was 'one of the first men of Science in Prussia', but Clark had forgotten his name.[18]

If the visit to Germany in May–June had been a revisitation of Albert's youth, the visit in high summer coincided with a cruel wrench. He was already feeling frustrated – 'cut to the heart' – that he would not be able to revisit Coburg. Then came news of a death. While the Queen was dressing, early one morning in their Berlin apartments, Albert came into the room holding a telegram. '*Mein armer Cart ist gestorben*,' he said. There were many tears from both of them. Albert's valet, Cart, an independent-minded Swiss, had suffered a heart attack in Morges. He had been a true friend to Prince and Queen, not only dressing Albert, but also performing secretarial duties such as the copying of letters. 'He was the only link my loved one had about him which connected him with his childhood, the only one with whom he could talk over old times. I cannot think of my dear husband without Cart! He seemed part of himself!'[19] Victoria did not say so, but it was especially sad for a man who did not, as such, have friends, to have sustained such a loss. There was Stockmar – but he was a father-figure, and now resident in Germany. There were many politicians, and others, with whom Albert had friendly dealings. But all his intimate relationships were with his family or with Cart. He was a Prince without an Horatio.

By the time they reached Babelsberg, the German royalties and military top brass were assembled, sweltering under their helmets and feathers. 'Large and dreadfully hot dinner,' the Queen's journal recorded. 'The whole Legation: Field Marshal Wrangel, Von Manteuffel, First Minister, Home Minister Von Massow, Obermarschall Graf Keller.' As well as being heavily uniformed, and very hot, the striking thing about them all was how very old they were. Here were throwbacks to a lost world – for example, the old Countess Blücher, a Scotswoman, who had married the nephew of the victor of Waterloo.

The next day, wilting with heat, they trooped into the Garrison Church in Potsdam, which was full of soldiers, whose combined voices were raised in 'a fine chorale'.[20] Behind the altar was that tomb which Albert had visited before. They made a special visit to the tomb once again a few days later.

Victoria remembered the trivial, though familially interesting, fact that her mother had been born on the very day that Frederick had died. The Duchess of Kent's grandmother – the Queen's and Albert's great-grandmother – had occupied the unlikely role of Frederick the Great's wife. Because it was a day of mourning, the old Duchess of Coburg would never allow Victoire's birthday to be celebrated.

For Albert, however, the tomb was eloquent of the differences between England and Germany, between the Enlightened Despot who had achieved so much for Prussia, and the Enlightened Prince Consort whose role was so circumscribed by the political and social rules of his adopted country. On a later day, Albert and Victoria were shown round Frederick's superb palace of Sanssouci, the silvery rooms, the festoons of creeper around the windows, the 400 orange trees in the orangery. 'Really one of the finest pieces of architecture that can be imagined!' exclaimed Victoria girlishly, 'but what an expense!'[21]

On 26 August, Albert's thirty-ninth birthday was celebrated. For the first time in their marriage, they were separated on this 'Blessed Day' from all their children save the firstborn. Vicky and Princess Louise of Baden arranged his present-table. The Queen gave him a life-size oil painting of Princess Beatrice by Horsley, and a collection of photographic views of Gotha, and a paperweight of granite and deer's teeth from Balmoral. Vicky gave him a portrait of herself by Hartmann, an iron chair for the garden at Balmoral and a drawing executed by herself. The other children had all sent letters. The birthday surprise was that Ernst arrived from Coburg. Outside the window, nature provided Balmoral weather, pouring rain, and the band played German hymns.[22] The next day, just before they left, they received another visitor from the past, Stocky, 'who promised to watch over our precious child'.[23]

The visit to the court at Berlin had confirmed pro-German sensibilities. A brief visit, earlier that summer, to Cherbourg – during which, it was said, Napoleon III had spent 25,000 francs on fireworks[24] – had preserved, superficially at least, the cordial relations between the Bonapartes and Victoria and Albert. Privately, Albert had been appalled by the rebuilding of the fortifications at Cherbourg, and by the display of naval force. It had heightened anti-French feeling in the British press, which felt that Britain was not doing enough to rearm. Albert agreed wholeheartedly. 'The war preparations in the French Marine are immense! Ours despicable! Our Ministers use fine phrases, but they do nothing. My blood boils within me.'[25]

The uneasy feelings between France and England had been the occasion, early in 1858, of yet another change of government. On 15 January, Palmerston's Cabinet heard of an attempted assassination attempt on Napoleon III outside the Paris Opera House. Three grenades were thrown, killing several people and injuring many bystanders. The ring-leader of the terrorists was an Italian named Felice Orsini, resident in England. He had travelled to Paris on a British passport, and the explosives had been manufactured in Birmingham. The French Foreign Minister, Count Walewski, sent a letter to Lord Clarendon, his counterpart, asking for assurance that the British would take more care in nosing out terrorists, and asking for an amendment in English Law which made it easier for such people to be apprehended and punished. There was outrage in England, though the Government grudgingly brought forward the Conspiracy to Murder Bill, which introduced no new law, but increased the penalties against proven conspirators. At first reading, it was passed. But there was chaos among the rag tag and bobtail whom Palmerston needed to cobble together to secure his Commons majority, and the Government was simultaneously running into trouble in the Lords securing the passage of the Government of India Bill, abolishing the East India Company. When it began to emerge that Palmerston had not given any formal answer to Count Walewski's dispatch, the Commons turned against the Bill. By their silence, a Radical MP named Milner Gibson averred, the Government had admitted that they had, in effect, sheltered assassins. His amendment of the Bill, censuring the Government, was supported by Lord John Russell, and Derby remarked to Disraeli, '*C'est le commencement de la fin*'.[26]

Gibson's motion of censure was passed by 234 to 215. Palmerston resigned. Who would be able to form an administration, other than Lord Derby who at the time was shooting wildfowl at Heron Court, staying with Lord Malmesbury? He let the shooting party know that he would be prepared to try to form a government, and on 20 February, he went to Buckingham Palace to kiss hands. Victoria's were the hands, but Albert stood beside her as the little ceremony took place.[27]

Since Peel's brave stand over the Corn Laws, the Conservatives were divided between old Tories and the Peelites. Derby could only cobble together a government by luring some of the Peelites back into the Tory fold. Gladstone was tempted, but in the event held back. Had he done so, he would almost certainly have become Chancellor of the Exchequer, rather than Disraeli. Malmesbury took the Foreign Office.

As far as foreign policy was concerned, the chief European problem was the future of northern Italy. Napoleon III conferred, very publicly, with Count Cavour, the Prime Minister of Piedmont, on 21 July 1858, and this was a clear signal that he was prepared to go to war, on behalf of the Italian nationalists, against Austria, thereby upsetting the whole balance of power in Europe as it had existed since 1815. As far as the British Empire was concerned, the Government's chief job was to ensure the safe passage of the Government of India Bill, also of 1858, and to guarantee the safety and stability of India after the winding-up of the East India Company. In both these areas, Albert was broadly in sympathy with Derby's approach. Equally, Albert sympathized with the speech Derby had made very shortly after taking office – 'There can be no greater mistake than to suppose that a Conservative ministry necessarily means a stationary ministry.'[28]

This belief was most strikingly demonstrated in the fact that it was Derby, rather than a coalition inclusive of the Radicals, who proposed and pushed forward an extension of the franchise. They spent most of 1858 unable, quite, to decide to what extent they would extend it, and how far reform could go. The diehards, such as the Home Secretary Spencer Walpole, believed that universal suffrage – a vote for every man, that is to say! – was 'a fateful step'. Derby realized he would never get such a measure past the more conservative back-benchers, and considered an £8 rating in the boroughs, so long as these voters only voted within their urban areas and were not, by some mistake, allowed to vote in country elections, where it was proposed that there should be a £25 rating. (That is, those who inhabited properties which would ensure a tax of these sums. The county voters would therefore all be figures of some substance, doctors, merchants and the like. They were a long way from giving the vote to the working classes; a measure which would eventually be brought in by Derby's third administration in 1867.)

It is during the discussion about reform of the voting franchise that one sees the first glimmerings of one of the great nineteenth-century alliances: between Benjamin Disraeli and the monarchy. But only a glimmering dawn. Albert and Victoria still, at this stage, saw Disraeli – Chancellor of the Exchequer by now – as the man who helped destroy the career of their hero Robert Peel. Both Disraeli and Victoria, however, saw that there was a symbiosis between monarch and people; that, far from an extension of parliamentary government being a threat to the monarchy, or monarchy – as in autocratic Russia and Austria – holding back political reform, the two went hand in hand: the popularity of the monarchy greatly bolstering the stability of political life for the Conservatives. Therefore, though neither Victoria nor Albert ever completely saw the point of Derby, being 'thrown' by his breezy, jokey manner, his concealment of high intelligence with facetiousness, his obsession with the horses and with Homer balancing his political career, they came round to him. Certainly, their reluctance, for example, to accompany him to Ascot races was not something with which later sovereigns would have sympathized.[29] They were both learning the ropes, and realizing that the success of constitutional monarchy depended on the sovereign's ability to get along with any administration.

Derby, for his part, adapted to the austerities of spending time in royal company. When the inevitable visit to Balmoral approached in late summer 1858, he recalled the discomfort of his stay there in 1852. He had been suffering badly from gout all summer, and had taken to his bed in the first week of September. Memory of the cramped, cold accommodation offered during his earlier visit to their Deeside Paradise urged Derby to wriggle out of his obligation. On 13 September, however, he went, and he was glad to have done so. Six years had brought great improvements to the house. His room was spacious and comfortable. And if deer-stalking in the rain – Albert stalked, Derby walked along to keep him company – was not Derby's ideal way of passing the time, these expeditions improved the gout, rather than the opposite. The unexpected bonus was the enjoyment of the evenings, which, for him, were hugely enlivened by the Prince of Wales.

Bertie was now nearly seventeen. A genial, though markedly non-academic, boy, he appeared at this stage to charm most people except his parents, who regarded him as a source of everlasting worry. Mr Gibbs, a barrister and Fellow of Trinity College, Cambridge, had, since 1852, been assigned the challenging role of private tutor to the two Princes, Bertie and Affie. Attempts to teach the Prince of

Wales mathematics had resulted in tantrums, with the Prince hurling his pencil across the room, hitting his tutor with a stick and pulling his younger brother's hair.[30] Things had calmed down a little since those early days, but neither the Queen nor Prince Albert thought highly of Gibbs's capabilities, and Bertie remained a tremendous disappointment.

Even in areas where you might have expected his parents to be pleased, they found reasons for censure. For example, much had been made of Bertie's confirmation in the April of that year, 1858. The boy was, after all, the future Supreme Governor of the Church of England, so the seriousness with which the whole ceremony was undertaken was entirely apt. It was surprising that, three months later, the Prince received an admonition from his father, a letter of many pages, when Albert heard of the Prince's intention to receive Holy Communion. Bertie, that July, was at Richmond. Albert had given him White Lodge, a Palladian villa in Richmond Park where he and Affie were to be 'crammed' by a Latin tutor named Reverend Charles Tarver, and by Gibbs himself. For company, they were to have three equerries of 'the highest character'. The Queen had decreed that no indiscipline was to be tolerated, such as slouching with hands in the pockets. 'Anything approaching a practical joke... should never be permitted.'[31] Tarver and Major Lindsay, one of the equerries, proposed that they should attend the Holy Communion Service at nearby Mortlake on the following Sunday and Bertie suggested that he join them. Far from this pleasing Albert, when he heard of the suggestion, he was disgusted. The sticking point for him was that this differed from his own, and the Queen's practice, with regard to taking the Sacrament.

> We have agreed upon taking it twice a year and have selected, as fixed periods, times at which the history of the Gospels and the Church festivals naturally prepare us and induce us to additional sanctity and at which we are sure not to be broken in upon by the gaieties of society or the demands of business (Christmas and Easter) as during these festivals everybody is

> at home with his family – We have chosen to take it away from, and undisturbed by the multitude who would stay for the show, if we were to remain in a public church after the service, and we have chosen the early morning as a time when the mind is still fresh and not diminished by the lengthy previous service.[32]

It is perhaps worth explaining, given the very different liturgical practices of the modern Church, what Albert means. In those days, it was almost unheard of for the Holy Communion to be the chief service on a Sunday morning in the Church of England. This service was either celebrated (as it still is) first thing in the morning, or as a 'stay behind service', as it was sometimes called, after the sung service of Morning Prayer.

Albert's concern, however, was clear. 'As our son, you would do well to keep to the example and practice of your parents.'[33]

Needless to say, the spell of cramming at White Lodge had little effect. The only new thing Bertie learned while he was there, after persistent questioning of Mr Tarver, was the process by which he, together with the rest of the human race, had been conceived. Albert and Victoria had deemed it prudent to keep the sixteen-year-old ignorant of the 'facts of life', but Bertie had winkled it out of the embarrassed clergyman.[34] Unlike some of the things he had learned or failed to learn from Mr Gibbs over the years, it was knowledge which he would put to very full use in the future.

Bertie, when he became seventeen in November, was handed a letter from Prince Albert which made him burst into tears, whether at the prospect of losing Mr Gibbs after five years, or simply at the momentousness of it all, was not clear. Instead of a tutor, he would have a governor, Colonel Robert Bruce, the brother of the Queen's close friend Lady Augusta Bruce, and the son of that Lord Elgin who acquired the Parthenon Marbles in the first decade of the century. Bertie was made a Lieutenant-Colonel, and awarded the Order of the Garter. He and Colonel Bruce were to travel to

Rome together in January to complete his education by acquainting him with the monuments of the classical past and the art treasures of the Renaissance. They reached Rome in January 1859, where they met Robert Browning. The poet was impressed. The Prince of Wales asked intelligent questions about modern Italian politics. But when commoners, even commoners as astute about human nature as Browning, meet royalties, do their critical faculties sometimes desert them? Bertie was already demonstrating his social skills and charm, but so far as Colonel Bruce was concerned, the Prince of Wales was a disappointment. In spite of, or perhaps because of, being dragged round on sightseeing tours, Bertie showed no interest in classical art or history, and when the governor laid on little dinners, so that Bertie could meet distinguished individuals such as Browning, the boy seemed only interested in dress, gossip and society.[35]

Albert's extreme optimism of temperament, which allowed him to dream of a liberal, united Germany, a peaceful Europe, a prosperous Britain ruled by Peelites, made him a stern father; for he was never able to give up hope that, with sufficient cajoling, Bertie would turn into a reflection of himself, and his own values, as Vicky had done.

Vicky joined in with her parents' desire to be judgemental about Bertie. When on his way to Italy, he had stopped off in Berlin, she sent home a thoroughly disobliging report:

> I am so very anxious he should speak a little better German, he has terribly forgotten it. It is difficult & perhaps hardly advisable to force lessons upon him here as he does not of his own accord wish for them. It is strange to see how much more he directs his attention and observation upon people rather than upon things and how quick he is at finding out their different qualities & what I sincerely rejoice in how [*sic*] he has given up laughing at people whom he does not know and only observing any little peculiarity or defect a person has; he does not do that at all now, which is a great point

> gained; people are so easily hurt by personal remarks & it is so dangerous a <u>habit wh. he used to have</u> to a great degree.[36]

Albert could place hopes in his other sons, but they were not, bar a calamity, going to become the King of England. Affie, by now fourteen, was destined for the Royal Navy and was already a 'middie' (midshipman). Albert was so proud of his examination results that he sent the marked papers to the Prime Minister reminding him that 'He solved the mathematical problems almost all without fault, and did the translations without a dictionary'. Derby, with the habitual facetious irony which Albert found so incomprehensible, replied, 'As I looked over them, I could not but feel very grateful that no such examination was necessary to qualify Her Majesty's Ministers for their offices, as it would very seriously increase the difficulty of framing an Administration.'[37]

Affie's younger brother Arthur, destined for a life in the army, was always his mother's favourite. He was still only a child of seven, however, and able to amuse his parents as much as to cause them anxiety.

'Arthur's legs are growing long, and he longs to be married,' Albert told Vicky, and recounted the child's words: 'It's too tiresome to have to wait so long when I shall be married and I see a pretty baby in the street, I shall buy it and give it to my wife.'[38] It was his younger brother Leopold who was beginning to cause alarm. The previous year, October 1857, Albert and Victoria had come home from Balmoral to find Prince Leopold, aged four, lame from a fall. He was still at this stage, on his mother's instructions, whipped for wrongdoing. Her mother used to protest that it was painful to hear a child crying after being whipped. 'Not when you have 8, mama,' Victoria had observed. By the time they had nine, and 'Baby' had arrived, Leopold was able, precociously, to write to his father, 'Dear Papa, I am so unhappy because you are not here. When will Vicky come back? Everything what I think, when I want to tell, I forget it. Baby makes such a noise, and when I am sitting opposite to Baby on the left side of the carriage, she kicks me, and she goes on saying oogly oogly. I have nothing else to say, but I send a Kiss & my love as well.'[39] At this stage they did not know that the boy was haemophiliac, though they had begun to have misgivings about him; they could tell that all was not well with the boy.

As for their sisters, Alice, at fifteen, was being lined up for her role in the dynastic marriage market, while Lenchen (twelve), Louise (ten) and Baby (two) awaited their turn.

Albert's preoccupation with the future of his own children was, of course, all part and parcel of his concern with education in general. When Derby had proposed the setting up of a boarding school, primarily for the education of the sons of soldiers, and as a potential training-ground for new soldiers, Albert had entered enthusiastically into every aspect of planning the new school – Wellington College. He and the Queen suggested that a Gold Medal should be presented annually to the boy who had exhibited most conspicuously the qualities the Prince most highly esteemed: 'It is not beyond the power of any boy to exhibit cheerful submission to superiors, unselfish good-fellowship with equals, independence and self-respect with the strong, kindness and protection to the weak, a readiness to forgive offences towards himself, and to conciliate the differences of others, and above all, fearless devotion to duty, and unflinching truthfulness.'[40] It is surely touching that Albert, with his high intellectual standards, should have wanted the medal to be awarded for moral qualities. And the reader who has not, by this stage of our story, come to weary of Albert's relentless virtuousness will readily see that he had all these qualities himself, even if his 'readiness to forgive offences towards himself' was not, in the marital context, matched by an empathy which would have helped him to understand why the storms and tempests occurred. But he did try. 'My love and sympathy are limitless and inexhaustible [*unerschöpflich*]', he told her after they had patched up one of their rows. 'As for your condition, I cannot, unfortunately, help you. You must fight it out for yourself with God's help.'[41] Rather chillingly, to us, but, we must hope, helpfully to her, he would write, 'I have noticed with delight your efforts to be unselfish, kind and sociable' and 'I willingly testify that things have gone much better during the last 2 years'.[42]

Albert accompanied the Queen to the opening of Wellington College on 29 January 1859. He spent 'thousands of pounds'[43] establishing a school library, and a comparable library at the military academy at Aldershot. The Queen diligently kept these libraries filled with appropriate new volumes. They were determined to introduce cleverness back into the British Army. While regarding Germany as their *Heimat*, and wishing to imitate the high standards of military training in the Prussian academies, they were also realistic enough to see that 'military science would be a chief, if not decisive agent, in any future European war'.[44] In Albert's eyes, the overwhelming threats to European security were the irresponsible military ambitions of France, and the sheer brute force and size of Russia. But the interest both he and the Queen took in Wellington and Aldershot showed that they wanted Britain – as it had so catastrophically *not* been in the Crimea – to be capable of showing itself a modern nation with a strong military capability. And that meant training up modern major-generals, with 'information animal, and vegetable and mineral'.

TWENTY

THE CARE AND WORK BEGIN

WRITING TO VICKY early one morning, not long after she was married, Albert said he was preparing for a meeting of the Duchy of Cornwall later in the day. 'I must rouse Mama and get dressed. Dann geht die Müh und Arbeit an!'[1] ('Then the Care and Work begin.') So it was every morning of his life. With each day, began Care and Work. And rousing Mother, choosing what clothes she would wear, and wondering what mood she was in. His close supervision of, and worry about, his nine children went hand in hand with constant attendance on the Queen. Management of the royal estates and palaces would, for another man, have been a full-time occupation. Nor did he ever rest simply with preserving the status quo. In 1859, for example, he drew up detailed plans to reorder the gardens at Kensington, designing an ornamental garden with colonnades with a winter garden with three '*hängenden Terrassen*', all to be paid for – it would cost £50,000 – by a public subscription of 2,500 Life Members; '£20 won't ruin you', as he argued.[2]

Throughout the year, and for the rest of his life, we find him complaining of illness, sometimes blaming his headaches, exhaustion and upset stomachs on 'worry about political affairs', as 'Sir Theodore Martin' tactfully prints the letter – irritation with Palmerston, as the original manuscript letter at Windsor makes clear.[3]

The speed and efficiency of modern means of global communication only increased his nervous sense of hurry. His heart beat with the agitation of a telegraphic machine. When he began his job in 1840, he had only Britain and Europe to worry about. Now, from America, and from every corner of the Empire and beyond came news – and news, to a temperament such as Albert's, meant information which appeared to beg for action. The United States was moving towards its Civil War. China was inflicting humiliation on the Royal Navy at sea. Nothing in Europe appeared to be following the blueprint which, taught by Stockmar, Albert had been expecting and hoping for the previous twenty years. In April 1859, Albert wrote to his stepmother:

> I can remember no period of equal confusion and danger. The ill-starred telegraph speaks incessantly from all quarters of the globe, and from every quarter a different language (I mean to a different purport). Suspicion, Hatred, Pride, Cunning, Intrigue, Covetousness, Dissimulation dictate the despatches, and in this state of things we cast about to find a basis on which peace may be secured! An agreeable occupation! At home we are now on the verge of a dissolution of Parliament, which is to take place on Tuesday; parties are broken up, and much embittered against each other; and with things in this state we are to find a sure basis for a Reform Bill which will satisfy the democrats, without driving monarchy and aristocracy to the wall! Also a pretty business.[4]

The young man who delighted in new inventions, such as the railroad, now regarded the telegraph as a thing of dread, not least because its dispatches increased his workload. The idealistic young German who had found in the British Parliament a beacon of political light was now a nearly forty-year-old who feared that monarchy might go to the wall – even though he remained, broadly speaking, in favour of parliamentary government, if only some sensible leaders, in the mould of Peel, could come forward. He dreaded the Radicals, and was impatient of the old men in a hurry, Russell and Palmerston, and their insatiable lust for power.

Certainly, this was an exceptional time in British politics, and June 1859 has been seen as 'a landmark in Victorian parliamentary politics'.[5] In the first half of the year, it looked as if Derby – who, of course, as an Earl, sat in the House of Lords – would be able to cobble together a Conservative majority in the House of Commons, provided the 'moderate opposition', the Whigs such as Lord John Russell, were prepared to vote with the Government. But Derby's proposed electoral reform – hinging on who, in the boroughs, would be eligible to vote – did not go far enough for Russell. Events abroad were also dividing British politicians, Russell's Whigs and the Radicals allying themselves to cheer on Cavour and the Italian nationalists, who wanted to throw out their Austrian overlords in northern Italy; Derby and some, but not all, Peelites inclining to support the status quo, not least because the 'champion' of Italian nationalists, in military terms, was Napoleon III. 'When, in '59, in spite of all our endeavours and warnings he [Napoleon III] made war in Italy against Austria and deceived us, Papa was most indignant and broke off all friendly, personal intercourse and had the worst opinion of him which was never removed', as the Queen recalled years later for her daughter Vicky.[6]

So, 1859 was a year of quite exceptional busyness. On 6 June, the 274 of the parliamentary Liberal Party met at Willis's Rooms in St James's Street. Palmerston, who did not really hold with political parties in the modern definition of the term, addressed them a little unwillingly, but he was raucously well received. The meeting, which was also attended by Russell, was called to demonstrate the unity of those who held 'Liberal' opinions, in social and foreign affairs. As a young man of twenty-five, already a minister in a Tory government, Palmerston had attended these premises when it housed a club called Almack's – known as 'the despair of the middle classes' – dressed in a bright-coloured coat and pantaloons and danced a quadrille with the Princess Lieven. Now, in his mid-seventies, wearing a black frock coat, shaking hands with Radicals such as John Bright, Palmerston was taking part in what was subsequently seen as the birth of the modern Liberal Party.[7] Five days later, this Liberal alliance helped carry a motion of no confidence in Derby's Government. The Prime Minister resigned.

The prospect of Palmerston's return as Prime Minister did little to improve Albert's dyspepsia. On her husband's advice, the Queen called for Lord Granville, leader of the Liberals in the Lords, to form an administration. He explained to the sovereign that Russell and Palmerston both regarded themselves as the *de facto* leader of the Liberals in the Commons[8] and that whereas Palmerston might serve under Granville, Russell most definitely would not. This spoke in Palmerston's favour. Albert and Victoria could see that, whereas Palmerston could hold together the coalition of Liberals and Radicals who made up the new party, Russell would not. With great reluctance, therefore, the Queen asked Palmerston to form a government. He offered Russell any job he wanted, and he took the Foreign Office. At the first Cabinet Meeting, Russell sat conspicuously as far away from the Prime Minister as possible. Genially, Palmerston called out, 'Johnnie, you'll find that place very cold. You had better come up here.'[9] The deeply conservative former Peelite, and deeply difficult to read, W. E. Gladstone was the Chancellor of the Exchequer. Unlike most of the Cabinet – Palmerston, Russell, the Dukes of Somerset, Argyll and Newcastle, the Earl of Elgin (Colonel Bruce's brother), Sidney Herbert et al., Gladstone was not an aristocrat. He came from the mercantile class, and his acceptance of the chancellorship a second time has been seen as 'the decisive moment in his public career'.[10] He would grow into the blackest of Victoria's bêtes noires, during her widowhood, but Albert and he always got on well – and indeed, in some respects were highly comparable characters, earnest, hard-working, scholarly and, while being conservative by temperament, both progressive by conviction.

Italy was moving inexorably, as the eyes of hindsight can now see, towards independent nationhood. While the French and Austrian armies fought the battles of Montebello, Palestro and Magenta, in Florence there had been a revolution. In defiance of Derby's considered aim to preserve the status quo of 1815, Europe was changing. Despite Albert's natural instinct to support Austria against the aggression of the French, there were too many aspects of the Italian situation for the status quo to be sustainable – among them the sheer unworkability, both of the Austrians holding the Italians within their Empire against their will; and of the Pope in modern times claiming temporal power, and territories which could only be defended by the medieval prospect in the nineteenth century of the Vicar of Christ raising an army.

Twelve days after Palmerston became Prime Minister, the armies of Napoleon III and Franz Josef of Austria met in the Battle of Solferino. Though it was a French victory, the losses on both sides were terrible. Napoleon approached the British Government to ask if they would act as brokers for a negotiated peace with Austria. Russell wrote to the Queen suggesting they accede to this request, and to create an Italian federation, with the cession of Lombardy to Piedmont, and the Veneto becoming an independent state. The Foreign Secretary said that he and the Prime Minister were 'humbly of the opinion that your Majesty should give to the Emperor of the French the moral support which is asked'. Victoria replied in no uncertain terms that Napoleon III 'made war on Austria in order to wrest her two Italian kingdoms from her, which were assured to her by the treaties of 1815, to which England is a party; England declared her neutrality in the war. The Emperor succeeded in driving the Austrians out of one of these kingdoms after several bloody battles. He means to drive her out of the second by diplomacy, and neutral England is to join him with her moral support in this endeavour. The Queen having declared her neutrality, to which her Parliament and people have given their unanimous assent, feels bound to adhere to it. She conceives Lord John Russell and Lord Palmerston ought not to ask her to give her "moral support" to one of the belligerents. As for herself, she sees no distinction between moral and general support'.[11]

Despite Palmerston's natural Francophile bias, and despite the fact the French and the British were still allies – for example, they were fighting a joint war against the Chinese while all these European convulsions were in progress – Albert felt that Britain should prepare for French treachery. He covered twenty manuscript sheets urging the Government to overhaul the defence of the south coast. 'At the Court of Napoleon,' he wrote, 'they play, they make love, enjoy themselves, dream, and between sleeping and waking, make decisions on matters of the greatest importance, *et la question ne s'examine qu'après* when things have happened.'[12]

As it happened, once Italy became an independent kingdom, the practically minded Queen was perfectly content to accept this as a fait accompli, but it would not happen until Albert was dead. Of much more immediate significance for Victoria and Albert was the extension of the franchise at home. Neither of them was in the strict sense of the word a democrat, but they were political realists who saw the inevitability, and indeed desirability, of representative, parliamentary government requiring an extended franchise. In fact, it was left to the Conservative Government, during Derby's third term of office, in 1867, to rectify the absurdity of five out of six adult males being without the vote. (The adult male population of England and Wales was about five million, the number of male voters was under a million.) The fact that it took longer than expected for this situation to be improved did not blind Victoria and Albert to the inevitability of it. Albert, a little cynically but surely accurately, saw that the motivation for a Reform Bill, at least in the minds of many in the political class, was not so much to bring in reform itself on any very conspicuous level, but 'to stop the agitation about Reform'.[13]

In the new world which was coming into being, the monarchy itself was to learn new roles. Not just the sovereign, but the entire Royal Family, in Albert's vision of things, needed to be exemplary. The family at Osborne, as now displayed in photographs and illustrated magazines, should be something to which the nation could look with respect, fellow feeling, affection. While the younger siblings, perhaps a majority of them, would be part of Albert's dynastic plan for the future of Europe and exiled, like poor Vicky, as soon as was practicable, the one child who must, by the very nature of his position, stay at home and represent the future of the new Albertian vision of monarchy was the Prince of Wales. Hence, the nervousness with which his entry upon manhood was greeted by both his parents.

In January 1859, Albert wrote to him, 'You are now standing on the threshold of the youthful period of life, and that makes me very, very anxious for you. For if you give the impression to others that you delight in giving them pain you will engender resentment and contempt where you absolutely require good will towards you.'[14]

As noted towards the end of the last chapter, though not everyone would like Bertie, the verdicts of those who met him were beginning to form a coherent picture. Vicky in Berlin noted the fact that he had lost much of his adolescent abrasiveness, and, while being totally unintellectual, he was a polite conversationalist who showed a genuine interest in other people. Robert Browning, a few weeks later in Rome, was struck by the Prince's pleasantness, and by his interest in politics. Albert, as well as being shocked by his eldest son's lack of interest in art and history, was chiefly terrified that history would repeat itself, and that the Prince of Wales, so rigidly force-fed with his parents' version of culture and morality, was going to become a throwback. The dreadful spectres – of Victoria's dissolute uncles, with their mistresses and their useless, extravagant modes of life – never ceased to haunt Bertie's parents; add to the memory of the late George IV, guzzling and swigging and womanizing, the unforgettable horrors of Coburg and Gotha, Duke Ernst I, hated by the people, openly mocking the marriage bed with a succession of women; while brother Ernst was no better.

Surely nurture could conquer nature? Surely the will of Albert, so benign in its intentions, would persuade the young man to follow in the footsteps, not of his appalling great-uncles, grandfather and uncle, but of his virtuous father, the Angel?

We last saw Bertie in Rome, having dinner with Browning. Some days after that, Bertie, together with his companions Colonel Bruce and Odo Russell,[15] later Lord Ampthill, had an audience with Pope Pius IX. With gormless attention to the things which caught his fancy in the Vatican, rather than to the important artefacts such as Greek manuscripts or the Belvedere torso, he wrote back to his father, 'The staircase was lined by the Swiss Guards who looked very picturesque in their peculiar and handsome dress; we had to pass through many rooms, before we reached that in which the Pope was; he came to his door and received me very kindly, I remained about ten minutes conversing with him, and then took my departure.'[16]

Albert did not find this satisfactory. 'Considering the Antagonism of Protestant England and the position of the House of Brunswick it may be said to form a moment in history – the first Prince of Wales visiting the Pope. I was eager to hear an account of your conversation and a description of the man.'[17] Likewise, Bertie's attempts to write to his father about the wars and uprisings in Italy were received with a long put-down, and an analysis of the whole international situation.

The dangerous situation in Italy made it impossible for Bertie to have a culture-crammed Grand Tour, even had he wanted it. His father bombarded him with letters explaining the situation in terms which would not have been agreeable to the pro-Italian Lord John Russell and Palmerston. Carried away by his passionate interest in the subject, Albert allowed his by now good grasp of English to go slightly haywire.

> The evacuation of Rome is our favourite scheme [*sic*] as most likely to remove one of the main elements of war, France pretended to wish it and gave out she would do so (evacuate) if she could prevail on Austria to evacuate the Legations. Austria's refusal to do so was the Cause of the French staying against their own wish at [*sic*] Rome and Austria is therefore answerable for

> foreign occupation and any misgovernment carried on under it. Austria says, we came only when asked and stay as long as we are asked by the power which wants us, France has no right to ask us to evacuate, if the Pope does, we shall go. The Pope has now done so and both are bound to stand by their promises. The Emperor Napoleon will be glad on one account, as the occupation of Rome is named as the cause of the intense hatred to [*sic*] him by the Italian liberals, by those who have at various times attempted to take his life. We still hope to be able to maintain and preserve the peace of the world, but Sardinia will not be pleased if the Italians were satisfied as it would deprive her of 'political capital', an American expression which Colonel Bruce can explain to you.[18]

When Bertie returned from Italy, there were still four months to be filled before the boy could be sent to Christ Church, Oxford, a place of learning where Albert was optimistic enough to suppose he would settle down to the detailed intellectual self-application which had characterized his own time of study at Bonn. Buckingham Palace, with its 'succession of gaieties', was an unsuitable place for the constant study which Albert required of his son. They sent him to Edinburgh, where he and his governor resided at Holyrood House and the young man attended university lectures on science and Roman history. Even when Bertie joined the family at Balmoral later in the year, he was required to spend two hours daily in study. Why not let the boy read a novel by Sir Walter Scott? asked Tarver – the clergyman who had so inappropriately suggested taking Bertie to the Holy Communion in Mortlake. Albert's testy reply was, 'I should be very sorry that he should look upon the reading of a novel (even by Sir Walter Scott) as a day's work.'[19] While Bertie's father contemplated with dismay the uselessness of his son's mind, the mother ruminated sadly on his unsatisfactory appearance. 'Bertie… is… grown and spread; but not improved in looks; the mouth is becoming so very large and he will cut his hair away behind and divide it nearly in the middle in front, so that it makes him appear to have no head and all face. It is a frightful coiffure.'[20]

In November 1859, a month into the Michaelmas term, Albert wrote optimistically to Stockmar, 'The Prince of Wales is comfortably settled in Oxford'. In the same letter, Albert added, 'I send you *Adam Bede*, the novel which has made a great sensation here, and which will amuse you by the fullness and variety of the studies of human character. By this study, your favourite one, I find myself every day more and more attracted.'[21] Albert was not the first, nor the last, father to take an interest in human nature while remaining absolutely blind to the qualities of his own son. The myopia, in a man so intelligent, is, however, very marked. Nothing would prevent Bertie from developing into a sensualist, but neither of his parents ever seemed capable of seeing his good sides – his geniality, his ability to rub along with different sorts of people, surely qualities which were useful in the modern kind of monarchy which Victoria and Albert were creating. Rigidly committed to making Bertie into a carbon-copy of Albert, which would never have been a possibility in any event – for who could match the Angel? – they doomed themselves to disappointment.

'The only use of Oxford,' Albert told him, 'is that it is a place for study, a refuge from the world.'[22] Why, in that case, did they send him to Christ Church, the most aristocratic of the colleges, where so many of the undergraduates liked to spend their time hunting, or womanizing, or going to the races? At Bonn, though cocooned with aristocratic or princely house-companions, Albert had been treated, more or less, like a normal undergraduate. This was not Oxford's way. Bertie resided at Frewin Hall in the Cornmarket, in private lodgings. Whenever he walked into a lecture hall or into the cathedral (which is Christ Church's college chapel) everyone stood up. The constant presence of Colonel Bruce was 'a good deal laughed at', according to Lord Clarendon.[23]

At the end of his first term, Bertie was examined by the Dean of Christ Church, Henry Liddell – father of Alice in Wonderland. The Prince's answers to questions about English history, from Alfred the

Great to Henry III, were 'ready clear and for the most part right', even if his 'pen is not so ready as his tongue'. This, on the whole, surprisingly good report was received as a total disappointment by Bertie's parents. When he returned to Windsor for Christmas, the Queen noted 'a growing listlessness and inattention and... self-indulgence... Let me never hear of your lying on a sofa or an armchair except you are ill or returned from a long fatiguing day's hunting or shooting.'[24]

Mixed with all the righteous longing to train his son to be a worthy monarch, there was an element of wistfulness, perhaps downright envy. Bertie had the one quality which Albert now found himself so shockingly to have lost entirely. Youth. When, in February 1860, Bertie reported from Oxford that he had been skating – formerly one of Albert's favourite pastimes – the forty-year-old father wrote to his eighteen-year-old son, 'I have lost it altogether and would not have been fit for it for some time even if the frost had continued. Fever, pain in all my limbs and violent cold in head and chest have made me suffer a good deal and pulled me down very much.' A forty-year-old in normal health would have wanted to skate with his son. Albert had become prematurely aged, by whatever the mystery illness was which would dog him for the rest of his life. The slender beautiful youth, whom Victoria had watched gliding across the ice during their idyllic Windsor Christmases less than two decades earlier, had become a bald, exhausted, paunchy, angry man. Illness was destroying him.

In 1859, Alfred Tennyson, the Poet Laureate and Albert's Isle of Wight neighbour, brought out the first four of his *Idylls of the King*. He ordered 40,000 copies to be printed, 10,000 of which were sold in the first week of publication. Prince Albert was one of the purchasers, and he read the poems with profound admiration and absorption. 'They quite rekindle the feeling with which the legends of King Arthur must have inspired the chivalry of old, whilst the graceful form in which they are presented blends those feelings with the softer tone of our present age,'[25] he wrote to the poet, asking him if he would personally sign the royal copy. Tennyson had already published, in 1842, the fragment entitled *Morte d'Arthur*, which described the wounded King, his Round Table-Fellowship dissolved, being placed on his barge and escorted across the misty lake by three mysterious queens. Like the *bal costumé* in which Victoria and Albert had appeared as Eleanor of Aquitaine and Edward III, like the Gothic Revival House of Lords interior by Pugin, like the historical murals of Dyce or G. F. Watts, Tennyson's Arthurian poems are statements about modernity clothed in historical costume. The four *Idylls* (Tennyson pronounced the word *Idles*) have two enormous themes which were of immediate appeal. One was the possibility of a whole society breaking up. The Round Table did so because of Guinevere's unfaithfulness to Arthur, and because of the distraction of the Grail Quest dissolving the knights' resolution to unite in the construction of a just society loyal to the King. The second underlying theme was men being undermined by their relationships with a woman. Enid thinks she is happy with Prince Geraint, but he feels emasculated by her, weakened by 'uxoriousness', and feels the need to go off and prove himself on a Quest. The King himself, lynch-pin of society, is weakened and humiliated by Guinevere's unfaithfulness. Two years after Albert's death, Tennyson dedicated the reprint of the *Idylls* to his memory. Leslie Stephen was not the only critic to compare Tennyson's Arthur to the Prince Consort,[26] and the wags christened the *Idylls*, when they were complete, *The Morte d'Albert*. Tennyson, in his dedicatory poem, advertised the *Idylls* at the same time, by reminding readers that Albert had loved his poem,

> Perchance as finding there unconsciously
> Some image of himself[27]

If Albert found an image of himself in the mythic figure of King Arthur, the '*I which has been*, with all its impressions, remembrances, experiences, feelings' had never really left the *Heimat*. Now that Vicky was married to the Crown Prince of Prussia, and Albert's epistolary contact with her was so frequent, he returned more and more in spirit to Germany, while he and Victoria planned a dynastic alliance for the next daughter in line, Princess Alice. Prince Louis of Hesse-Darmstadt joined the Royal Family for a torrentially wet Ascot week in June. Alice, seventeen, and he, twenty-three, dutifully 'fell in love'. Lord Clarendon, who saw that Alice was clever and original, regretted her betrothal to a dull boy with a dull family in a dull country.[28]

The Queen was more preoccupied by the political crisis than she was, at that moment, by Alice's marriage prospects. Palmerston and Lord John Russell dined and stayed at Windsor during that Ascot week. Lord John alarmed the Queen by telling her about Garibaldi and his advance through Italy. Palmerston spoke of the proposed Reform Bill. The question in their minds was whether Gladstone would throw in his lot with the Radicals, or whether, as Chancellor, he would be too concerned with raising taxes to pay for the Chinese war to risk his career on a resignation. Palmerston, who was always attentive to the political ambition of his rivals, could both sense Gladstone's extreme ambition and the fact that he planned a 'lurch to the left' in order to secure his future, and the future direction of the Liberal Party. Palmerston told the Queen 'it was evident every impediment would be put in its [the Reform Bill's] way to prevent its passing & they were obliged to get on with the Estimates, also to propose new taxes for the Chinese War. This he thought would prevent Mr Gladstone from resigning, — besides the sense of the ruinous position he would place himself in, — thrown together with Bright, who was against everything in the Constitution; Crown, Lords, Commons (as at present established), Aristocracy, Navy, Army — & in whose "vocabulary, the word honour was not to be found".'[29]

A different form of populism was closer to home. At the end of the Oxford year, at the Encaenia ceremony in the Sheldonian Theatre, Bertie led in the procession of professors and doctors and Heads of Colleges to the riotous cheers of the undergraduates in the galleries.[30] Could it be that his father's repeated injunctions, that royal popularity had to be earned, was not true? Could it be that the crowds loved him simply for being royal, regardless of how uninterested he was in Cimabue or the housing conditions of the working classes?

To avoid the hazards resultant upon Bertie having too much time on his hands during the university vacation, Albert arranged for him to visit Canada and the United States. It was entirely Albert's idea that this innovation should take place – a royal prince on a significant foreign visit – and Albert had done his best to stage-manage the tour from afar, drafting all Bertie's speeches for each stopping place in the journey and getting his travelling companion, the Duke of Newcastle, to put them into workable English prose. They arrived at St John's, Newfoundland, one of the oldest European settlements in North America, on 23 July. In the evening, there was a ceilidh, at which the Prince whirled about the dance floor until 2 a.m., and, when the Newfoundlanders seemed hesitant about the dance steps, Bertie acted as a caller. The fact was picked up by the New York press, to Albert's horror. 'Never forget how constantly you are watched observed and described,' he wrote. Bertie justified himself, 'I am quite aware that I am closely watched and must be careful in what I do'. He denied bossing the dancers, but added, 'besides if I did it would have been not to be wondered at, as I never saw so many people who knew so little about dancing.'[31]

The *New York Herald* reported that Bertie had 'whispered sweet nothings to the ladies' as he directed them. When he crossed the border into the United States, there was intense interest and enthusiasm. He was received at the White House by President Buchanan. The bachelor-president's thirty-year-old niece acted as hostess. 'I thought Miss Lane a particularly nice person and very pretty,' Bertie told his mother.[32] It was reported that when he reached New York, a crowd of 300,000 showered him with flowers, and that when he appeared at a Grand Ball at the Opera House, 5,000 crammed into a space which could only accommodate 3,000. It was not exactly what Albert had expected, and both parents, who were so troubled by Bertie's lack of seriousness and, to his mother's eyes, peculiar appearance, searched in their minds for an explanation. Victoria decided that 'Bertie was received in the United States as no one has ever been received anywhere, principally from the (to me incredible) liking they have for my unworthy self'.[33] Whatever the reason for such adulation, the Queen would not permit Vicky to use the phrase 'those horrid Yankees'.[34]

While Bertie wowed the Americans, Albert and Victoria had been in Germany, accompanied by his sister Alice, and the entourage of Lady Churchill, Miss Bulteel, General Grey, Sir Charles Phipps and Colonel Ponsonby – this is the hardcore of the old Whig court. Mary Bulteel would marry Henry Ponsonby and they would be largely responsible for sustaining the monarchy during one of its worst crises, which, little did they realize, was looming, namely the death of the Prince Consort and the nervous collapse of the Queen.

The Ponsonbys, in particular, were sustained by humour. Ponsonby's son recalled a family anecdote about a small shooting party at Frogmore. The Duchess of Kent, who lived there, objected to the pheasants and other game being shot. Prince Albert and a party were shooting there one day when they saw her carriage coming. The Prince ordered everyone to hide behind the palings but her carriage came unexpectedly close and 'she was inexpressibly surprised' to see the Prince and a dozen others crouching in their hiding place. We may doubt if the Prince laughed at his mother-in-law's surprise. Humour was not his strong point. In fact in later years Ponsonby (in reminiscence of his time as equerry to 'the great and the good') says he had 'unmistakable marks of the Snark'.[35] (Of the Snark's characteristics, 'The third is its slowness in taking a jest./Should you happen to venture on one,/It will sigh like a thing that is deeply distressed:/ And it always looks grave at a pun'.[36]) Princess Beatrice, writing to George VI shortly after this anecdote reached print, complained about Ponsonby's indiscretion. She considered it disloyalty in a courtier to use the safety valve of laughter.

Not that there were many laughs to be had on that last German visit. No sooner had they reached Belgium than a telegraph message informed them that Albert's stepmother (and cousin) Marie was dying, and by the time the train reached Verviers, they heard by telegram that she had died. Coburg, when finally reached, was deep in mourning. Ernst and Fritz (the Crown Prince of Prussia) were standing at the railway station at five in the afternoon in their mourning clothes. The carriage rumbled through the beautiful arch of the Schloss Ehrenburg, and there were Ernst's long-suffering Duchess and Vicky in deepest mourning too, 'long black veils with a point'.[37]

The funeral took place at Gotha on 27 September; his stepmother was buried at seven in the morning. The next day, back in Coburg, Albert walked with his brother, the Queen, Duchess Alexandrine, Vicky, Fritz and Alice to see the mausoleum, which he and Ernst had designed in the Italianate for their birth-parents. Their mother Luise had been moved from her grave in Sankt Wendel to lie beside the husband who had humiliated, tortured and divorced her. Closure, of a kind, had, however, been found for the two sons.

So much for the past. The future, Albert and Victoria had met upon their arrival at Schloss Ehrenburg – Little Willy, their grandson, who came stomping towards them, Mrs Hobbs, the English nurse, holding his limp hand to disguise the fact that the left arm was useless. 'He has Fritz's eyes and Vicky's mouth, and very fair curly hair. We felt so pleased to see him at last!' the Queen wrote.[38] She was destined to die in his malfunctioning arms, forty-one years later on the Isle of Wight.

Once they had completed the rounds of visits, they called on Stockmar, who had retired to his native Coburg, frail and old. He had three more years to live. His furious cousin Karoline Bauer claimed that he died lonely and entirely unloved by his wife and family, having given his life to the royalties he served. If true, he would be neither the first nor the last royal servant whose family life was destroyed by the demands of the work.

In their remaining time in Coburg, Albert and Victoria developed a routine. He went to the Schloss Kalenberg, some four or five miles out of town, each day, to hunt. The drive of wild boars made it impossible for any of them actually to miss, though spears were carried, in order to finish off the animals which had only been wounded by the rifle cartridges. Fritz, Lord John Russell and Colonel Ponsonby all tactfully shot merely one boar, allowing Albert to kill three. The next day, being a Sunday, they all attended Divine Worship – at which one of Ernst's hymns was sung – and the Grand Duke and Grand Duchess of Weimar (the son and daughter-in-law of Goethe's patrons) came over to spend the day with them. Each day, Victoria and her daughters did sketching.[39]

Albert's emotions were overwrought. On his last visit to the Rosenau with Ernst, he burst into tears, and declared that he was sure he would never see the place again.[40] It was a strange intimation for a man who had just turned forty-one. Perhaps it was the heightened tension of thoughts and feelings which caused a potentially serious accident. Albert was riding alone in a four-horse open carriage, driven by a coachman, about three miles out of town when they approached a railway crossing. There was a bar drawn across the road, to prevent carriages crossing the railway line when a train was coming. Albert noticed it too late, and, being unable to turn the carriage or to stop the horses, he jumped. One of the horses, crashing into the barrier, hit its head and was killed. The others somehow broke out of their harnesses and rushed on into Coburg. Upon recognizing the horses as they galloped into town, Colonel Ponsonby himself jumped into a two-horse carriage and rode off on the road which he knew would have been Albert's route home. The coachman was stunned, but Albert sustained almost no injury, beyond a few cuts to his face. The next day he was well enough to visit Ernst's Museum of Natural History and Mineralogy, and to walk in the Hofgarten. But the accident had taken the gilt off their visit and the return journey was melancholy.

At Frankfurt, they were met by the Prince Regent of Prussia, who accompanied them as far as Mainz. 'All seemed so different from the breakfasts at Coburg,' Victoria told her journal in the soulless Rheinischer Hof hotel. Victoria was going down with a fever, and by the time they reached Coblenz, where they took leave of Fritz, Vicky and Little Willy, Victoria said she had not felt so ill since she nearly died of pneumonia in Ramsgate in 1835. She had a raging sore throat and the fever was rising. But unlike her husband, she was robust. By the time they reached Brussels, although she was confined to her bed, the sore throat had gone and she felt much better. Albert energetically explored the Trade Exhibition in the Belgian capital that October, and on the anniversary of their betrothal, their *Verlobungstag*, he bought her a pretty bracelet. 'God grant we may see many more anniversaries!'[41] In fact, they would see just one.

No sooner had they returned to London than *Dann geht die Müh und Arbeit an*!

Albert found an England gripped with anti-German and pro-Italian feeling. *The Times* published some disobliging article about Prussia almost every week, much to his displeasure ('What abominable articles *The Times* has against Prussia!'[42]) and the national mood was sentimentally in favour of Cavour and Garibaldi. Albert was torn. On the one hand, so advanced an anti-Catholic as he was could not happily contemplate siding with Catholic Austria, and the Pope, in favour of keeping the Italian status quo. As he remarked in a letter to Vicky, 'To your question, whether it would ever be right or useful for a State to seal a Concordat with the Pope, I would answer a decisive No!'[43] To the Prince Regent of Prussia, Albert wrote:

> I personally still fail to believe that the corporate state in Italy, brought into being by revolution, civil war, treachery, and invasion, can prove a success. Here, however, enthusiasm is so high that such a view is regarded as high treason. It is a peculiar phenomenon which Statesmen may do well to study and to lay to heart. The deep horror at the long misgovernment of Rome and Naples, which Austrian pressure alone has made possible ever since 1816, is the thing which causes the nation to shout aloud for Liberty, and dulls every other feeling or consideration.[44]

From his perspective, Albert could not possibly have seen the very short distance into the future in which Prussia, upon which he set such hopes, would wage a brutal war, in 1866, not only against Austria, but also against the southern German states, including that of Darmstadt. The forthcoming marriage of Alice to Prince Ludwig, the future Grand Duke of Hesse-Darmstadt, far from making peace, either in the family or in Europe, would in fact emphasize how disastrously they were at war. But no one, even in Prussia, could have foreseen the speed and success of Bismarck's ratcheting up of military power. Accordingly, on 30 November, Albert sealed the arrangement and the pair were betrothed, 'a most touching, and to me most sacred moment',[45] thought the Queen.

Two weeks later, their old friend Lord Aberdeen died. 'I was too miserable yesterday to hold my pen,'[46] Albert wrote to Vicky in Berlin. It was not simply grief which laid him low, however. He was suffering from shivering fits and violent attacks of vomiting. Something was clearly amiss. The likeliest explanation which has yet been proposed for Albert's condition was made in Helen Rappaport's 2011 book *Magnificent Obsession*, in which she suggested the diagnosis of Crohn's disease, the symptoms of which are abdominal pain, vomiting, diarrhoea and fever. Another possibility is abdominal tuberculosis, a disease from which many died in the nineteenth century, undiagnosed because the condition did not even have a name until 1882. Clearly, Albert, who had always had a weak digestion, had become seriously ill by the time he was forty/forty-one and his whole appearance had altered.[47]

Work had by now become an addiction from which there was no escape. Had the balance of power in the marriage been different, Victoria could have simply insisted on Albert taking three months of total rest. Had he lived in the age of therapists or analysts, he might well have been advised that he was working with such a frenzy in order to disguise from himself painful realities – the loss of his children, through their maturity, the storminess of the marriage, the agony of childhood memory, stirred up by the sad visit to Coburg. He devoted the autumn to detailed and lengthy correspondence with the Duke of Somerset, First Lord of the Admiralty, about the necessity of building up the navy with ironclads. And, as the old King of Prussia failed in strength, Albert conducted what was in effect a correspondence course in constitutional history with Vicky, pages-long disquisitions on the role of Cabinet Ministers, and their relationship with a constitutional monarch; how the monarch, restricted as he or she might be to the role of simply advising, can and should exercise political influence.

The family assembled at Windsor for a true German Christmas. The thermometer dropped. It was the coldest Christmas recorded for fifty years – in Staffordshire as low as minus 15 degrees Celsius (5 degrees Fahrenheit), and in Edinburgh minus 14 degrees. Twenty-two degrees of frost. Every branch, every twig, every window pane crystalline with ice. From Germany came welcome presents – a bust of Little Willy from Vicky; from brother Ernst a boar's head. To Stockmar, Albert wrote, 'The poor birds miss your kind and sympathetic hand, which used to scatter bread-crumbs for them in the days that are gone!'[48]

He and the Queen listened to the distant bells of Datchet. 'It reminded him of Home and Germany. Otherwise he disliked, as much as I did, our ugly and melancholy so called merry peal. At church our chapel, which was decorated with holly, his beloved Christmas Hymn, composed in 1842, was sung, and as always on Christmas Day, Palm Sunday and Easter Day, by his express wish, 4 wind instruments accompanied the organ and voices as is the custom in Germany'.[49] The iller he became, the less at home in England he felt. And he was, by now, 'very ailing and tired'.[50]

They left Windsor for Osborne on 2 January 1861. Affie had left them for Berlin, to visit Vicky. Just as the rest of the family were setting out for the Isle of Wight, they received the news that the old King had died. His brother, the Prince Regent, succeeded as King Wilhelm I, and Vicky was now the Crown Princess of Prussia, in line to become the Queen.

The previous winter, Albert had written to Bertie that he was now too old and ill to skate. The perfect skating conditions in January 1861, however, made it irresistible and Barton pond was hard and smooth as marble. Even during their short holiday on the Isle of Wight, however, Albert could not truly relax, and made visits of inspection to Portsmouth and Gosport to see the progress of the fortifications which were being built. France could not be trusted. Nothing could be left to chance. Nor could anything quite be trusted to function properly, without Albert's eye being cast over it.

If this was true of the defence of the realm, and of the right governance of Germany, and the future of Italy, how much truer was it of the career of the Prince of Wales. Bertie had been hoiked out of Oxford and been established at Madingley Hall, just outside Cambridge, where he could benefit from the purer intellectual atmosphere of a university of which his own father was the Chancellor. Bertie was now an undergraduate at Trinity College. The Reverend Charles Kingsley, Regius Professor of History, was engaged as his tutor. He was nineteen years old. Albert gave him permission to smoke cigars. And now began the serious business of finding a suitable wife for his son.

TWENTY-ONE

SHE WAS THE STRONG ONE

QUEEN VICTORIA'S SEVENTY-FIVE-YEAR-OLD mother, the Duchess of Kent, had cancer. For some years, she had been suffering from a variety of painful complaints. There was a skin infection, erysipelas, which would not go away. There was persistent pain in a swollen right arm. When this was operated upon, Sir James Clark warned Prince Albert that the tumour they discovered was almost certainly malignant. Neither the Duchess nor the Queen were told of the danger.

Blindness about the seriousness of an illness affecting our nearest relatives or loved ones is a phenomenon too common to require explanation. In Queen Victoria's case, the myopia was all the more extreme because of the psycho-history between her and her mother. Her actual girlhood in Kensington Palace, though unusual by any standards – she was, after all, being prepared to become the Head of State – had been nothing like as miserable as her memory had created, nor as lonely. At first, she had had the companionship of her half-siblings, Charles and Feodore. Uncongenial as the Conroys later became to her, she had the Conroy children to play with. She had a doting governess in the Baroness Lehzen who pandered to her every whim, and who was forbidden to chastise her royal charge. She had over a hundred dolls. Visits to the theatre and, later to the ballet and the opera, were frequent. Famous actors and musicians and singers came to the house. Moreover, Victoria enjoyed robust health.

The fly in the ointment was Conroy, and, specifically, the Duchess of Kent's subservience to this rogue. We now know far more than Victoria herself knew – certainly more than she knew as a girl – both about Conroy and about the Duchess's awkward position. The discovery of Conroy's diaries in the twentieth century revealed that he was of the insane belief that his wife, the former Miss Fisher, was the natural daughter of the late Duke of Kent (to whom Conroy was aide de camp), possibly, even, that Miss Fisher's mother had been married to the Duke. For this, there exists not a scintilla of evidence, and, as for his fathering Miss Fisher, he was in Canada by the time she was conceived and the dates do not fit at all.

Nevertheless, such manias are difficult to dislodge and they supply the clear motive for Conroy's unscrupulous behaviour as the chief adviser to the Duchess of Kent about the upbringing of the infant Alexandrina Victoria. The so-called Kensington System, which had been nominally devised to keep Victoria out of the eye of the Courts of George IV or William IV, were chiefly designed to prevent the Duchess, with her very poor grasp of English and her total absence of friends in England or acquaintance at Court, from being welcomed as part of the Royal Family, and thereby seeing just how fiendish Conroy was being. (Apart from anything else, he conned the Duchess of Gloucester, the Duchess of Kent's elderly neighbour in Kensington Palace, out of £18,000, with which to buy himself an estate.) The gossips believed that Conroy and the Duchess were lovers. There is no evidence for this, but Victoria might well have subsequently suspected it. The surviving papers of the Duchess of Kent, full of Lutheran piety and care for her little daughter, make no psychological sense had she been Conroy's lover. This in many ways foolish, vain woman, who had been a pawn in the dynastic marriage game in her youth, gave herself up entirely, in middle age, to doing her best for her child.

In Victoria's teens, the child changed. The tempestuous nature of Victoria, which has been on display at various intervals during this book, became very marked. Her half-siblings had married and gone to live in Germany. She was more and more alone with Mama and Conroy, whom she loathed. Things came to a head in Ramsgate in 1835, when she was ill with influenza, turning to pneumonia. Conroy used her physical weakness to try to force her to sign a document making the Duchess regent even beyond Victoria's majority. With the iron will which would, later in life, be a match for even the more formidable of her Prime Ministers, the young Victoria refused, even when – as it would seem – he physically tried to force her hand. Conroy was never forgiven for this incident, and nor was Victoria's mother for condoning it. As soon as she became Queen, at the age of eighteen, Victoria moved into Buckingham Palace with Lehzen, and the Duchess was in effect banished. Of course, 'dear Mama' or 'dearest Mama' was often trailed around as part of the royal entourage. She frequently came over to Buckingham Palace from the house in Belgravia where Victoria had dumped her, and at Windsor she was given a house at Frogmore. But there was no real intimacy, and sometimes, when, for example, she had spent two weeks on the Isle of Wight staying at Osborne, the Duchess would be reduced to tears at the thought that she had seen Albert, and spent time with her grandchildren, but spent almost no time alone with her daughter.

When the Duchess was dying, however, Victoria, still unaware, at least with her surface-consciousness, of how serious the condition was, reverted to her old childhood love of her mother. For the first two months of the new year, 1861, the great issues of the day, in the Queen's mind at least, receded into a misty unimportance. The business of government kept her in Buckingham Palace, but her mind was with Mama, and whenever she could she travelled to Windsor to see her. She wrote a formulaic letter to the French Emperor, rejoicing in the joint success of the British and French armies against the Chinese. She fired off the occasional note or memorandum to the Prime Minister or to the Foreign Secretary signalling the difference between the royal attitude to the Italian crisis and that of HM Government,[1] and she refused Lord John Russell's requests to be allowed to write to express solidarity with 'General Garibaldi'. She encouraged Palmerston to go ahead with commissioning twenty-six or even twenty-seven ironclad ships. But by the end of February, none of these things mattered much to her. She had withdrawn to Frogmore, to sit beside the bed of her mother and, in so doing, to revisit her personal *Kinderszenen*. Having actively disliked her mother for most of her own grown-up life, Victoria now fell in love with the old lady. In the absence of Lehzen, an absence now of nearly twenty years, Victoria became, as it were, a child again, her mother's little friend in Kensington Palace. The bedroom in Frogmore, with its pale lilac walls, was crammed with mementos of those Kensington days. The face on the pillow, which had now almost ceased to eat, grew ever thinner, gaunter, paler. The original swelling had been under her right arm – it was presumably a lymph node. Now the swellings grew beneath her left. Her hands were covered in sores.

Albert was absent much of the time. He had thrown himself into work after the Christmas holidays, committees of the Fine Arts Commission, and meetings at Trinity House, headquarters of the Lighthouse Service; he had been elected Master. 'Poor dear Papa has been suffering badly with toothache since three days,' she wrote, slightly lapsing into a German idiom as she told Vicky the state of things, with the breezy unsympathy of the healthy when discussing the chronically sick, '... which wears and worries him dreadfully, and seems particularly obstinate; it comes from inflammation of the tooth; he had a little of it last autumn on the journey to and at Coburg, and also at Babelsberg in '58, but not near so bad as now; I hope, however, it is a little better today, but dear Papa never allows he is any better or will try to get over it, but makes such a miserable face that people always think he's very ill.'[2] It was difficult for her not to be impatient with the way in which Albert was 'so completely overpowered by everything'.[3] She blamed him for working so hard, 'staying up talking so long and to too many people'.[4]

The illness of the two individuals she most relied upon was going to remove Victoria's two emotional props, her mother and her husband, in the space of ten months. By the beginning of March, Victoria could not hide from herself the seriousness of her mother's illness, and 16 March was what she called 'the most dreadful day of my life'.[5] The poor Duchess died.

'I feel so verwaist [orphaned],' she wrote to her uncle Leopold, while recognizing, 'How this will grieve and distress you!'[6] The grieving process was to be an especially painful one. Victoria went through all her mother's papers and private belongings. She discovered, what must have been clear to those about her for the previous thirty years, that the Duchess of Kent had been the most devoted of mothers. The idea that her mother had been her enemy had been a delusion, born in Victoria's adolescent brain, and never properly laid to rest until it was all but too late. Here was an abundance of evidence! Every little note which they had exchanged when Victoria was a child, every little love letter the Duchess had written to her – often simply bidding her good night – some tiny note, written on pink papers in the 'dear German language' and trimmed with lace, had been carefully preserved. Here, too, was the sad record of how rejected and unloved the Duchess had felt after Victoria's accession. Here was evidence of her having been a doting grandmother, who had never forgotten a birthday or an anniversary, and who had patiently come to Court, or to Osborne and Balmoral, to spend happy times with her grandchildren, only to suffer coldness and snubs from her daughter. As Victoria's sad recollections, and her letters to Albert, so often revealed, she had the most painful capacity for self-criticism and remorse. 'Not a scrap of my writing or of my hair has ever been thrown away – and such touching notes in a book about my babyhood.'[7] The cruel evidence that she had been an ungrateful cold daughter to the most doting of mothers was almost more than she could bear, and she descended into a nervous collapse.

The cruel clarity with which Victoria saw the situation emboldened Vicky to express the hope that, having estranged her mother, Victoria and Albert might not be about to do the same to their eldest son. Bertie, Vicky told the Queen, 'loves his home and feels happy there and those feelings must be nurtured, cultivated for if once lost they will not come again so easily. I admire dear Papa's patience and kindness and gentleness to him so much I can only hope and pray that there may never be an estrangement between him and you – it would be the source of endless misery to you both.'[8]

These were brave words to write. The Queen took to her bed and seldom came downstairs. Meals were sent to her in her room, and it was left to Albert to make the necessary funerary arrangements with the courtiers. 'LADIES: Mourning dress with crape veil, no bonnet. GENTLEMEN: Black evening Coat, with trousers and boots: white cravat, round hat. Stars and Ribands will be worn. Mourning scarf &c will be provided in Wolsey's Chapel, Windsor. A special train will leave the Paddington terminus of the Great Western Railway for Windsor at 9.25 on the morning of Monday the 25 Inst and will return after the ceremony'.[9]

The Duchess had originally hoped that her mortal remains would be conveyed from England, where she had never felt at home, and 'come to rest in my beloved fatherland [*in lieben Vaterland*]. But I have given up this dear wish... because it would be troublesome and cause a stir'. She asked instead that she should be placed in the vault of St George's Chapel next to her husband the Duke of Kent. It was placed very near the coffin of Princess Charlotte.[10]

She would not rest there for long. Relatively recently, the Queen had expressed disapproval of the Coburg habit of entombing their relatives in a mausoleum, but, such was the extravagant weight of grief, and so deeply had the Coburg mausoleum for Duke Ernst I impressed Victoria, that she commissioned an elaborate mausoleum for her mother at Frogmore, designed by Albert and Grüner between them. There was no shortage of money to pay for it. The old lady had £8,722 10s. 5d. in her account at Coutts' Bank. Four judicious life assurance policies, taken out for her by Sir John Conroy, with the Guardian, the Equitable, the Sun and the Provident, meant that *in toto* her estate was worth a very respectable £34,294 2s. 3d.[11] This is the inflationary equivalent in 2019 of nearly £4 million.[12] She distributed some of her wealth, leaving £700 to Conroy's son Sir Edward, and between seven and seventeen shillings per annum to the gardeners at Frogmore.

Uncle Leopold wrote to the Queen, begging her, 'Don't let yourself too much be taken possession of sorrow.'[13] But, as Victoria admitted to her eldest child, 'You are right… I do not wish to feel better.'[14] To Albert, she promised that she would still do her work with the boxes, and keep up her correspondence with the politicians, but she implored him not to make her any public duties. 'As regards myself, my most precious Angel, I can assure you that I will shrink from no work – but I must remain in retirement. I can give no parties this year. And I pray that I may remain in the country, for some time at least.'[15] She reminded him that when his father had died they had remained for four weeks in complete seclusion at Windsor, and she wanted to do the same, escaping directly to Osborne and not returning to London until the autumn.

He did his best to dissuade her, and to come out of herself a little, if only to quell the rumours, which quickly flew around, that she had become insane in her grief. She cautiously and timorously replied, 'If levees must be held – perhaps you cd hold a few – and then later we cd see which was best done… Work I will hard [*sic*] but I can't see people at present who don't feel my immense grief nor can I have festivities.'[16]

The spring, summer and autumn of 1861 were full of incident – both in the family, and in the political sphere, especially in international affairs. Albert was deeply and wearily involved with them all. Everything that happened, however – his worsening relationship with his son, his painful feeling of sadness that Vicky was so far away, his interest in the creation of the new Kingdom of Italy, his worry over the Schleswig-Holstein duchies, his concern for the foundation of the new military academy at Sandhurst, his design of the Duchess of Kent's mausoleum, his planning of a visit with the Queen to Ireland, his passionate advocacy of rearmament in the army and navy, and the strengthening of fortifications on the south coast of England, his worry about the American Civil War and its impact on the United

Kingdom – all these things happened against a background of the Queen's often uncontrolled fits of tears, or rage, anger directed most usually at him. When she was able to control herself, he would write to her, in English, in a mode which was more like that of a fierce psychiatrist or a schoolmaster:

> I can give you a very good certificate this time, and am pleased to witness with your own improvement. What I can do to contribute to your getting over the painful sensations which a return to Windsor under such sadly altered circumstances will be readily and cheerfully done. My advice to be less occupied with yourself and your own feelings is actually the kindest I can give, for pain is felt chiefly by dwelling upon it, and can thereby be heightened to an unbearable extent. This is not hard philosophy but common sense supported by general experience. If you take increased interest in things unconnected with personal feelings, you will find the task much lightened, of governing those feelings in general which you state to be your great difficulty in life. God's assistance will not fail you in your endeavours.[17]

Albert was not an especially religious person, but, while fully accepting the March of Mind and the advancement of scientific ideas, he did accept the orthodoxies, albeit with 'modern' interpretation. He was probably very close in spirit to Charles Kingsley, whom he selected as the most eligible tutor for Bertie at Cambridge – the category of churchmanship which no longer exists but which used to be called Broad. Like most people in the nineteenth century, Christian and not, he was preoccupied by the future life. Tennyson agonized about it in his greatest poetic sequence, *In Memoriam*. G. F. Watts, the most intensely spiritual of all the nineteenth-century painters, and one of the least Christian, expressed their yearnings for immortality in paint. It was the time of spiritualism, of séances, and of *Mr Sludge, 'the Medium'*. It paved the way for the era of the Society for Psychical Research (founded 1882), when even non-believers and scientists united with the orthodox in seeking some tangible proof of what there was 'behind the veil, behind the veil' (*In Memoriam*, LVI). Tennyson knew that he owed the laureateship to Albert's deep admiration, both for the *Idylls* and for *In Memoriam*, and when the Queen was widowed, she marked and scored her copy of the latter poem, and would ask the poet for private interviews. Tennyson found these embarrassing – 'I am a shy beast and like to keep in my burrow' – but he knew what was required: 'Talk to Her as you would to a poor woman in affliction, that is what She likes best.'[18] One of the lyrics from *In Memoriam* which she found 'spoke' to her most purely was section XXV.

I know that this was Life – the track
    Whereon with equal feet we fared;
    And then, as now, the day prepared
The daily burden for the back.

But this it was that made me move
    As light as carrier-birds in air;
    I loved the weight I had to bear,
Because it needed help of Love:

Nor could I weary, heart or limb,
    When mighty Love would cleave in twain
    The lading of a single pain,
And part it, giving half to him.

Beside these lines, in her copy of the poem, the Queen wrote, 'So it was for 22 years'.[19] Although the last year of Albert's marriage was, from his point of view, the most difficult, from Victoria's it was the time when their love showed itself strongest. From the moment she lost her mother, her need-love for him became absolutely dominant. (The other great love poem which this year suggests is Shakespeare's sonnet 'Let me not to the marriage of true minds/Admit impediments…') It was when she needed him most, and when she was at her most abject, that, as far as she was concerned, the marriage was strongest.

One April day in 1861, walking back together from the Swiss Cottage at Osborne, 'talking of Eternity', she said to him, 'I feel not quite composed and full consolation [*sic*] and care not how it is; but I feel sure all was right. She, Mama, is happy, is near us and we shall certainly meet again, – & there must be identity or else it would be annihilation.' Albert entirely agreed.

> He said that there could be no doubt as to the future state, and its being one of great happiness – but that how it could be was something we could not know, and need not plague oneself about. That the real, tender, strong affections must continue for they formed the fairest and purest parts of our Souls. He also said, on my observing, on that very mistaken notion of God punishing us for loving those dear to us too much by taking them away, that that was a mistaken interpretation – that what was meant was – that without some of those near and dear being taken – we should not feel or properly understand that there was another and a better world. How soothing and true this is.[20]

Some days later, on 28 April, as they were walking together in the gardens of Buckingham Palace she begged him '(as I had frequently done before) pardon for crying and so often appearing sad – then I was so happy with him!'[21] The moments of intimacy were becoming fewer. The sad truth was, he could no longer bear her unhappiness, and he kept apart from her as much as possible. Queen Victoria, in a candid moment, once admitted to Henry Cole that in the last year of Prince Albert's life 'she had had scarcely anything of his company'.[22]

Of the many events overseas which preoccupied Prince Albert during 1861 – the proclamation of the Kingdom of Italy (recognized by Britain almost immediately, though not by most of Europe), the death of Count Cavour, the outbreak of Civil War in America, the continuation of the crisis in China – what was uppermost was the fate of two Baltic duchies, Schleswig and Holstein, which since the Middle Ages had been part of the Kingdom of Denmark. The Danish crown could pass to a female, but the dukedom of Schleswig-Holstein could not. This was one element of the conundrum. The other was the demographic fact that Holstein and south Schleswig were German-speaking. Up to and including the Treaty of Vienna in 1815, the Germanness of the duchies was not of great political significance, since there was no national entity called Germany which could have laid claim to the culturally German duchies. But since the establishment of the German *Zollverein*, and the growth of the power of Prussia, and the uncertainty about a potential heir to the Duchy coming through a female line, things were more complicated. In 1852, the European powers agreed that the succession of the duchies should pass to the heirs of Frederick VII of Denmark, but at the same time, the Danish agreed not to incorporate the duchies into the kingdom. It was therefore technically possible for there to be a Duke of Schleswig-Holstein who was not the King of Denmark. The 1852 treaty had promised them to Christian of Glücksburg, Frederick VII's nearest heir in the female line. Though Christian foreswore his claim, his son, Christian of Augustenburg, came forward as a claimant. From Albert's point of view, he had two very strong claims. First, he was, officially, and in the eyes of international law since 1852, the legitimate claimant. Secondly, he had married Queen Victoria's niece, the daughter of Feodore of Hohenlohe-Langenburg.

It is likely that the reader, unless addicted to the pedigrees of northern monarchies and the by-ways of Scandinavian history, has allowed the eye to glaze over the previous paragraph. Palmerston, in 1863, would make the joke that passed into legend: 'Only three people have ever really understood the Schleswig-Holstein business—the

Prince Consort who is dead—a German professor, who has gone mad—and I, who have forgotten all about it.'[23]

In life, however, the Prince Consort believed that the intractable problems of the duchies could be resolved peaceably. He had two solutions which he hoped would be pursued simultaneously. First, the Germanies, with Prussia as their leader, would lay aside the easy option of military success and the acquisition of territory as a means of 'diverting attention from the state of things at home'. In a long letter to the King of Prussia (Vicky's father-in-law) Albert spelt out what he deemed to be Prussia's great opportunity. This was to avoid striving either after 'democratic absurdities' or 'reaction'. Instead, Prussia, and eventually, a united Germany, should strive for what existed, at present in Europe, only in Britain and Belgium, a government which was 'in harmony with the wants and wishes of the people'.[24] Albert proposed 'a slow, well-thought-out, persistent, courageous, truly German and thoroughly liberal policy'.[25] He wrote at such length, he told the King, because Wilhelm knew of 'my friendship for you, and my German feelings'.[26]

At the same time, the King of Prussia would have been aware of Albert's feelings for Denmark. Visitors to the Schloss Friedenstein at Gotha can see that Albert's family links with the Danish Royal House went back to the Schmalkaldic League in the sixteenth century. There had been several cases of intermarriage between Denmark and Saxe-Gotha, and many of Albert's forebears were members of the Danish chivalric Order of the Elephant, as you can see from the number of heraldic elephants adorning the picture frames and furniture at Gotha. The first dynastic marriage which Albert had arranged, between his daughter and the Hohenzollerns, would help to further the liberalization of Prussia, the creation of a government in harmony with the wants and wishes of the people. A second marriage, between the Prince of Wales and the apparently very beautiful Alexandra, daughter of the future King Christian IX of Denmark, would surely bring about harmony between Denmark and the Prussians.

That was the idea, and Albert would be dead before the vanity of the hope was made clear to him, before the Prussians had bullied Austria into forming an alliance against Denmark, and before the war of 1863–4. By then harmony had been forgotten. Fritz, Vicky's husband, had gone to war, as a cavalry officer, to help take possession of the duchies, and the German armies had also invaded Jutland. Palmerston's instincts had been to side with the Prussians; the public mood in Britain, partly whipped up by the popularity of the beautiful young Princess Alexandra, sided with the Danes. The Schleswig-Holstein war was the first indication that none of Albert's ideals or plans for a happy, peaceful Europe would be followed. By 1866, the Prussian armies were crushing the southern German duchies and kingdoms and vanquishing the Austrians. By 1870, they had defeated the French at Sedan, Napoleon III would be driven into exile, and the Kingdom of Prussia had been proclaimed an Empire, its King a Kaiser.

Vicky entered enthusiastically into the idea that Bertie's betrothal to Princess Alexandra of Denmark would solve two problems at once. It would cement Denmark and Prussia in bonds of love; and it would force Bertie, still only nineteen, to settle down. After the Cambridge term finished, Bertie had been allowed to go to Ireland, to attend the military camp at the Curragh, where he was supposed to train as an officer. It was hoped that, by the time his parents came to visit him in August 1861, Bertie would have become proficient enough to command a company in front of his parents. Colonel Henry Percy VC, a Crimean veteran, was his commanding officer, and, unfortunately for the young Prince, this professional soldier, a great personal favourite of the Queen's, decided, 'You are too imperfect in your drill, and your word of command is not sufficiently loud and distinct.'[27]

It was yet another example, as far as his parents were concerned, of Bertie failing to make the grade. 'The Prince of Wales has acquitted himself extremely well in the camp,' Albert wrote to Stockmar,[28] but the wish was father of the thought. It was a wet but, on the whole, jolly trip to Ireland. Alice, Lenchen and Affie accompanied their parents, and the royal party met many old friends, few of whom went so far as to be actually Irish, except in the most notional sense, but who were nonetheless very welcoming – the Marquess and Marchioness of Kildare, Lady Charlemont, the Duke of Leinster and Lord Carlisle among them. Excursions to Killarney with Lady Castlerosse provided Victoria with many of the things she most liked – lakeside views to sketch, peasanty lunches in cottages. 'Much of this, and indeed of the scenery in general, reminds one of the Highlands', as she scarcely needed to write to her uncle on 2 September.[29]

It was four days after the Queen posted this letter that Bertie, in the Curragh Camp, underwent the decisive experience, recorded in his engagement diary as 'N.C. 1st time'.[30] Some of the more high-spirited of his brother-officers had imported a companion from London called Nellie Clifden. Upon their recommendation, Bertie escaped his minders by climbing through a window of his quarters, and into the bedroom of one of these young blades. There Nellie introduced him to the activity which would occupy so many of his remaining forty-nine years on this planet. His career as 'Edward the Caresser' had begun.

Four days after he had written 'N.C. third time' in his engagement book, Bertie crossed the Channel and took a train to Coblenz, where the Moselle meets the Rhine, and attended the manoeuvres of the Prussian army. He was met there by Vicky and Fritz, who had been delegated by their parents to arrange the meeting between the Prince of Wales and Princess Alexandra of Denmark.

Vicky was able to report back to Queen Victoria that 'I think General Bruce must have been struck by her', but it was not General Bruce who was going to marry her. When Bertie had been shown Princess Alexandra's photograph back in June, he had exclaimed, 'From that photograph I would marry her at once.'[31] When he met the sixteen-year-old, however, it was clear that he found her disappointing. Her mother said to Vicky, 'If something came of this marriage it would not be political, thank God, but personal,' but this was one of those diplomatic occasions when the precise opposite of the truth seems to be the right thing to say. From a political point of view, the Germans, most of them wound up to believe that the invasion of Schleswig-Holstein was the simplest solution to the problem of those two duchies, believed that a dynastic match between Britain and Denmark was highly undesirable. Albert's brother Ernst, when he met Vicky, 'was very violent', as she told her father, 'and said he would do what he could to dissuade you from such a marriage; if you were to stay with him some weeks, he should not despair of succeeding'.[32] Had he done so, he would have spared Alexandra much future misery. No one doubted that the slender, rather flat-chested, innocent Danish girl was charming, and kind and virtuous, but Nellie Clifden – 'much run after by the Household Brigade', as Charles Carrington put it[33] – she was not. When Bertie returned to England, Nellie became his mistress, and in clubland and the raffish world of officers and young rakes the rumours flew round that she was already styling herself the Princess of Wales.

Prince Albert as yet knew nothing of this. It was still possible for him to write to Stockmar that Bertie had covered himself with military esteem in Ireland and was now returning to Cambridge to pursue his scholarly endeavours. The moment he and the Queen had returned from Ireland, Albert had thrown himself, as usual, into a multitude of concerns, and although they were at Balmoral from the beginning of September until 22 October, scarcely a day passed without the Prince Consort penning letters on the European situation to the King of Prussia (whose coronation took place at Königsberg on 18 October), to Lord Clarendon (who had represented the Queen at that ceremony), to the Prime Minister and to the Foreign Secretary. Wilhelm's coronation was not at all what Albert would have hoped or expected. Indeed, the new King of Prussia, far from espousing the values of Sir Robert Peel, representative parliamentary government and constitutional monarchy, appeared to assert, in his speech from the throne, a belligerent belief in the Divine Right of Kings. 'I receive this crown from the hand of God!' he proclaimed, waving his sword aloft. 'The speech of the King has alarmed and rather distressed us,' Victoria confided in Vicky.[34]

He was also much concerned with the question of training volunteer reserve troops, in addition to the Regular Army, and with the re-equipment of the army with breech-loading rifles. 'I am glad to see from your letter of the 7th inst., that you keep up the steam about our defences,' he wrote, with a slightly odd idiom, to Palmerston. 'Shorncliffe sadly wants a drill-ground, and I am happy to hear that three-hundred acres of land adjoining the camp are to be had.'[35] His complete loss of faith in France as a useful or possible ally, and his incredulity about the viability of the new Kingdom of Italy made Albert all the more anxious that Britain should be fully armed and ready for the next international emergency. In the same spirit of enthusiasm, later in the year, when the weather was foul and his health had collapsed, and he had many other things on his mind, he went over to inspect the building of the new Staff College and Royal Military Academy at Sandhurst.[36] Almost to the very end of his life, Albert was demonstrating how right Wellington had been to want him to become Commander-in-Chief of the army.

On 22 October, having left Balmoral, the Royal Family moved southwards, pausing in Edinburgh to give Albert time to lay two foundation stones, one for the new post office and one for the Industrial Museum. It had been a strange parting from his Highland retreat. John Brown, his favourite gillie, had disturbed them with his valediction. For ten years now, the Prince had been struck by Brown's 'magnificent physique, his transparent honesty and straightforward, independent character'.[37] Victoria and Albert both loved Brown's directness, his lack of sycophancy, the fact that he in no way resembled the stuffy courtiers of Windsor. 'My Angel and I both observed to one another that these Scotch people were so superior – and he said to me, "England does not know what she owes to Scotland".'[38] So it was particularly disturbing when Brown blurted out, upon parting from them, that he hoped 'we should all be well through the winter and return all safe, and above all, that you may have no deaths in the family'.[39] There was a difference, of course, between believing in Brown's common sense, and attributing to him the gift of second sight, but the words stayed with the Queen.

Albert was already wilting, with his by now too-familiar symptoms of sleeplessness, indigestion and consequent exhaustion, when they came south, and although the Queen begged his Private Secretary, Sir Charles Phipps, to put a limit on his frequent journeys between Windsor and London, there appeared to be no stopping his addiction to ever-heavier workloads.

It was Lord Torrington, whose turn it was as a Lord-in-Waiting to do a stint at Windsor, who, on 12 November, brought to Albert's ears what had been the gossip of the clubs ever since they reopened after the summer break. Nellie. Bertie. The Curragh. 'Oh!' the Queen would remember, 'that face, that heavenly face of woe and sorrow which was so dreadful to witness!'[40]

It is easy for a twenty-first-century reader to mock Prince Albert for his reaction to the story, which he was told at the very point when they were going to formalize the betrothal of Bertie to the angelically pure Alexandra of Denmark. If such a reader discards modern assumptions – boys will be boys, wild oats, such things happen, better to laugh them off – he or she will recognize that this incident represented the destruction of all Albert's unrealistic hopes. For nearly twenty years, he had devoted himself to ensuring that Bertie should be like himself, and not like his brother Ernst or his father Ernst; that Bertie should be interested in art history, and the condition of the working classes, and yearn to build up a happy stable family with a virtuous young wife. Albert's entire picture of what a constitutional monarch should be, fashioned before the idea was conceived by Stockmar, and developed by himself through the virtuous and often pleasurable experience of family life, was that the monarch of the modern age should be someone who commanded the respect of the emergent middle-class voters, the £10 borough electors, the sort who might have smiled indulgently at the rakish stories of old Lord Palmerston's antics with parlour maids, but who would never sully their own suburban hearths with such sordidities. Albert and Victoria, by their frequent self-exposure, by photography and painted portraiture, as models of family rectitude were all part of the package, and Bertie's antics showed all too clearly how little the son and heir intended to follow in the father's footsteps.

Consumed with rage and sorrow, and sickness and disappointment, Albert sat down and penned a letter to his son which was eleven pages of quarto (203 × 254 millimetres). 'Nellie Clifden,' he wrote, 'boasts even of having been down here at Windsor to meet you when you last stayed here, with us for your birthday! I knew you were thoughtless and weak, and often trembled at the thought that you might get into some scrape through ignorance and a silly love of doing what you might have heard of others having done without further reflections or enquiring whether it was in itself right or wrong, but I could not think you depraved!'[41]

The awful memory of the ribald laughter which had been provoked by the memoirs of La Belle Grecque were etched into Albert's brain. He knew how publicly ridiculous his father's depravity had appeared. He knew how diminished his uncle Leopold was by all the rumours concerning himself and Stockmar's cousin, Karoline Bauer. As for Victoria's uncles, their antics with the equivalents of Nellie Clifden were the stuff of cartoons by Gillray and Rowlandson. Albert was not going to mince his words. He wanted Bertie to know that he knew precisely that the young man had climbed 'through the windows of your hut and had sexual intercourse with her'. With this one stupid, sordid act, he had, as it were, ripped into shreds the Holy Family-like icons of Winterhalter. Remembering how La Belle Grecque had behaved in relation to his own father, Albert told Bertie, 'she will probably have a child or get a child and you will be a reputed father! If you were to try to deny it, she can drag you into a Court of Law to force you to avow it. Passing over the grief which you have caused us by the crime committed towards us and the country for the sake of which you have been invested with rank and wealth and which sees its own honour defiled in that of its Royal Family...'

Albert concluded his letter by saying that he did not have it in his heart even to see his son, but that in future they should communicate through General Bruce. Albert's biographer, perhaps more than that biographer's readers, must feel more sympathy for the father than the son at this point. If it had not been for Nellie Clifden, Albert's death at the age of forty-two would be a story of pure tragedy. Bertie's undignified attachment to Nellie threatens, however, the ending of our tragedy – and it is a quite genuine tragedy – with the ribald gusto of a Feydeau farce. Or, it is as if, in a theatre about to bring down the curtain on a run of *Hamlet* before the coming of Christmas demands a month of pantomime, a confusion of stage-management has occurred; the low drumbeat, and the preparation of Fortinbras and Horatio to bid the soldiers shoot, and flights of angels bid the Prince to his rest, are interrupted by the strains of Offenbach; the sable costumes and pale faces of the tragic players are suddenly mingled with the rouged and pomaded smiles of a chorus-line of girls dancing the can-can.

No wonder that 'Sir Theodore Martin', who kept 'his' tragic buskin on until the end, omitted any reference to the reason why, on 25 November, Albert should have gone down to Madingley Hall, near Cambridge, 'on a visit to the Prince of Wales'.[42] He does, however, tell us that the Prince's diary recorded, 'Am full of rheumatic pains and feel thoroughly unwell. Have scarcely closed my eyes at night for the last fortnight.'[43] He was weak and he could feel the cold grip of death upon him.

And in his last political act, his intervention in the crisis over American seizure of the British merchant ship *Trent*, Prince Albert demonstrated that his political work was sometimes effective, and benignly so.

Albert did indeed go to Madingley Hall to talk with Bertie, and they were reconciled. The son was full of contrition. 'Parents' hearts are prone to forgive their children when they see them sincerely penitent,' Albert wrote to Bertie on 20 November, four days before his visit to Madingley.[44] All Albert's symptoms, sleeplessness, indigestion, cold turning to influenza, were worsening by the day. The visit to Bertie did not help, partly because it so exacerbated the father's nerves, partly because the weather was foul, and pacing about at Madingley with the Prince of Wales in the rain was a bad idea. The walk took much longer than either man intended, because they took a wrong turning. When he returned to Windsor in wet clothes, Albert was exhausted.

He should have taken complete rest. Not that this would have saved his life if, as some now suspect, he was in the advanced stages of stomach cancer. His own doctors were to fear that he was suffering from typhoid fever. At this very moment, however, there arose the danger that Britain might become embroiled in war with America.

Since March, the President of the United States, Abraham Lincoln, had determined that the war which had begun two months earlier in Charleston harbour, South Carolina, should restore the State of the Union. Many in Britain, however, supported the Confederacy of the six Southern States. Two of the Confederate envoys, Mason and Slidell, had boarded a British packet ship, the *Trent*, in Havana. Their aim was to come to London and to persuade the Government of Lord Palmerston to recognize the Confederate States of America, an idea for which there was strong support, especially among British Liberals, and in the cotton towns of Lancashire, whose livelihoods were threatened by the belligerence of the Yankee North. Three Southern deputies were already installed in London, so Mason and Slidell would not have been the first to arrive.

Palmerston's emergency meeting with Gladstone (who supported the Confederates), with a Judge of the Admiralty, Dr Stephen Lushington, with First Lord of the Admiralty, the Duke of Somerset, and with Sir George Grey, Chancellor of the Duchy of Lancaster, established the fact that 'according to the law of the nations as... practised and enforced by England in the war with France, the Northern Union being a Belligerent is entitled by its ships of war to stop and search any neutral merchantman, and the West Indian packet is such, and to search her, if there is reasonable suspicion that she is carrying enemy's dispatches, and if such are found on board to take her to the Port of the Belligerent, and there to proceed against her for condemnation'.[45]

It was obvious, given Palmerston's long track record of gunboat diplomacy, how he would naturally proceed, but he and the Cabinet could see that this would be an extremely dangerous path to follow. An American Federal steamer of war had lately arrived in Southampton harbour, just over the water from Osborne, with the clear intention of intercepting the *Trent* before she came into her English harbour.

In fact, the *Trent* had already been held up by *another* American warship, the *San Jacinto*. By the end of November, the news reached London. The *Trent* had been stopped, an American captain, Wilkes, came on board, and apprehended the two Confederates. The American public felt that they had 'given to Great Britain a dose of her medicine in the previous era'.[46] The Governor of Massachusetts publicly congratulated the captain who had made the arrests and said he should be awarded a gold medal.

In Britain, as the American Ambassador, Henry Adams, made clear, a very different patriotism took a very different view of the *Trent* episode. When the *Trent* reached Southampton there was a public upsurge of anti-Americanism. 'The people are frantic with rage, and were the country polled, I fear 999 men out of a thousand would prepare for immediate war.'[47] Meanwhile, the Cabinet was told by Lord John Russell that he was drafting a dispatch demanding the release of Messrs Mason and Slidell, and an apology.[48]

Albert, though feeling so deeply ill, felt he must intervene to prevent an escalation of the crisis. All sorts of wild talk was flying about, including the claim by General Winfield Scott, a Northern general then in Paris, that Napoleon III was preparing to join forces with the North in a war against England.

Albert realized that the Northern States were in no mood for such a war. Having suffered a heavy defeat at the hands of the Confederates at the Battle of Bull Run on 21 July, they were not likely, in November, to risk a war against the greatest maritime power in the world.[49] Moreover, as the submissions of Mrs and Miss Slidell made clear – Slidell's wife and daughter, who were waiting for him in London – the captain who made the arrests explained he did so off his own bat, and not under instruction from the US Government or the President. In other words, there was room for diplomatic manoeuvre if they drafted the dispatch correctly.

On Sunday, 1 December, Albert rose early as usual, and forced himself to his desk, confessing himself later that morning to be 'very wretched', and scarcely capable of holding a pen. The Albertian draft was the one which was, in the event, dispatched. It asked for confirmation that Captain Wilkes did not act under the President's instruction, and made clear that the British Government could not allow its flag to be insulted. Lord Palmerston considered Albert's alterations 'excellent'. The Americans agreed to release the two delegates and their two companions. The crisis was averted.

The same could not be said of the crisis in Albert's health. All that day, 1 December, he shivered, and was unable to eat. The next day, his tongue was brown and dry. One of the doctors, William Jenner, assured the Queen that there was no cause for alarm. Sir James Clark came to examine the Prince and found him lying on his sofa in a dressing gown. On the 3rd, the Prime Minister wrote to the Queen suggesting that they seek further medical advice. Victoria wrote a snubbing reply:

> The Queen is very much obliged to you for the kind interest displayed in your letter received this day. The Prince has had a feverish cold the last few days, which disturbed his rest at night, but Her Majesty has seen His Royal Highness before similarly affected and hopes that in a few days it will pass off. In addition to Sir James Clark, the Queen has had the advantage of the constant advice of Dr Jenner, a most skilful Physician, and Her Majesty would be very unwilling to cause

> unnecessary alarm, where no cause exists for it, by calling in a Medical Man who does not upon ordinary occasions attend at the Palace.[50]

Colonel Phipps, Albert's equerry, wrote in confidence to Palmerston that the Queen was in a very nervous state and Albert was already irritable to have three medical attendants, the two doctors named, and Mr Brown, the Windsor apothecary. Phipps believed that if another medic were to arrive it would frighten Victoria. Her Majesty had been 'upset and agitated' at Palmerston's sensible suggestion.

The next day, the doctors decided that Albert was suffering from a gastric fever and warned that the symptoms would last for a month. 'The Queen is at present perfectly composed… but I must tell you, most confidentially, that it requires no little management to prevent her from breaking down altogether,'[51] Phipps warned Palmerston.

For the next day or two, Albert, his temper frayed, his tongue and throat furred, and still suffering from diarrhoea, was capable of walking, but he was too weak for much. At night he moved from one room to another, incapable of sleep. On Sunday, 8 December, he asked to be moved into the large Blue Room, called the King's Room, since it was here that William IV had died. A piano was brought into the neighbouring chamber and Princess Alice played 'Ein feste Burg ist unser Gott'. On that day, too, Clark and Jenner realized that they should consult other doctors. Sir Henry Holland, one of the physicians-in-ordinary to the Prince, and Dr Watson, one of the physicians-in-ordinary to the Queen, were brought in. Lord Clarendon remarked that he 'would not trust Sir James Clark and Sir Henry Holland to look after a sick cat'.[52]

Everyone noticed, and none were reassured by, the Queen's unnatural cheerfulness. On 9 December, the carapace cracked, and she got into 'a great state of agony' at the prospect of so many doctors possibly causing Albert alarm. Dr Watson warned the Prime Minister that Albert was 'very ill', and Palmerston instructed him to remain at the Castle overnight – an instruction which infuriated the Queen, who regarded it as 'most improper interference'. Nevertheless, she allowed Watson to stay overnight. The Prince was now in a confused state, but always affectionate when Victoria was near him, laying his weak head on her shoulder. On the 12th, he coughed up large amounts of mucus, a fit which was followed by uncontrollable shivering. By the 13th, the doctors were sufficiently satisfied by his condition that they said the Queen could safely take a short walk. While Victoria was out, Dr Jenner ran in to Lady Augusta Bruce to say that they had feared Albert was going to die there and then.

When Lady Augusta broke the news to the Queen that Albert was close to death, Victoria became almost incoherent, muttering, 'The country... oh the country... I could perhaps bear my own misery, but the poor country.' She nevertheless regained her composure when she went back into the sick-room. He was his old self. He kissed her affectionately and called her his '*gutes Frauchen*'. Every half hour they gave him brandy. Sir James Clark told her that the Prince's condition had taken a turn for the worse.

At 1, 2 and 3 a.m. messages were taken to the Queen that the Prince was bearing up, and at 6 a.m. Mr Brown said, 'I think there is ground to hope the crisis is over.' Victoria came into the Blue Room at 7 a.m. on Sunday, 14 December. The sun was rising. It was a bright morning. Jenner detected a 'decided rally' in the patient, but Albert's breathing was unnaturally rapid. The Queen asked if she could go out on the terrace, which she did, accompanied by Alice. The military band was playing in the distance, and Victoria burst into tears. One by one, the children were led into the room, and at half past nine that morning, she summoned the Dean of Windsor, the Hon. Gerald Wellesley.

By now the death-room – there could be no doubt whatever that this was what it had become – was growing crowded. Bertie and Lenchen were at the end of the bed. Alice was by her father's side. Later in the day, they were joined by Louise and Arthur. Beatrice was too young for such a scene, Affie was away at sea and Leopold was in Cannes for his health. Here too were the Prince and Princess of

Leiningen, Miss Hildyard the governess, General Bruce, the doctors and Colonel Phipps. At some point that afternoon, Alice calmly said, 'That is the death rattle.' The Queen exclaimed, 'Oh, yes, this is death. I know it. I have seen it before.' She leaned over him, whispered, '*Es ist dein kleines Frauchen*.' He whispered that he wanted her to give him '*ein Kuss*'. The Queen started to scream when the full horror of the situation became clear. She ran to fetch Beatrice from the nursery, so that Baby could be with her father when he died. Since Victoria had plainly lost control, it is difficult to know how accurate her later memories of the tragic scene actually were. Emotion recollected in tranquillity is not necessarily truthful. As she later remembered, 'Two or three long but perfectly gentle breaths were drawn, the hand clasping mine and… All, all was over… I stood up, kissed his dear heavenly forehead & called out in bitter and agonising cry: "Oh, my dear Darling!" and then dropped on my knees in mute, distracted despair, unable to utter a word or shed a tear! Ernest Leiningen & Sir C. Phipps lifted me up, and Ernest led me out… Then I laid down on the sofa in the Red Room, & all the gentlemen came in and knelt down & kissed my hand, & I said a word to each. Distantly, the Castle clock chimed the third quarter after ten.'[53]

The body lay in the Castle until 23 December, the day of the funeral. The Queen was too distraught to attend this ceremony, which took place in St George's Chapel with the rest of the Royal Family, the Gentlemen and Lords-in-Waiting, the Heralds, led by Garter King of Arms, a guard of honour of the Grenadier Guards, the Cabinet – all but Palmerston who was suffering from gout – the Queen's Household, and a special train-load, brought down by the Great Western Railway – of bishops, dukes and the like. It was all done in a seemly, dignified, Church of England manner. Yet the Chapel was filled by those who had either been slow to see the Prince Consort's virtues, or who had never done so. Those who had most cause to mourn him, and those with whom he felt instinctive empathy, were those who made the nineteenth century a glory-age for Great Britain: the engineers, the scientists, the university reformers, the museum curators, the art historians, the social reformers, the city planners, the philanthropists, the choral societies and orchestras, the librarians.

The real twist, the surprise-denouement of the Victoria and Albert story, is one which would surprise any but him, if they did not know the ending: namely, she was the strong one. Few except Sir James Clark had witnessed the extent of the Queen's emotional vulnerability and the toll it took upon the husband. Had anyone apart from Clark seen it, however, they would surely have imagined that it was Albert who was shoring up Victoria, and that without his support she would collapse. In the short term, this would clearly be the case, and in the immediate aftermath of his death, immediate in so grievous a bereavement being within the next twelve months, it did indeed look as if the Queen would be unable to continue her work. She herself feared she was losing her reason, let alone her ability to function as a Head of State. In an earlier book, I traced her slow conquest of the demons, of grief and self-doubt, which plagued her in those dark days. I came to see her life, which was destined to outlast Albert's by forty years, as, in a way, a cruel liberation, a liberation through heartbreak, in which she learned how to live alone, and without him on whom she had depended for everything.

It was 1861, two years after Charles Darwin had published *On the Origin of Species*, and perhaps there was no more bizarre example in England of a surprisingly powerful life-force, in Victoria, showing itself to be much more robust than that of Albert. He remarked to her at about this time – and it is a truly chilling remark for a man of only forty-two – 'I do not cling to life. You do; but I set no store by it. If I knew those I love were well cared-for, I should be quite ready to die tomorrow… I am sure if I had a severe illness, I should give up at once, I should not struggle for life. I have no tenacity of life.'[54]

Sir James Clark, a misogynist old doctor, felt that the monarchy itself could not long survive the calamity of Albert's departure. Four years after it, in 1865, Clark wrote in his diary, 'By his premature death, the Prospects of the Monarchy are bad indeed. No one who knows the character of the Queen and of the Heir Apparent can look forward to the future without seeing troubles in that quarter before us.'[55]

Gladstone, when he had become Prime Minister and was having his notorious difficulties with the Queen, made it very clear what was lost when Albert died. It was Albert's influence, and the influence, on a larger scale, of the Court. Albert provided patronage, in so many senses of the term, to science, technology, all branches of the arts, to the cultural life of the nation, in a way which Victoria never wished to do. During his lifetime, there was a sense of all these things being at the heart of life, of the Victorian economic and political success story being chiefly of value because they enabled these things. Hindsight makes us see that his visions for Europe were doomed to fail. As for his political ambitions at home, and his having been 'king in all but name', this too was probably illusory. England did not want or need an enlightened despot such as might have flourished in Germany during the *Goethezeit*. But this was not to say that it did not need orchestras, chambers of commerce, art galleries, libraries, universities, museums, societies to further decent housing for the poor, and all the many good causes to which he gave such active encouragement.

Albert lay in the vault of St George's for a year, the time it took to commission, design and build the mausoleum at Frogmore. Naturally, the Queen turned to Ludwig Grüner. Who else could have been the architect for such a shrine? It is designed in the form of a Greek cross, with a diameter of 69 feet. Albert was taken there in December 1862. For nine years, as his body lay there, the work of embellishment continued around him, appropriately enough, since he loved building projects and would surely have approved of Grüner's pan-European vision. It does not feel like England. The red Portuguese marble with which the walls would eventually be covered mingles with other coloured marbles from Italy, Greece and France. The walls are decorated with reproductions of the works of Raphael, using Albert's own photographic collection of the Italian master.[56] The tomb itself in which he would eventually be laid to rest was hewn from a single piece of Aberdeenshire granite. Atop lie the two young lovers, carved in white marble by Baron Carlo Marochetti. It is his last work.[57]

Rightly is the mausoleum opened to the public. But equally rightly, the days when this happens are few, and the place is a private monument to the private love of two very different, very complicated and very fallible people. For all the storms and battles of their marriage, it was in itself something of a monument. Marochetti's figure of the Queen is of the young woman, lying beside her lover. The real Victoria must wait forty years after her husband's death before she could join him, and it is impossible not to be moved, as one thinks of the extreme age of the woman who would eventually be laid beside her Angel's side. Frogmore contains and embodies their private love, of which we are only allowed a glimpse. The imperfections of the marriage, which a modern biography is intrusive enough to reveal, will only enhance the observer's admiration. Both Victoria and Albert were full-time busy public figures, steering the institution of monarchy into something which could be sustained in the era of universal suffrage and parliamentary government. They were also the parents of nine children, and they were very strong characters. It would not have been possible for them to live together for twenty-two years without friction. Albert wanted the Royal Family, almost as an idealized icon captured by Winterhalter or by carefully chosen photographers, to inspire suburban families to lives of rectitude. Twenty-first-century readers, with their long life-expectancy, perhaps find more to admire in the Victoria and Albert we discover from that evidence Princess Beatrice so wanted to destroy – two fallible beings, whose egos clashed, but who remained lovers to the end, and who were in every sense partners.

That is the pair, enclosed in the almost fantastical, but always moving, setting of the mausoleum. Albert's more public memorials remain in Kensington. In 1868, the Queen laid the foundation stone of the Royal Albert Hall, loosely based on Gottfried Semper's Dresden Opera House. When one sits through the glory of the Promenade

Concerts each summer (they moved here after the Queen's Hall was bombed in the Second World War) one senses most powerfully Albert's connectedness with the arts. Opposite the Royal Albert Hall, in Kensington Gardens there rises the spindly Gothic of George Gilbert Scott's memorial. The 14-foot gilded statue of Albert sits, holding the catalogue of the Great Exhibition of 1851. The symbolic statuary ascending from its base speaks of the multiplicity of the Prince's concerns and interests, Agriculture, Manufacture, Engineering and Commerce symbolized in marble. The frieze of 169 figures shows the painters, the architects, the musicians, the poets, the sculptors, while on the pillars are symbolic representations of Geometry, Chemistry, Geology and Astronomy. In the late twentieth century, the memorial fell into such decay that there was even talk of pulling it down. Such was the distaste for Victorian design, and the low esteem, among the liberal establishment, for the Royal Family, that the idea met with some enthusiasm. It was largely owing to the energy of Jocelyn Stevens, Chairman of English Heritage, that the Albert Memorial was spared, and restored. Now it is seen to be what it always was, one of the glories of London. It is a splendid reminder of Prince Albert's greatness. In British history, no other public figure of comparable ability, breadth or benign influence even touches him. There is certainly no comparable royal figure in British history. In the museums of the Albertopolis, his good influence bears fruit to this day. We are right to honour his memory.

Prince Albert never learnt to play the English courtly, or upper-class, game of wearing talent lightly. He never saw the point of facetiousness, which is a quality so English that other languages do not really possess a word for it. He was disliked at Court, and in the Royal Household, because he was a know-all, who really did know best. He was the butt of aristocratic society. He did not especially understand or take an interest in gossip. Yet when he met academics, writers, musicians, artists, mill-owners, inventors, scientists, servicemen and artisans, they found an unpompous, genial person who shared their interests and could talk to them without 'side' or frivolity.

Eight years after Albert died, W. E. Gladstone had become the Prime Minister, and came down to the Isle of Wight from London for an audience with the Queen. Henry Ponsonby, then an equerry-in-waiting, but soon to be Victoria's most inspired and helpful Private Secretary, was sent to Cowes to escort the Prime Minister back to Osborne House. Ponsonby embarked on one of his favourite themes, the similarity in character between Albert and King William III.

Gladstone 'entirely disagreed. William, though a great statesman and able warrior winning campaigns although losing battles, was hard, cold and cruel. The Prince whose only resemblance was his shyness was warm and kind. The Prince could never have effected the revolution: he had not the indomitable will of William nor could he ever have perpetrated the Glencoe massacre.' Ponsonby said they were both unpopular with the English. 'But Gladstone denied the Prince was. He was unpopular with smart society but the people liked him'.[58]

His dreams for Germany, and for the future of Europe, were pathetically unrealistic. The dynastic marriages he planned for his children, far from spreading peace and constitutional monarchy on a British liberal model, did nothing to check the rise of militarism and nationalism. The family quarrels to which these dynastic marriages gave rise could even be seen as an ingredient in the tragedy of the First World War. It was not until Europe had undergone the blood-lettings of two world wars, and bloody revolutions and civil wars, that it came close, in the decades post-1950, to realizing the Albertian vision of German federalism and a *Zollverein* keeping the peoples of Europe in peace and prosperity.

The defining, the central, event of his public life was the Great Exhibition of 1851. It called forth all the virtues he brought to public life – practicality, intelligence, a belief in commerce, a passion for technology and invention. It enabled him to work with those he liked best – intelligent males with practical skill. His colleagues on the Royal Commission of 1851 signalled the arrival of a new world order, urban, commercialized, industrial. Albert's presence among them, and the midwife skills which enabled him to encourage Victorian trade and industry, placed that ancient institution, the British monarchy, at the heart of nineteenth-century change. In other parts of Europe, the monarchy was rightly seen as the ally of reaction. In Britain, it performed a Janus function. The Royal Family, of which Albert was the youthful patriarch, were part of the new order. But their hereditary status and their ritual public function helped to ensure that Britain would embrace neither the turbulence of revolution nor the illusions of reaction. Albert was not the sole architect of British constitutional development, but he undoubtedly played a vital role in the evolution of that strange hybrid: a monarchy held in check by a representative parliament; a democracy whose ultimate power wore a crown. The personal loss, when he died so young, was irrecoverable, and no one in British public life has ever replaced or rivalled him. The legacy, however, did not die with him.

BIBLIOGRAPHY

Aronson, Theo, *Queen Victoria and the Bonapartes*, London, Bobbs-Merrill, 1972

Ashley, Evelyn, *The Life and Correspondence of Henry John Temple, Viscount Palmerston*, 2 vols., London, Bentley, 1879

Auerbach, Jeffrey A., *The Great Exhibition of 1851: A Nation on Display*, New Haven, Yale University Press, 1999

Auerbach, Jeffrey A. and Hoffenberg, Peter H. (eds.), *Britain, the Empire and the World at the Great Exhibition of 1851*, Aldershot, Ashgate, 2008

Bauer, Karoline, *Memoirs*, London, Remington and Co., 1885

Beem, Charles and Taylor, Miles (eds.), *The Man Behind the Queen: Male Consorts in History*, Basingstoke, Palgrave Macmillan, 2014

Beier, A. L. (ed.), *The First Modern Society: Essays in English History in Honour of Lawrence Stone*, Cambridge, Cambridge University Press, 1989

Bennett, Alan, *The History Boys*, London, Faber and Faber, 2004

Benson, A. C. and Esher, Viscount (eds.), *The Letters of Queen Victoria 1837–1861*, 3 vols., London, John Murray, 1908

Boeckmann, Daniel, 'Ludwig Gruner, Art Adviser to Prince Albert', MA dissertation, School of World Art Studies and Museology, University of East Anglia, 12 September 1996

Bosbach, Franz, *Die Studien des Prinzen Albert an der Universität Bonn (1837–1838)*, Munich, K. G. Saur, 2010

Bosbach, Franz and Davis, John, *Prinz Albert – Ein Wettiner in Grossbritannien*, Munich, K. G. Saur, 2004

Bosbach, Franz and Davis, John et al., *Die Weltausstellung 1851 und ihre Folgen*, Munich, K. G. Saur, 2002

Briggs, Asa, *Victorian Things*, London, Batsford, 1988

Brindle, Steven (ed.), *Windsor Castle: A Thousand Years of a Royal Palace*, London, Royal Collection Trust, 2018

Cannadine, David, 'The Last Hanoverian Sovereign? The Victorian Monarchy in Historical Perspective 1688–1988', in A. L. Beier, David Cannadine and J. M. Rosenheim (eds.), *The First Modern Society: Essays in English History in Honour of Lawrence Stone*, Cambridge, Cambridge University Press, 1989

—*Victorious Century*, London, Allen Lane, 2017

Cecil, David, *Lord M*, London, Constable, 1954

Chadwick, Owen, *The Victorian Church*, 2 vols., London, A. & C. Black, 1966

Chamberlain, Muriel E., *Lord Aberdeen: A Political Biography*, London, Longman, 1983

Checkland, S. G., *The Gladstones: A Family Biography, 1764–1851*, Cambridge, Cambridge University Press, 1971

Cole, Henry, *Fifty Years of Public Work of Sir Henry Cole accounted for in his deeds, speeches and writings*, 2 vols., London, G. Bell and Sons, 1884

Collier, Carly, *British Artists and Early Italian Art, 1770–1849: The Pre-Raphaelites*, Warwick, University of Warwick, 2014

Colquhoun, Kate, *A Thing in Disguise: The Visionary Life of Joseph Paxton*, London, Fourth Estate, 2003

Colvin, Howard, *A Biographical Dictionary of British Architects*, New Haven, Yale University Press, 1995

Cope, Charles, *Reminiscences of Charles West Cope, RA*, London, Richard Bentley and Son, 1891

Crook, J. Mordaunt, *The Rise of the Nouveaux Riches*, London, John Murray, 1999

Davenport-Hines, Richard and Sisman, Adam (eds.), *One Hundred Letters from Hugh Trevor-Roper*, Oxford, Oxford University Press, 2014

Davis, John R., *The Great Exhibition*, Stroud, Sutton, 1999

Dickens, Charles, *Dombey and Son*, London, Chapman and Hall [n.d.]

Eastlake, Lady, *Contributions to the Literature of the Fine Arts with a Memoir*, London, 1870

Eckermann, Johann, *Gespräche mit Goethe*, part III, Leipzig, F. A. Brockhaus, 1848

Eissenhauer, Michael and Wiebel, Christiane, *Eine adlige Kindheit in Coburg*, Coburg, Kunstsammlungen der Vester Coburg, 2000

Eyck, Frank, *Prinzegemahl Albert von England, eine politische Biographie*, Erlenbach/Zurich/Stuttgart, Eugen Remtsch Verlag, 1961

Ferguson, Sarah, HRH Duchess of York, and Stoney, Benita, *Victoria and Albert: Life at Osborne House*, London, Weidenfeld & Nicolson, 1991

Ffrench, Yvonne, *The Great Exhibition*, London, Harvill, 1950

Field, Leslie, *The Jewels of Queen Elizabeth II: Her Personal Collection*, London, Weidenfeld & Nicolson, 1987

Foster, R. F., *Modern Ireland, 1600–1972*, London, Allen Lane, 1988

Fraser, Antonia, *Perilous Question: The Drama of the Great Reform Bill 1832*, London, Weidenfeld & Nicolson, 2013

Fulford, Roger, *The Prince Consort*, London, Macmillan, 1949

—(ed.), *Dearest Child: Letters Between Queen Victoria and the Princess Victoria, 1858–1861*, London, Evans Bros., 1964

—(ed.), *Dearest Mama: Letters Between Queen Victoria and the Crown Princess of Prussia 1861–1864*, London, Evans Bros., 1968

—(ed.), *Your Dear Letter: Private Correspondence of Queen Victoria and the Crown Princess of Prussia, 1865–1871*, London, Evans Bros., 1971

Gash, Norman, *Sir Robert Peel: The Life of Sir Robert Peel after 1830*, London, Longmans, 1963 (referred to in text as Gash, Vol. II)

Gieger, Etta K., *Die Londoner Weltausstellung von 1851: im Kontext der Industrialisierung in Grossbritannien*, Essen, Blaue Eule, 2007

Greville, Charles, *The Greville Memoirs, 1814–1860*, ed. Lytton Strachey and Roger Fulford, 8 vols., London, Macmillan, 1938

Grey, Lieutenant-General the Hon. C., *The Early Years of His Royal Highness the Prince Consort, compiled under the direction of Her Majesty the Queen*, London, Smith, Elder & Co., 1867

Grunewald, Ulrike, *Luise von Sachsen-Coburg-Saalfeld (1800–1831)*, Cologne/Weimar/Vienna, Böhlau Verlag, 2013

Hare, Augustus, *The Story of My Life*, 6 vols., London, G. Allen, 1896–1900

Hawkins, Angus, *The Forgotten Prime Minister: The 14th Earl of Derby*, 2 vols., Oxford, Oxford University Press, 2007, 2008

Henker, Michael and Brockhoff, Evamaria, *Ein Herzogtum und viele Kronen. Coburg in Bayern und Europa*, Augsburg, Haus der Bayerischen Geschichte, 1997

Hewison, Robert, *John Ruskin*, Oxford, Oxford University Press, 2007

Hill, Rosemary, *God's Architect*, London, Allen Lane, 2007

Hilton, Boyd, *A Mad, Bad, and Dangerous People? England 1783–1846*, Oxford, Clarendon Press, 2006

Hobhouse, Hermione, 'The Monarchy and the Middle Classes: The Role of Prince Albert', in *Bürgertum, Adel und Monarchie/Middle Classes, Aristocracy and Monarchy*, Munich/London/New York/Paris, K. G. Saur, 1989

—*Thomas Cubitt, Master Builder*, London, Macmillan, 1971

Hodder, Edwin, *Life of Lord Shaftesbury*, London, Cassell, 1886

Jagow, Kurt (ed.), *Prinzgemahl Albert: Ein Leben am Throne: Eigenhändige Briefe und Aufzeichnungen, 1831–1861*, Berlin, Verlag Karl Siegismund, 1937

Kelly, J. N. D., and Walsh, M. J., *Oxford Dictionary of Popes*, revised edn, Oxford, Oxford University Press, 2010

Kennedy, A. L. (ed.), *My Dear Duchess: Social and Political Letters to the Duchess of Manchester, 1858–1869*, London, John Murray, 1956

Kilvert, Francis, *The Diaries of Francis Kilvert*, 3 vols., ed. William Plomer, London, Jonathan Cape, 1939

Lawrence, D. H., *Kangaroo*, London, William Heinemann, 1950

Leapman, Michael, *The World for a Shilling: How the Great Exhibition of 1851 Shaped a Nation*, London, Headline, 2001

Longford, Elizabeth, *Wellington: Pillar of State*, London, Weidenfeld & Nicolson, 1972

McCloskey, Deirdre, *Bourgeois Equality: How Ideas, Not Capital or Institutions, Enriched the World*, Chicago, University of Chicago Press, 2016

Marsden, Jonathan (ed.), *Victoria and Albert: Art and Love*, London, Royal Collection Enterprises Ltd, 2010

Martin, Sir Theodore, *The Life of the Prince Consort*, London, I. B. Tauris, 2012 (originally published between 1877 and 1879 by Smith, Elder & Co.)

Miers, Mary, *Highland Retreats: The Architecture and Interiors of Scotland's Romantic North*, New York, Rizzoli, 2017

Millar, Delia, *The Victorian Watercolours and Drawings in the Collection of Her Majesty the Queen*, London, Philip Wilson Publishers, 1995

Montagu, Jennifer, 'The "Ruland/Raphael Collection"', *Visual Resources*, Vol. III, 1986

Morley, John, *The Life of William Ewart Gladstone*, 3 vols., London, Macmillan and Co., 1912

Murphy, Paul Thomas, *Shooting Victoria: Madness, Mayhem and the Modernisation of the British Monarchy*, London, Head of Zeus, 2012

Netzer, Hans-Joachim, *Albert von Sachsen-Coburg-Gotha. Ein deutscher Prinz in England*, Munich, Verlag C. H. Beck, 1988

Pakula, Hannah, *An Uncommon Woman: The Empress Frederick, Daughter of Queen Victoria, Wife of the Crown Prince of Prussia, Mother of Kaiser Wilhelm*, New York, Simon & Schuster, 1995

Pattison, Mark, *Memoirs*, London, Macmillan and Co., 1885

Pevsner, Nikolaus and Lloyd, David, *The Buildings of England: Hampshire and the Isle of Wight*, Harmondsworth, Penguin, 1967

Plunkett, John, *Queen Victoria: First Media Monarch*, Oxford, Oxford University Press, 2003

Ponsonby, Arthur, *Henry Ponsonby. Queen Victoria's Private Secretary. His Life from His Letters*, London, Macmillan and Co., 1942

Ponsonby, D. A., *A Prisoner in Regent's Park*, London, Chapman and Hall, 1961

—*The Lost Duchess*, London, Chapman and Hall, 1958

Quiller-Couch, Sir Arthur (ed.), *The Oxford Book of English Verse*, Oxford, Clarendon Press, 1939

Rappaport, Helen, *Magnificent Obsession: Victoria, Albert and the Death that Changed the Monarchy*, London, Hutchinson, 2011

Read, Benedict, 'Sculpture and the New Palace of Westminster', in David Cannadine (ed.), *The Houses of Parliament: History, Art, Architecture*, London, Merrell, 2000

Rhodes James, Robert, *Albert, Prince Consort: A Biography*, London, Hamish Hamilton, 1983

Ridley, Jane, *Bertie: A Life of Edward VII*, London, Chatto & Windus, 2010

Ridley, Jasper, *Lord Palmerston*, London, Constable, 1970

Robinson, John Martin, *James Wyatt, Architect to George III*, New Haven/London, Yale University Press, 2012

Röhl, John C. G., *Wilhelm II: die Jugend des Kaisers 1859–1888*, Munich, C. H. Beck, 1993

Sotnick, Richard, *The Coburg Conspiracy*, London, Ephesus Books, 2008

Stephen, Leslie and Lee, Sir Sidney (eds.), *Dictionary of National Biography*, London, Smith, Elder & Co., 1885–1900

Strachey, Lytton, *Queen Victoria*, London, Chatto & Windus, 1921

Strong, Roy, *Painting the Past: The Victorian Painter and British History*, London, Pimlico, 2004

Tasler, Angelika, *Macht und Musik*, Cologne/Weimar/Vienna, Böhlau Verlag, 2017

Tennyson, Alfred Lord, *The Poems of Tennyson*, ed. Christopher Ricks, 3 vols., London, Longman, 1969

Tennyson, Charles, *Alfred Tennyson*, London, Macmillan, 1950

Turner, Michael, *Osborne House*, London, English Heritage, 1989

Urbach, Karina (ed.), *Royal Kinship: Anglo-German Family Networks 1815–1918*, Munich, K. G. Saur, 2008

—*Bismarck's Favourite Englishman: Lord Odo Russell's Mission to Berlin*, London, I. B. Tauris, 1999

Vickers, Hugo, *Alice: Princess Andrew of Greece*, London, Hamish Hamilton, 2000

Weintraub, Stanley, *Albert, Uncrowned King*, London, John Murray, 1997

Wilson, A. N., *Victoria: A Life*, London, Atlantic, 2014

Winter, Emma, 'Prince Albert, Fresco Painting and the New Houses of Parliament, 1841–51', in Bosbach and Davis, *Prinz Albert – Ein Wettiner in Grossbritannien*

Woodham-Smith, Cecil, *Queen Victoria: Her Life and Times. Volume I, 1819–1861*, London, Hamish Hamilton, 1972

Woodward, E. L., *The Age of Reform, 1815–1870*, Oxford, Clarendon Press, 1938

Wordsworth, William, *The Poetical Works of Wordsworth*, London, Oxford University Press, 1950

Young, G. M., *Victorian England: Portrait of an Age*, London, Oxford University Press, 1936

Young, Paul, *Globalization and the Great Exhibition: The Victorian New World Order*, London, Palgrave Macmillan, 2009

NOTES

1 PRINCESS BEATRICE'S WAR WORK

1 RA AEC/GG/012/FF2/13.
2 Ibid.
3 http://www.queenvictoriasjournals.org/home.do.
4 RA VIC/MAIN/Z/261.
5 RA AEC/GG012/FF2.
6 RA AEC/GG/012/FF2/16.

2 HIS MOTHER

1 Michael Turner, Chapter 27, 'Victoria and Albert, 1840–1861', in Steven Brindle (ed.), *Windsor Castle: A Thousand Years of a Royal Palace*, p. 369.
2 Delia Millar, *The Victorian Watercolours and Drawings in the Collection of Her Majesty the Queen*, catalogue number 3991.
3 Hugo Vickers, *Alice: Princess Andrew of Greece*, p. 259.
4 In this book, I spell German first names as they would themselves have spelt them – so, Luise, not Louise; Ernst not Ernest. Albert was christened Albrecht but from an early age known as Albert.
5 D. A. Ponsonby, *The Lost Duchess*, p. 159.
6 Ulrike Grunewald, *Luise von Sachsen-Coburg-Saalfeld (1800–1831)*, p. 39.
7 Ibid., p. 229.
8 D. A. Ponsonby, *The Lost Duchess*, p. 19.
9 Ibid., p. 30.
10 Ibid., p. 34.
11 Ibid., p. 35.
12 Grunewald, pp. 58–9.
13 RA VIC/MAIN/M2/CFP/8. Prince Albert has labelled the file with the note, 'These papers have been brought together from the scattered correspondence of the Duchess of Kent found after her death. Letters concerning the Marriage of the Duke and Duchess of Kent', 16 March 1861.
14 RA VIC/MAIN/M2/CFP/20.
15 Quoted Robert Rhodes James, *Albert, Prince Consort: A Biography*, p. 7.
16 RA VIC/MAIN/Y/154/Z.
17 RA VIC/MAIN/Y/154/4.
18 RA VIC/MAIN/Y/154/2.
19 D. A. Ponsonby, *The Lost Duchess*, p. 109.
20 Richard Sotnick, *The Coburg Conspiracy*, p. 179.

21 D. A. Ponsonby, *A Prisoner in Regent's Park*.
22 D. A. Ponsonby, *The Lost Duchess*, p. 110.

3 THE WETTINS

1 Grunewald, p. 84.
2 Staatsarchiv Gotha LATh f. 8.
3 Lieutenant-General the Hon. C. Grey, *The Early Years of His Royal Highness The Prince Consort, compiled under the direction of Her Majesty the Queen*, p. 16.
4 RA M40 C FP 15.
5 Grey, p. 17.
6 Grunewald, p. 89.
7 D. A. Ponsonby, *The Lost Duchess*, p. 116.
8 Ibid., p. 117.
9 Grunewald, p. 121.
10 Ibid., p. 118.
11 Ibid.
12 Ibid., p. 119.
13 Ibid., p. 92.
14 D. A. Ponsonby, *The Lost Duchess*, p. 131.
15 Ibid., p. 139 but quote from actual book BL.
16 Grunewald, p. 136.
17 D. A. Ponsonby, *The Lost Duchess*, p. 160.

4 CHILDHOOD: '*TOUT RAPPELLE L'HOMME À SES DEVOIRS*'

1 Johann Eckermann, *Gespräche mit Goethe*, part III, 23 October 1828, p. 682.
2 Angelika Tasler, *Macht und Musik*, p. 25.
3 Hans-Joachim Netzer, *Albert von Sachsen-Coburg-Gotha. Ein deutscher Prinz in England*, p. 66.
4 Staatsarchiv Coburg LAA 6866 f. 2 v.
5 Grey, p. 24.
6 All quoted Netzer, p. 64.
7 Ibid., p. 62.
8 RA VIC/MAIN/QVJ/1845, 20 August.
9 Tasler to the author.
10 'Traktat von Christoph Florschütz um 1824/5', Staatsarchiv Coburg, LAA 7667.
11 RA VIC/ADDA10/81/9.
12 All the above, Windsor, RA VIC/ADDA10/81/9.
13 RA VIC/ADDA10/81/12.
14 Grey, p. 4.
15 D. A. Ponsonby, *The Lost Duchess*, p. 161.
16 RA VIC/MAIN/M/40/4.
17 Grunewald, p. 199.
18 Karoline Bauer, *Memoirs*, p. 64.

5 'THESE DEAREST BELOVED COUSINS': ALBERT'S FIRST VISIT TO ENGLAND

1 Michael Eissenhauer and Christiane Wiebel, *Eine adlige Kindheit in Coburg*, p. 49.
2 Fürstlich Leiningensches Archiv, Amorbach.
3 Eissenhauer and Wiebel, p. 50.
4 RA VIC/ADDA10/81/9.

5 RA VIC/MAIN/Y/154.
6 *Oxford Dictionary of National Biography*, 'Charles Cecil Cope Jenkinson', Vol. X, p. 747, W. A. J. Archbold, revised H. C. G. Matthew.
7 Antonia Fraser, *Perilous Question*, p. 242.
8 RA VIC/MAIN/Y/54/61.
9 Ibid.
10 Netzer, p. 94.
11 Staatsarchiv Gotha Vol. I, Loc. I, Til. 2, Nr. 33, p. 76, f. 9.
12 Grey, p. 130.
13 Cecil Woodham-Smith, *Queen Victoria: Her Life and Times. Volume I, 1819–1861*, p. 157.
14 RA VIC/ADDA6/1.
15 Charles Greville, *The Greville Memoirs, 1814–1861*, ed. Lytton Strachey and Roger Fulford. 25 June 1837. Vol. III, p. 279.
16 Ibid.
17 Woodham-Smith, p. 160.
18 Ibid., p. 161.
19 Grey, p. 216.
20 Woodham-Smith, p. 163.

6 EUROPEAN JOURNEYS

1 Netzer, p. 96.
2 Franz Bosbach, *Die Studien des Prinzen Albert an der Universität Bonn (1837–1838)*, p. 122.
3 Stanley Weintraub, *Albert, Uncrowned King*, p. 57.
4 Grey, p. 171.
5 RA VIC/ADDA6/3.
6 Ibid.
7 RA VIC/ADDA10/81/8.
8 Ibid.
9 RA VIC/MAIN/Y/154/29.
10 Ibid.
11 RA VIC/MAIN/Y/188/33.
12 RA/VIC/MAIN/Y/188/42.
13 RA VIC/ADDA6/5.
14 Grey, p. 149.
15 RA VIC/MAIN/Z/141/22.
16 Grey, p. 154.
17 Sir Theodore Martin, *The Life of His Royal Highness The Prince Consort*, Vol. I, p. 38.
18 Grey, p. 154.
19 RA VIC/ADDA6/6.
20 Bosbach, p. 120.
21 Ibid., p. 80.
22 *Daily News*, No. 2385.
23 Martin, Vol. I, p. 21.
24 Bosbach, p. 124.
25 RA VIC/ADDA6/4.
26 Ibid.
27 Kurt Jagow (ed.), *Prinzgemahl Albert: Ein Leben am Throne: Eigenhändige Briefe und Aufzeichnungen, 1831–1861*, p. 18.
28 Martin, Vol. I, p. 29.
29 RA VIC/ADDA6/7.

30 RA VIC/ADDA6/10.
31 Ibid.
32 Martin, Vol. I, p. 29.
33 RA VIC/ADDA6/13.
34 Ibid.
35 RA VIC/ADDA6/14.
36 Ibid.
37 Ibid.
38 Grey, p. 197.
39 RA VIC/ADDA6/17.
40 RA VIC/ADDA6/18.
41 Ibid.
42 J. N. D. Kelly and M. J. Walsh, *Oxford Dictionary of Popes*, p. 307.
43 Grey, p. 200.
44 RA VIC/ADDA6/20.
45 Jonathan Marsden (ed.), *Victoria and Albert: Art and Love*, p. 18.
46 Weintraub, p. 71.
47 Grey, p. 212.

7 A SOMEWHAT ROUGH EXPERIENCE

1 RA VIC/ADDA6/6.
2 Grey, p. 246.
3 Ibid.
4 Queen Victoria's Journal (hereafter Journal), 10 October 1839.
5 Ibid.
6 Journal, 11 October 1839.
7 Journal, 12 October 1839.
8 Journal, 13 October 1839.
9 Journal, 15 October 1839.
10 Ibid.
11 Grey, p. 244.
12 Martin, Vol. I, p. 40.
13 Grey, p. 238.
14 Ibid., p. 255.
15 Martin, Vol. I, p. 53.
16 Quoted Woodham-Smith, p. 74.
17 Journal, 27 January 1840.
18 Ibid.
19 Journal, 28 January 1840.
20 Woodham-Smith, p. 259.
21 Ibid.
22 Seymour Diaries, BL Additional MS 60302 ff. 6–7.
23 Grey, p. 303.
24 Greville, Vol. IV, pp. 239–40.
25 RA VIC/MAIN/Z/141/194.
26 Journal, 10 February 1840.
27 Journal, 11 February 1840.
28 Woodham-Smith, p. 266.
29 RA VIC/ADDA6/25.
30 John Morley, *The Life of William Ewart Gladstone*, Vol. I, p. 17.
31 Martin, Vol. I, p. 87.

8 UNBOUNDED INFLUENCE

1 Boyd Hilton, *A Mad, Bad, and Dangerous People? England 1783–1846*, p. 23.
2 Ibid., p. 20.
3 Ibid., p. 15.
4 Tennyson, 'Locksley Hall', *The Poems of Tennyson*, ed. Christopher Ricks, Vol. I, p. 128.
5 *Pace* the online journal which says that this entry was written at Windsor Castle, the visit to the Chapel Royal and the evening in Buckingham Palace make it clear that they were in London.
6 Journal, 23 February 1840.
7 Martin, Vol. I, p. 71.
8 A. C. Benson and Viscount Esher (eds), *The Letters of Queen Victoria 1837–1861*, Vol. I, p. 224.
9 Woodham-Smith, p. 278.
10 Hannah Pakula, *An Uncommon Woman: The Empress Frederick*, p. 34.
11 Ibid.
12 Martin, Vol. I, p. 101.
13 Stockmar to Albert, 25 January 1854; Martin, Vol. I, p.76.
14 Greville, Vol. IV, p. 385.
15 Benson and Esher, *Letters*, Vol. I, p. 292.
16 Greville, Vol. IV, p. 387.
17 Journal, 14 June 1841.
18 Ibid.
19 Augustus Hare, *The Story of My Life*, Vol. V, p. 358.
20 *Oxford Dictionary of National Biography*, Vol. XXV, p. 123, Nigel Ashton.
21 Greville, Vol. IV, p. 398.
22 Benson and Esher, *Letters*, Vol. I, p. 295.
23 Greville, Vol. V, p. 39.
24 Benson and Esher, *Letters*, Vol. I, p. 268.
25 All the above, Fulford, p. 70.
26 Greville, Vol. IV, p. 383.
27 Gash, Vol. II, p. 263.
28 Liverpool Papers, BL Additional MS 38,303, f. 110 v.
29 Greville, Vol. IV, p. 408.
30 Peel Papers, BL Additional MS 40,432, f. 67 v.
31 Ibid.
32 Peel Papers, BL Additional MS 40,432, f. 1 r.
33 Peel Papers, BL Additional MS 40,432, f. 219 r.
34 Peel Papers, BL Additional MS 40,432, f. 192 r.
35 Peel Papers, BL Additional MS 40,432, f. 348 r.
36 Peel Papers, BL Additional MS 40,432, f. 349 r.
37 Greville, Vol. IV, p. 420.
38 Ibid., p. 421.
39 Peel Papers, BL Additional MS 40,344 f. 61 v.
40 Greville, Vol. IV, p. 420.
41 For all statistics, Gash, Vol. II, pp. 295–6.
42 Greville, Vol. V, p. 7.
43 Peel Papers, BL Additional MS 40,433, f. 89, r.
44 Peel Papers, BL Additional MS 40,433, f. 89 (translation).
45 Benson and Esher, *Letters*, Vol. I, p. 366.
46 Ibid.
47 Peel Papers, BL Additional MS 40,432, f. 291 r.
48 Benson and Esher, *Letters*, Vol. I, p. 361.
49 Ibid.

50 Ibid., p. 362.
51 Ibid., p. 357.
52 Ibid., p. 359.
53 Ibid., p. 367.
54 Ibid., p. 375.
55 Ibid., p. 363.
56 Netzer, p. 198.
57 Weintraub, p. 128.
58 Ibid., p. 129.
59 Peel Papers, BL Additional MS 40,513, f. 88 r.
60 Gash, Vol. II, p. 690.
61 Greville, Vol. V, p. 39.
62 Journal, 22 September 1842
63 Martin, Vol. I, p. 137.
64 Ibid., p. 140.
65 Ibid., p. 143.

9 PUBLIC ART, PUBLIC LIFE

1 Roy Strong, *Painting the Past*, p. 74, a reprint of the book published in 1978 (Thames and Hudson) as *And When Did You Last See your Father? The Victorian Painter and British History*.
2 Eckermann, 7 April 1829, p. 336.
3 Rosemary Hill, *God's Architect*, p. 128.
4 Peel Papers, BL Additional MS 40,432 f. 219 r.
5 Peel Papers, BL Additional MS 40,432, f. 266 r.
6 Emma Winter, 'Prince Albert, Fresco Painting and the New Houses of Parliament, 1841–51', in Franz Bosbach and John Davis, *Prinz Albert – Ein Wettiner in Grossbritaannien*, p. 150.
7 Peel Papers, BL Additional MS 40,432, f. 257 r.
8 Peel Papers, BL Additional MS 40,432, f. 257 v.
9 Winter, p. 157, quoting Memoirs of Baron von Bunsen, pp. 366–7.
10 Benedict Read, 'Sculpture and the New Palace of Westminster', in David Cannadine (ed.), *The Houses of Parliament: History, Art, Architecture*, p. 19.
11 Peel Papers, BL Additional MS 40,432, f. 259 r.
12 Charles Cope, *Reminiscences of Charles West Cope, RA*, p. 226.
13 *Westminster Review*, 1842.
14 Daniel Boeckmann, 'Ludwig Gruner, Art Adviser to Prince Albert', MA dissertation, School of World Art Studies and Museology, University of East Anglia, 12 September 1996, p. 4.
15 *Athenæum*, 25 September 1841.
16 Boeckmann, p. 6.
17 *Athenæum*, 6 July 1844, p. 627.
18 M scr. Dresden App. 1, 20, 30.
19 Hermione Hobhouse, 'The Monarchy and the Middle Classes: The Role of Prince Albert', in *Middle Classes, Aristocracy and Monarchy*, p. 59.
20 Lady Eastlake, *Contributions to the Literature of the Fine Arts with a Memoir*, pp. 171–2.
21 RA VIC/MAIN/Y/195/10.
22 Martin, Vol. I, p. 181.
23 Journal, 3 September 1843.
24 Martin, Vol. I, p. 181.
25 RA VIC/MAIN/Y/204/40.

26 Martin, Vol. I, p. 195.
27 RA VIC/MAIN/Y/186/12.
28 Ibid.
29 Martin, Vol. I, p. 202.
30 RA VIC/ADDA10/82.
31 RA VIC/MAIN/Y/186/13.
32 Ibid.
33 Ibid.
34 Seymour Diaries, BL Additional MS 60303 f. 88.
35 Martin, Vol. I, p. 209.
36 Ibid., p. 211.
37 Ibid., p. 212.
38 Ibid.
39 RA VIC/MAIN/Y/204/77.
40 Jennifer Montagu, 'The "Ruland/Raphael Collection"', *Visual Resources*, Vol. III, 1986, pp. 167–83.

10 NEPTUNE RESIGNING HIS EMPIRE TO BRITANNIA

1 Martin, Vol. I, p. 181.
2 Ibid., p. 343.
3 Journal, 7 September 1843.
4 Journal, 4 October 1837.
5 Journal, 9 January 1839.
6 Journal, 23 September 1839.
7 Journal, 8 February 1845.
8 John Martin Robinson, *James Wyatt, Architect to George III*, p. 232.
9 RA VIC/MAIN/Y/240/11, Prince Albert's diary, 23 April 1845. Albert never did master the correct forms of English address among the aristocracy, calling Lady Isabella Blachford, the daughter of a duke, 'Lady Blachford'.
10 Benson and Esher, *Letters*, Vol. II, p. 36.
11 Journal, 22 June 1845.
12 RA VIC/MAIN/F/21/53, memorandum by Prince Albert, 21 October 1844.
13 Journal, 29 March 1845.
14 Hermione Hobhouse, *Thomas Cubitt, Master Builder*, p. 384.
15 Howard Colvin, *A Biographical Dictionary of British Architects*, p. 101.
16 Greville, Vol. V, pp. 235, 431.
17 Nikolaus Pevsner and David Lloyd, *The Buildings of England: Hampshire and the Isle of Wight*, pp. 756–9; Michael Turner, *Osborne House*, p. 20.
18 HRH Duchess of York and Benita Stoney, *Victoria and Albert: Life at Osborne House*, p. 32.
19 Marsden, p. 18.
20 Journal, 8 September 1844.
21 Journal, 1 September 1846.
22 Netzer, p. 216.
23 Benson and Esher, *Letters*, Vol. II, p. 37.
24 Gash, Vol. II, p. 560.
25 Ibid., p. 533.
26 Ibid., p. 526.
27 Ibid., p. 561
28 'Pairing' is an arrangement whereby individual members of the House from opposite sides agree that, if one of them has to be absent from a debate, the other will abstain from voting.

29 Benson and Esher, *Letters*, Vol. II, p. 89.

11 MALTHUSIAN CALAMITY

1 RA VIC/MAIN/Y/186/188/21.
2 Cutting preserved by Albert in RA VIC/MAIN/Y/186.
3 E. L. Woodward, *The Age of Reform, 1815–1870*, p. 118.
4 S. G. Checkland, *The Gladstones: A Family Biography, 1764–1851*, p. 416.
5 Jagow, p. 155.
6 RA VIC/MAIN/188/30.
7 Aberdeen Papers, BL Additional MS 43,043 f. 9 r, Victoria to Aberdeen, 28 October 1845; RA VIC/MAIN/B8/13, Aberdeen to Victoria.
8 Muriel E. Chamberlain, *Lord Aberdeen: A Political Biography*, p. 417.
9 Greville, Vol. IV, p. 235.
10 Weintraub, p. 165.
11 Martin, Vol. I, p. 402.

12 CAMBRIDGE

1 RA VIC/MAIN/Y/204/34, Prince Albert's diary for October 1843.
2 Martin, Vol. I, p. 187.
3 Ibid.
4 See below.
5 My sense of these things is quickened by reading *One Hundred Letters* from Hugh Trevor-Roper, especially pp. 295 ff.
6 Rhodes James, p. 174.
7 Ibid., p. 175.
8 Journal, 27 February 1847.
9 *The Poetical Works of Wordsworth*, p. 493.
10 In the RA most of the documents relating to his work as Chancellor date from his first three years in office.
11 RA VIC/MAIN/F/32/20.
12 RA VIC/MAIN/F/32/4.
13 RA VIC/MAIN/F/32/21.
14 RA VIC/MAIN/F/32/23, 13 November 1847.
15 RA VIC/MAIN/F/32/20.
16 RA VIC/MAIN/F/32/29.
17 RA VIC/MAIN/Y/240/181.
18 RA Cambridge Papers/Vol. 1, ff. 14–15.
19 RA VIC/MAIN/F/32/92.

13 THE YEAR OF REVOLUTIONS

1 RA VIC/MAIN/Y/186/49 ('*Wir haben gestern unsere Revolution gehabt die in Rauch aufgegangen ist*').
2 RA VIC/MAIN/Y/186/61.
3 RA VIC/MAIN/Y/186/60.
4 R. F. Foster, *Modern Ireland, 1600–1972*, p. 316.
5 RA VIC/MAIN/Y/188/48.
6 Journal, 7 May 1848.
7 RA VIC/MAIN/Y/188/48.
8 Journal, 2 May 1848.
9 Journal, 12 May 1848.
10 Journal, 4 May 1848.

11 Journal, 7 May 1848.
12 Journal, 9 May 1848.
13 Journal, 10 May 1848.
14 Edwin Hodder, *Life of Lord Shaftesbury*, p. 371.
15 Martin, Vol. II, p. 48.
16 Ibid., p. 49.
17 RA VIC/MAIN/Y/186/58.
18 Evelyn Ashley, *The Life and Correspondence of Henry John Temple, Viscount Palmerston*, Vol. I, p. 139.
19 Ibid., p. 141.
20 Ibid., p. 142.
21 Shaftesbury, when still Ashley, married Lady Emily Cowper, daughter of Earl and Lady Cowper; upon the Earl's death, Lady Cowper married Palmerston.
22 Benson and Esher, *Letters*, Vol. II, p. 203.
23 RA VIC/MAIN/Y/204/19.
24 Ibid.
25 RA VIC/MAIN/Y/204/198.
26 RA VIC/MAIN/Y/204/203.
27 RA VIC/MAIN/Y/204/204.
28 RA VIC/MAIN/Y/204/211.
29 RA VIC/MAIN/Y/204/213.
30 RA VIC/MAIN/Y/204/214.
31 RA VIC/ADD MS MISCELLANOUS 1867.
32 *The Literary Gazette and Journal of Belles Lettres, Arts, Sciences &c.*, 1848, p. 714.
33 Weintraub, p. 210.
34 RA VIC/ADDJ/1867.
35 *St James's Gazette*, 11 January 1893 (write-up of whole affair).
36 RA VIC/MAIN/Y/204/254.

14 BALMORAL

1 Mary Miers, *Highland Retreats: The Architecture and Interiors of Scotland's Romantic North*, p. 105.
2 *Oxford Dictionary of National Biography*, Vol. VIII, p. 201, Muriel E. Chamberlain.
3 Miers, p. 105.
4 Martin, Vol. II, p. 135.
5 *Oxford Book of English Verse*, ed. Sir Arthur Quiller-Couch, p. 778.
6 J. Mordaunt Crook, *The Rise of the Nouveaux Riches*, p. 219.
7 Journal, 5 September 1848.
8 Benson and Esher, *Letters*, Vol. II, p. 194.
9 Journal, 8 September 1848.
10 Journal, 13 September 1848.
11 Journal, 10 September 1848.
12 Journal, 17 September 1848.
13 Journal, 25 September 1848.
14 Journal, 27 September 1848.
15 Journal, 28 September 1848.
16 Journal, 30 September 1848.
17 Journal, 1 October 1848.
18 Jagow, p. 150.
19 RA VIC/ADDA6/31.
20 Ibid.
21 Martin, Vol. II, p. 159.

22 Quoted Paul Thomas Murphy, *Shooting Victoria: Madness, Mayhem and the Modernisation of the British Monarchy*, p. 280.
23 Martin, Vol. II, p. 207.
24 Ibid., p. 211.
25 Owen Chadwick, *The Victorian Church*, Vol. I, p. 284.
26 Martin, Vol. III, p. 208.
27 Chadwick, Vol. I, p. 303.
28 Jagow, p. 175.
29 Benson and Esher, *Letters*, Vol. II, p. 281. And see Vol. II, p. 275 where she wrote, 'Dr Arnold said very truly, "I look upon a Roman Catholic as an enemy in his uniform; I look upon a Tractarian as an enemy disguised as a spy".'
30 Martin, Vol. II, p. 327.
31 Benson and Esher, *Letters*, Vol. II, p. 249.
32 Hansard, series 3, Vol. 114, col. 83 (3 February 1852).
33 Ibid., p. 252.
34 Jagow, p. 213.
35 Journal, 27 June 1850.
36 Murphy, p. 319.
37 G. M. Young, *Victorian England: Portrait of an Age*, p. 79.
38 Martin, Vol. II, p. 291.
39 Ibid., pp. 291–2.
40 Ibid., p. 307.
41 Benson and Esher, *Letters*, Vol. II, p. 264.
42 Ibid., p. 277.

15 THE GREAT EXHIBITION OF 1851: THE MAKING OF THE MODERN WORLD

1 Robert Hewison, *John Ruskin*, p. 81.
2 D. H. Lawrence, *Kangaroo*, p. 284.
3 Charles Dickens, *Dombey and Son*, p. 337.
4 Mark Pattison, *Memoirs*, p. 244.
5 Hilton, p. 629.
6 John R. Davis, *The Great Exhibition*, p. 12.
7 Ibid., pp. 20–21.
8 Ibid., p. 21.
9 Ibid., p. 24.
10 Ibid., p. 26.
11 Yvonne ffrench, *The Great Exhibition*, p. 27.
12 RC/H/1/2/9 (archive of the Royal Commission for the Exhibition of 1851).
13 *The Times*, 18 October 1849.
14 Ibid.
15 Jeffrey Auerbach, *The Great Exhibition of 1851: A Nation on Display*, p. 33.
16 RC/H/1/4/131.
17 Quoted Paul Young, *Globalization and the Great Exhibition*, p. 46.
18 Martin, Vol. II, p. 248.
19 RA VIC/MAIN/F/24/34.
20 Auerbach, p. 33.
21 RC/H/1/2/43.
22 RC/H/1/2/50.
23 Hansard, series 3, Vol. 112, cols. 901–4 (4 July 1850).
24 RC/H/1/3/123.
25 Ibid.

26 Auerbach, p. 45.
27 '*L'illustre défunt*'. RA VIC/MAIN/F/24/45.
28 Quoted Henry Cole, *Fifty Years of Public Work of Sir Henry Cole*, Vol. I, p. 163.
29 *The Times*, 29 June 1850.
30 Michael Leapman, *The World for a Shilling*, p. 77.
31 RC/H/1/4/22, pasted pages from *The Times*. There are no dates but it appears to be June–July 1850 from the context.
32 RC/H/1/4/28.
33 For the previous two pages, ffrench, passim, and Kate Colquhoun, *A Thing in Disguise: The Visionary Life of Joseph Paxton*, Part 2, 'Air'.
34 RC/H/1/4/98.
35 Colquhoun, p. 173.
36 RC/H/1/5/133.
37 Martin, Vol. II, p. 349.
38 RC/H/1/5/19.
39 Ibid.
40 RC/H/1/5/27.
41 Martin, Vol. II, p. 333.
42 Ibid., p. 334.
43 RC/H/1/4/1.
44 Ffrench, p. 61.
45 RC/H/1/4/109.
46 RC/H/1/4/141.
47 RC/H/1/4/115.
48 *Catalogue of the Great Exhibition Volume III. Official Descriptive and Illustrated Catalogue. Foreign States*, pp. 1426–7.
49 Franz Bosbach and John R. Davis et al., *Die Weltausstellung von 1851 und ihre Folgen*, p. 277.
50 Ibid., p. 180.
51 Franz Selmeier, 'Eisen, Kohle und Dampf' in Etta K. Gieger, *Die Londoner Weltausstellung von 1851 im Kontext der Industrialisierung in Grossbritannien*, p. 55.
52 Deirdre McCloskey, *Bourgeois Equality: How Ideas, Not Capital or Institutions, Enriched the World*, p. 649.

16 THE GREAT EXHIBITION: 'SO VAST, SO GLORIOUS'

1 RA VIC/MAIN/F/24/89.
2 RA VIC/MAIN/F/24/183.
3 RC/H/1/6/2.
4 Ffrench, p. 217.
5 *Gothaisches Tageblatt*, 5 February 1851.
6 RC/H/1/6/107. '*Du kannst dir gar nicht denken, was es für Mühe brauchte, manche Leute dazu bewegen, nur ETWAS zu schicken.*'
7 RA VIC/MAIN/F/24/125.
8 RC/H/1/6/115.
9 RC/H/1/6/14, *Gothaisches Tageblatt*, 5 February 1851.
10 RC/H/1/6/15 and 16, 14 and 16 February 1851.
11 RC/H/1/6/235.
12 RC/H/1/6/81.
13 RA VIC/ADDA15/1728, quoted Marsden, p. 88.
14 Martin, Vol. II, p. 369.
15 When Queen Elizabeth wore this at the Coronation Ball in 1937, Chips Channon called it an 'ugly spiked tiara'. Princess Elizabeth (later Queen Elizabeth II) wore it

as 'something borrowed' at her wedding to Prince Philip of Greece in 1947. It was also worn by Princess Anne on her wedding to Captain Mark Phillips in 1973. See Leslie Field, *The Jewels of Queen Elizabeth II.*

16 Martin, Vol. II, p. 365.
17 Ibid., p. 366.
18 Davis, p. 133.
19 Marsden, p. 313.
20 1851 GREAT EXHIBITION, METROPOLITAN DISTRICT, Class IV, machines for direct use. Prince Albert's copy, RA.
21 Quoted Asa Briggs, *Victorian Things*, p. 32.
22 Ibid., p. 65.
23 Ibid., p. 53.
24 RC/H/1/5/111.
25 RC/H/1/6/8.
26 RC/H/1/8/42.
27 RA VIC/MAIN/F25/5.
28 RC/H/1/B/207.
29 RC/H/1/C/256.
30 RC/H/1/C/223.
31 RC/H/1/C/231.
32 RA VIC/MAIN/F/26/33.
33 RA VIC/MAIN/F/26/34.
34 RA VIC/MAIN/F/26/38.
35 Ibid.
36 RA VIC/MAIN/F/25/85.
37 Francis Kilvert, *The Diaries of Francis Kilvert*, Vol. I, p. 21.

17 STORMY WEATHER

1 Journal, 16 September 1852.
2 Martin, Vol. II, p. 468.
3 RA VIC/MAIN/Y/206, Sir James Clark's diary, 28 February 1850. Copy of Sir James Clark's diary in Sir Theodore Martin's hand. Original in Royal College of Physicians.
4 Greville, Vol. V, p. 267.
5 Jagow, p. 249; Martin, Vol. I, p. 491.
6 Bonham's Sale Catalogue of Redgrave's sketch, 21 June 2005.
7 Elizabeth Longford, *Wellington: Pillar of State*, p. 404.
8 Leslie Stephen and Sir Sidney Lee (eds.), *Dictionary of National Biography*, Vol. XXIV, p. 345.
9 RA VIC/MAIN/Z/140/18.
10 RA VIC/MAIN/Z/140/15.
11 RA VIC/MAIN/Z/140/22.
12 RA VIC/MAIN/Z/261/16.
13 RA VIC/MAIN/Y/188/61.
14 Jasper Ridley, *Lord Palmerston*, p. 396.
15 Ibid., p. 398.
16 Graham Papers, BL Additional MS 79599 f. 91.
17 Jasper Ridley, *Lord Palmerston*, p. 421.
18 Benson and Esher, *Letters*, Vol. III, p. 3.
19 RA VIC/MAIN/Y/194/48.
20 RA VIC/MAIN/Y/194/54.
21 Benson and Esher, *Letters*, Vol. III, p. 67.

22 Martin, Vol. III, p. 120.
23 Ibid., p. 109.
24 Ibid., p. 117.
25 Ibid., p. 109.
26 Ibid., p. 125.
27 Jasper Ridley, *Lord Palmerston*, p. 371.
28 Ibid., p. 372.
29 *Encyclopaedia Britannica*, Vol. VII, p. 453.
30 Website of the Royal Victoria Patriotic Building (http://www.rvpb.com), *passim*.
31 RA VIC/MAIN/Z/140/61.
32 RA VIC/MAIN/Z/261/52.
33 RA VIC/MAIN/Z/39/18 and 19.
34 Woodham-Smith, p. 503.
35 RA VIC/ADDA6/33.
36 Ibid.
37 John Plunkett, *Queen Victoria: First Media Monarch*, p. 144.

18 THE PRINCE CONSORT AT LAST

1 RA VIC/MAIN/Y/206.
2 RA VIC/ADDA6/36.
3 See Charles Beem, 'Why George of Denmark did not become a King of England', in Charles Beem and Miles Taylor (eds.), *The Man Behind the Queen: Male Consorts in History*, pp. 81–91.
4 RA VIC/MAIN/Y/154/63.
5 David Cannadine, 'The Last Hanoverian Sovereign? The Victorian Monarchy in Historical Perspective, 1688–1988' in A. L. Beier et al. (eds.), *The First Modern Society: Essays in English History in Honour of Lawrence Stone*, p. 127 ff, quoted Urbach, 'Prince Albert the Creative Consort', in Beem and Taylor (eds.), p. 152.
6 RA VIC/MAIN/Y/154/64.
7 Quoted Martin, Vol. IV, p. 65.
8 RA VIC/MAIN/Z/140.
9 A. N. Wilson, *Victoria*, p. 203.
10 RA VIC/MAIN/Z/140, 5 November 1856.
11 RA VIC/MAIN/Z/261/25.
12 Martin, Vol. IV, p. 82.
13 Alan Bennett, *The History Boys*, p. 55.
14 Benson and Esher, *Letters*, Vol. III, p. 160.
15 *London Gazette*, 5 February 1856, pp. 410–11.
16 Press Release, The Lord Ashcroft Gallery, 'Extraordinary Heroes', exhibition at the Imperial War Museum, 9 November 2010.
17 Angus Hawkins, *The Forgotten Prime Minister: The 14th Earl of Derby*, Vol. II, p. 127.
18 Plunkett, p. 147.
19 Martin, Vol. IV, p. 35.
20 Ibid., p. 37.
21 Ibid., p. 63.
22 Ibid., p. 62.
23 Ibid., p. 145.
24 *Your Dear Letter*, p. 297.
25 Ibid.
26 Ibid., p. 299.
27 Ibid.

28 Theo Aronson, *Queen Victoria and the Bonapartes*, p. 69.
29 Martin, Vol. IV, p. 66.
30 Jasper Ridley, *Lord Palmerston*, p. 535.
31 Aronson, p. 83.
32 Ibid., p. 81.
33 Ibid.
34 Journal, 17 August 1857.
35 RA VIC/MAIN/Y/194/69 v.
36 Journal, 14 October 1857.
37 Ibid.
38 Ibid.
39 RA VIC/MAIN/Y/191/70.
40 RA VIC/MAIN/NOTES ON THE BIRTH OF THE CHILDREN 12 March 1857.
41 RA VIC/MAIN/NOTES ON THE BIRTH OF THE CHILDREN/38–39.

19 KING IN ALL BUT NAME

1 John C. G. Röhl, *Wilhelm II: die Jugend des Kaisers 1859–1888*, p. 106.
2 Martin, Vol. IV, p. 177.
3 RA VIC/MAIN/Z/261/23 *Remarks, Conversations, Reflections*.
4 Roger Fulford (ed.), *Dearest Child: Letters between Queen Victoria and the Princess Victoria 1858–1861*, p. 14.
5 Ibid., p. 31.
6 RA VIC/MAIN/Z/262/47.
7 RA VIC/MAIN/Z/261/22 *Remarks, Conversations, Reflections*.
8 *Dearest Child*, p. 66.
9 Ibid., p. 68.
10 Ibid., p. 94.
11 Ibid., p. 91.
12 Martin, Vol. IV, pp. 208–9.
13 RA VIC/MAIN/Y/195/25.
14 Martin, Vol. IV, p. 240.
15 *Dearest Child*, p. 112.
16 RA VIC/MAIN/Z/140/60–62.
17 RA/VIC/MAIN/Y/206, Sir James Clark's diary, 13 August 1858.
18 Ibid.
19 Martin, Vol. IV, p. 281.
20 Ibid., p. 287.
21 Ibid., p. 292.
22 Ibid., p. 298.
23 Ibid., p. 299.
24 Ibid., p. 273.
25 Ibid., p. 278.
26 Hawkins, Vol. II, p. 156.
27 Ibid., p. 159.
28 Ibid., p. 166.
29 Ibid., p. 43.
30 Jane Ridley, *Bertie*, p. 26.
31 Ibid., p. 41.
32 RA VIC/MAIN/Z/141/28.
33 Ibid.
34 Jane Ridley, *Bertie*, p. 41.

35 Ibid., p. 42.
36 RA VIC/MAIN/Z/1/51.
37 Martin, Vol. IV, p. 301.
38 RA VIC/MAIN/Y/195/17, 28 July 1858.
39 RA VIC/MAIN/M/19/67.
40 Martin, Vol. IV, p. 114.
41 RA VIC/MAIN/Z/140/46–51.
42 RA VIC/MAIN/Z/140/63, undated but from context it must be 1858–9, since they post-date some words of 1857, and they could not have been written after 1859–60 when things went from bad to worse.
43 Martin, Vol. IV, p. 387.
44 Ibid.

20 THE CARE AND WORK BEGIN

1 RA VIC/MAIN/Y/195/13.
2 RA VIC/MAIN/Y/195/32.
3 Martin, Vol. IV, p. 484; RA VIC/MAIN/195/36.
4 Martin, Vol. IV, p. 425.
5 Hawkins, Vol. II, p. 220.
6 *Your Dear Letter*, p. 299.
7 Jasper Ridley, *Lord Palmerston*, p. 488.
8 Palmerston, though a Viscount, sat in the Commons, because his title was an Irish viscountcy. Lord John Russell, as son of a Duke, was called 'Lord' but did not hold a peerage in his own right. When he was later granted an earldom he sat in the Lords as Earl Russell.
9 Woodward, p. 166.
10 Ibid.
11 Benson and Esher, *Letters*, Vol. III, p. 353.
12 Quoted Aronson, p. 98.
13 RA VIC/ Z//MAIN/141/44.
14 RA VIC/MAIN/Z/461/21.
15 See Urbach, *Bismarck's Favourite Englishman*.
16 RA VIC/EVIID, quoted Ridley, *Bertie*, p. 43 – quoted by Albert in his letter RA VIC/MAIN/Z/141/43.
17 RA/VIC/MAIN/Z/141/43.
18 RA VIC/MAIN/Z/141/44.
19 Jane Ridley, *Bertie*, p. 44.
20 *Dearest Child*, p. 208.
21 RA VIC/MAIN/Y/187/64.
22 Jane Ridley, *Bertie*, p. 44.
23 Ibid.
24 Ibid., p. 45.
25 Martin, Vol. V, p. 90.
26 *The Poems of Tennyson*, ed. Christopher Ricks, Vol. III, p. 264.
27 Ibid., p. 263.
28 A. L. Kennedy (ed.), *My Dear Duchess, Social and Political Letters to the Duchess of Manchester, 1858-1869*, p. 197.
29 Journal, 9 June 1860.
30 *The Times*, 21 June 1860.
31 RA VIC/MAIN/Z/461/113.
32 Jane Ridley, *Bertie*, p. 49.
33 *Dearest Child*, p. 279.

34 Ibid.
35 Arthur Ponsonby, *Henry Ponsonby. Queen Victoria's Private Secretary. His Life from His Letters*, p. 27.
36 Lewis Carroll, *The Hunting of the Snark*, London, Macmillan, 1876.
37 Martin, Vol. V, p. 196.
38 Ibid.
39 Ibid., pp. 200–209.
40 Rhodes James, p. 262.
41 Martin, Vol. V, p. 216.
42 Ibid., p. 229.
43 RA VIC/MAIN/Y/195/51.
44 Jagow, p. 352.
45 Journal, 30 November 1860.
46 Martin, Vol. V, p. 255.
47 Helen Rappaport, *Magnificent Obsession: Victoria, Albert and the Death that Changed the Monarchy*, p. 258.
48 Martin, Vol. V, p. 271.
49 RA VIC/MAIN/Z/491/33.
50 RA VIC/MAIN/Z/491/31.

21 SHE WAS THE STRONG ONE

1 Benson and Esher, *Letters*, Vol. III, p. 427.
2 *Dearest Child*, p. 308.
3 Ibid.
4 Journal, 21 February 1861.
5 Benson and Esher, *Letters*, Vol. III, p. 435; *Dearest Child*, p. 317.
6 Benson and Esher, *Letters*, Vol. III, p. 435.
7 *Dearest Child*, p. 319.
8 Ibid., p. 318.
9 Liverpool Papers, BL Additional MS 38,303, f. 292 r.
10 Arthur Ponsonby, *Henry Ponsonby*, p. 64.
11 RA VIC/MAIN/M/10/72.
12 UK Inflation calculator.
13 RA VIC/MAIN/Z/140/43.
14 *Dearest Child*, p. 319.
15 Martin, Vol. V, p. 324.
16 RA VIC/MAIN/Z/90/122.
17 RA VIC/MAIN/Z/140/29.
18 Charles Tennyson, *Alfred Tennyson*, p. 337.
19 Ibid., p. 336.
20 RA VIC/MAIN/Z/491/41.
21 Ibid.
22 Ffrench, p. 288.
23 Lytton Strachey, *Queen Victoria*, p. 185.
24 Martin, Vol. V, p. 312.
25 Ibid., p. 346.
26 Ibid., p. 314.
27 Jane Ridley, *Bertie*, p. 54.
28 Martin, Vol. V, p. 385.
29 Ibid., p. 383.
30 Jane Ridley, *Bertie*, p. 54.
31 Ibid., p. 53.

32 RA VIC/MAIN/Z/141/89.
33 Jane Ridley, *Bertie*, p. 57.
34 *Dearest Child*, p. 359.
35 Martin, Vol. V, p. 403.
36 Ibid., p. 416.
37 *Herald and Weekly News Press*, 31 March 1883.
38 RA VIC/MAIN/Z/491/13.
39 *Dearest Child*, p. 365.
40 Roger Fulford (ed.), *Dearest Mama: Letters between Queen Victoria and the Crown Princess of Prussia, 1861–1864*, p. 132.
41 RA VIC/MAIN/Z/141/f.94.
42 Martin, Vol. V, p. 417.
43 Ibid.
44 RA VIC/MAIN/Z/141/95.
45 Woodham-Smith, p. 531.
46 Woodward, p. 296.
47 Ibid., p. 297.
48 Benson and Esher, *Letters*, Vol. III, p. 469.
49 Woodham-Smith, p. 535.
50 Ibid., p. 536.
51 Ibid., p. 537.
52 Ibid., p. 539.
53 Martin, Vol. V, p. 442.
54 Ibid., p. 415, '*Ich hange gar nicht am Leben; du hängst sehr daran…*'
55 RA/VIC/MAIN/Y/206, Sir James Clark's diary, December 1865.
56 Montagu, p. 179.
57 Indeed, he did not live to see the figures in place, since the mausoleum was finished in August 1871, and Marochetti died in 1867.
58 Arthur Ponsonby, *Henry Ponsonby*, p. 250.

INDEX